How-to Build & Scale Robotics Startups

TRIBE OF ROBOTS

Hard-Won Lessons from 25 Robotics Startups

By Prathamesh Khedekar

TRIBE OF ROBOTS

preserve confidentiality while maintaining the integrity of the lessons.

Image credits and third-party attributions appear in a dedicated section of this book.

ISBN (Paperback): 979-8-9942966-2-2

ISBN (Hardcover): 979-8-9942966-3-9

Library of Congress Control Number: 2026906914

Published by Boring Sage

Lake Forest, Illinois

www.boringsage.com

Printed in the United States of America

Dedication

To my Aae, Dipti Khedekar, and my Baba, Deepak Khedekar,

who devoted their lives to my education.

To Lutfulla & Rikhsi Agzamov, for all that you are.

To Arun Khedekar, Alka Khedekar, Prakash Khedekar, Ashok Khedekar, Munavvar Shakirova, Baqo & Makhmuda Agzamov for all that you are, and all that you have been.

To my sister, wife, son, and family,

for their unwavering support throughout this journey.

Why I Wrote This Book

After more than 100 demos, countless sleepless nights debugging our robots, and over a dozen launches of our robot-led safety and security services for startup and enterprise clients across the United States, the United Kingdom, and Australia—for a diverse spectrum of clients ranging from Apple to Amazon, Slack to DoorDash, Merck to Toyota, and General Motors to Ally Financial—I heard the words that no founder or engineer wants to hear.

The words I heard that day still echo in my mind:

"Despite all our efforts and execution across the board—from funding to operations and engineering—we have not been able to reach the mark needed to take this venture forward. Our journey here has come to an end. We will need to shut down operations. Effective immediately, we will need to hand over control to a private firm. I tried my best, but this is where we need to draw the line. This is as hard for me as the founder as it is for you, the people who helped bring this company this far."

Those words stayed with me.

It was our founding team speaking on a video call. The dream with which I had landed in the Valley had ended in failure. This was the hardest exit of my career—and the one that hurt the most.

For years, I had worked tirelessly orchestrating countless demos and launches across continents, making enormous sacrifices on many fronts that are well beyond the scope of this book. To see the journey end this way was devastating.

That experience left me with one question:

Why?

Why did a venture blessed with an exceptional team, a strong set of investors, and a product engineered with extraordinary precision, one backed by clients across verticals spanning pharmaceuticals, technology, finance, automotive, and e-commerce—still fail?

That question led me to spend the next several years studying robotics startups across the world: those that succeeded and those that faced the same fate as ours. I started studying the trajectory of these startups and their founders, tracing their roots across the board—from inception to funding, decision-making to partnerships, the whole DNA that shaped their trajectory. Over time, what started as a personal quest for answers turned into a long-term mission to understand the true forces that shape robotics startups, leading to much deeper and more systematic research into this landscape.

This book is a culmination of that research.

From the academic corridors of MIT, Carnegie Mellon, ETH Zurich, the University of Oslo, UCLA, UC Berkeley, the University of Electronic Science and Technology of China, and other leading institutions to the frontlines of entrepreneurship, it traces the journeys of twenty-five robotics founders and the startups they built across three decades—from 1992 to 2026. Their stories unfold in cities and towns across the world—from Ji'An in China to Toulouse in France and Odense in Denmark—within university laboratories, research groups, accelerators, factory floors, and entrepreneurial ecosystems where many of the foundations of modern robotics were forged.

Through these journeys, you will learn the distinct approaches adopted by each of these founders to conceptualize their startups, fund them, scale them, and, most importantly, navigate the risks inherent in this path, defying the odds that often derail even the most well-funded and talented teams in this arena.

You will also discover the strategies and business models they adopted to overcome challenges along the way, from securing early adopters with MVP-grade products, to raising capital from mission-aligned investors, identifying strategic hubs to scale production, and ultimately building companies that create lasting value for their teams, investors, and customers.

At its heart, this is a story about founders and the ecosystems around the world that shaped their journey—cities, universities, and communities that helped give rise to many of the robotics startups defining the twenty-first century. By examining their decisions, the risks they took, their breakthroughs, and their failures, this book distills hard-earned lessons for the next generation of entrepreneurs and researchers building the future of robotics.

How To Use This Book

This book is organized into seven core segments, each covering the journeys of founders and their startups operating across distinct classes of robots—from industrial mobile robots and warehouse systems to consumer robots, delivery bots, and humanoids. Depending on the arena in which you plan to build your startup, or are already building one, the lessons across these segments—along with the business models they culminate in—are intended to equip you with a series of approaches you can adopt to address the bottlenecks inherent along your path and ultimately bring your mission to life.

My goal here is to ensure that through these case studies you are able to distill the decisions that helped these companies navigate some of the most challenging stages of their journeys and apply those insights to your own.

Below is a simple framework you can use to apply the insights from this book to your startup journey.

01. Map Your Bottlenecks

Take a piece of paper, ideally an A4 sheet, and make a list of your A+ bottlenecks across the full spectrum of your startup's lifecycle: funding, customers, scale, partnerships, manufacturing, product–market fit, and early adopters. Keep these in mind as you start studying the trajectories of the startups and founders in this book.

02. Start With Your Segment

Based on your primary class of robots, deeply study the trajectories of the startups and their corresponding founders operating in that space that are covered in the respective segments of this book. Segment I is applicable across the board, so start there and then proceed to your primary segment from Segments II–VIII based on the class of robots you are building. Whether you are building humanoids, warehouse robots, quadrupeds, social robots, or delivery robots, start here.

03. Decode Founder Decisions

After reading each chapter, reflect on and document the specific decisions the founders made, along with the specific parameters of their business models that apply to your startup and can help you address the bottlenecks across the areas you listed in Step 01.

04. Learn From Adjacent Segments

Now repeat this process for segments that may not directly apply to you but are still important, since the underlying decisions founders made and the driving factors behind them are often domain-agnostic.

05. Consolidate Insights — Build Your Founder Playbook

Now reflect on the lessons, decisions, and parameters of the business models that you have documented as you advanced through this book, listing solutions to each of the bottlenecks you identified in the areas that matter most to you and your team today. At this point, you will have distilled these insights into a founder playbook that captures the approaches most relevant to your startup and its mission.

06. Unblock Your Bottlenecks & Bring Your Mission to Life

With your founder playbook now in place, start applying these insights to systematically address the most critical bottlenecks in your startup today. Your goal here is to study these founder journeys and translate their decisions into actions that you can execute today—actions that will help you move at a rapid pace, scale in a resource-efficient manner, and ultimately bring your mission to life: one that you started this journey with in the first place and one that your team has put all their hopes, efforts, time, and dreams into.

Table of Contents

Part I — The Rise of Robotics Startups

1. From Silicon Valley to Shenzhen — A Road Less Traveled ..1

Part II — North American Industrial Robotics Leaders

2. Boston Dynamics — DARPA Origins to Hyundai Scale.......13
3. Amazon Robotics — Kiva Acquisition to 1M+ Robots..........39
4. Kiva Systems — Webvan Roots to Amazon Exit.................65
5. Fetch Robotics — Willow Garage Lab to Zebra Exit............87
6. 6 River Systems — Six Years to Shopify Exit....................117
7. Symbotic — The F1 Engine Powering Walmart.................143
8. Clearpath — University of Waterloo Lab to Rockwell Exit..165

Part III — European Industrial Robotics Powerhouses

9. AutoStore — Norwegian Roots to Whole Foods Scale......205
10. Universal Robots — Danish Cobots to Teradyne Exit........239
11. ANYbotics — ETH Zurich Labs to Shell and Vale.............267
12. Exotec — French Engineering to UNIQLO Scale..............299

Part IV — The Consumer Robotics Wave

13. iRobot — MIT Lab to Roomba Scale.................................325
14. Social Robots — Why They're So Hard 339
15. Astro, Anki, Jibo, & Mayfield — Vision to Market Reality ...347

Part V — Last Mile Delivery Robots

16. Delivery Robots — Core Vectors Driving Adoption 375
17. Starship — Estonia to 60+ U.S. Campuses with Sodexo...383
18. Serve Robotics— Postmates Lab to NASDAQ with Uber..391
19. Kiwibot — Bogotá to UC Berkeley with SkyDeck...............401

20. Coco Robotics— UCLA Dorms to Helsinki with DoorDash 415

Part VI — Humanoid Startups: Lessons from Pioneers

21. Dawn of Humanoid Robots 423
22. Figure AI — Illinois Farm to BMW Factories 431
23. 1X — Oslo Labs to Norway's Humanoid Moonshot 447
24. Agility Robotics — Oregon State Labs to Amazon Scale...461

Part VII — Humanoid Startups: Lessons from Challengers

25. UBTECH — Shenzhen Roots to Airbus and Audi Fleets ...473
26. Apptronik — NASA Labs to Mercedes-Benz Factories 485
27. AgiBot — Chengdu Origins to Thailand Scale 499

Part VIII — Lessons & Next Steps

28. Hard-Won Lessons from 25 Robotics Startups..................513
29. How Founders & Engineers Can Apply This Book.............533

Image Credits

Notes on Sources, Author Experience, and Educational Use

References

About the Author

Chapter 01: From Silicon Valley to Shenzhen — A Road Less Traveled

On a cold, foggy morning in December 2018, our pilot, navigating through dense fog, harsh winds, and roaring thunderstorms, somehow mustered the courage to land the plane at San Francisco International Airport. My heart was pounding at 110 miles per hour. I had barely slept the night before. Within minutes of landing, I rushed through the airport, got into an Uber, and was soon on my way to the address I had been given on El Camino Real in Silicon Valley.

About half an hour later, the young, tech-savvy driver from the Valley dropped me in front of a quiet building.

A Knock on a Door in Silicon Valley

As I gazed through the glass wall at the robots, a few people entered the room, and soon I was fielding questions ranging from the philosophy of life to computer networks, program management, robotics, and mission-critical systems. After six hours of warm, well-thought-out, and carefully structured interview sessions, crafted with the same level of care and precision as the robots themselves by the team that built them, I did okay and was selected to join this robotics startup, that was freshly funded by YC, Bloomberg, and Sequoia.

This was around the time I had returned from Europe, having designed mission-critical systems for the governments of the Netherlands and Norway. Building the equivalent of private versions of Google Maps and WhatsApp for these agencies, along with the geo-redundant data centers needed to host them, ensuring end-to-end encryption, 99.9999 percent reliability, and the ability to withstand reasonable scales of earthquakes and tsunamis, helped me learn the A to Z of software, systems design, networking, and the telecom space.

With that expertise honed, robotics and AI felt like the next natural step. The reason being: with the widespread adoption of robots and the maturity of the underlying tech stack, it did not take long to realize that the same data that was landing at these data centers and transmitted through the cell phones and laptops of public safety professionals would one day be transmitted by robots and AI-enabled safety and security systems around the world.

That realization, along with the conversations that had started transpiring from devices to robots in the arena, were the early signals that made it increasingly clear that robots, in particular, were steadily gaining adoption across the industry.

The Turning Point

Curious to understand where this shift was emerging from and where it might lead, I started spending more time studying robotics before and after work. After months of self-study sessions aimed at decoding the underlying technology and business stack, I turned to researching the journeys of emerging startups in this field, ranging from the incubation of the legged robotics lab at Carnegie Mellon University that gave birth to Boston Dynamics, to Amazon's $775 million acquisition of Kiva Systems, whose warehouse robots now form nearly half of the workforce layer of the e-commerce giant.[1,2]

Inspired by the journeys of these startups, their founders, and their impact, I decided to venture into robotics. After having self-studied what I believed to be the A-to-Z of the engineering and business stack of this arena, starting with ROS; to perception and navigation algorithms, and the corresponding AI stack ranging from YOLO to transformers and more, I started reaching out to a number of robotics startup founders.

To each, I would craft a DM followed by an email stating my understanding of their mission, their potential A+ problem, and an approach that I would naively think of as a solution to that problem, proposing it in a one-pager. One of those conversations eventually translated into an interview and a role that allowed me to contribute to the development and launch of commercial robots. I remain deeply grateful to these founders for placing their faith in me and for inspiring others in the field through their grit and actions.

The Brutal Reality of a Robotics Startup

Over the period of the next few years, after orchestrating hundreds of demos and launches of our robot-led safety and security services for clients across the startup arena like Slack, DoorDash, Yelp, and Pivotal, to enterprise giants ranging from Apple to General Motors, Toyota to Amazon, and Western Digital to Merck across the U.S., U.K., and Australia, I came to terms with the reality of robotics startups and learned some of the hardest lessons in the arena—lessons that came at an extreme cost of time, health, and capital.

By then, we had exhausted our financial arteries and, despite landing Fortune 100 clients across these countries and having made our share of mistakes, some avoidable and some inevitable, our journey in the robotics startup arena came to an abrupt end.

During this time, I learned that building a robotics startup is brutally hard.

Years of hard work, sleepless nights, and a steadfast belief in our mission had brought us to that moment. That day, the founders, over a video call, explained the decision that had been made and the underlying reasons to sell the assets and the company to a private firm.

That night, as I went back to bed, my mind raced through all the demos I had done, the deployments, conversations with clients who had promised to expand our service, those long nights, rehearsals to get every bit of the demos and launches right, and the sheer effort that went into demonstrating the value of our product to our clients.

This was the hardest exit of my career, and the one that hurt the most.

I could not digest the fact that a product of such caliber, designed by a team of some of the most inspiring minds in the space, could not find its place in the market.

That question stayed with me.

While this journey ended on that day, the question that kept me awake at night was: Why?

The Question That Wouldn't Let Me Sleep

Why did a product of such caliber, engineered with surgical precision, designed with the finest levels of Japanese craftsmanship, and led by founders who brought a rare combination of rigor, discipline, and a deeply humane approach to life and startups, end this way?

Why do some robotics startups succeed while others fail—even when they appear to have the right team, the right technology, and sometimes even the right investors?

Why did some startups with only a fraction of the resources we had manage to achieve successful exits to companies like Amazon, Shopify, and Teradyne, while others with remarkable products and strong investor backing—companies such as Anki, Jibo, and Mayfield Robotics' Kuri—struggled to survive?

These were the questions that stayed with me.

These questions eventually led me down a path of research that would span several years. I started studying robotics startups around the world—their founders, their investors, their decisions, their partnerships, the whole DNA that shaped their trajectory. I wanted to understand the patterns behind success and failure in this industry.

Studying 25 Robotics Startups — From Ji'An, China to Nedre Vats, Norway

Having already sacrificed a great deal of health, time, and capital on our first robotics startup venture, I decided to spend my nights researching startups across both ends of the spectrum—successes and failures—their founders, and the ecosystems that shaped them. Over time, what started as a personal quest for answers turned into a long-term mission to understand the true forces that shape robotics startups, leading to a much deeper and systematic research into this landscape.

This book is a culmination of that research.

Over the course of this journey, I studied dozens of robotics companies around the world and closely examined the paths taken by 25 of them, along with the trajectories of their founders. These are some of the most extraordinary journeys of individuals from diverse backgrounds and distant corners of the world who have gone on to build successful robotics companies.

From an invention created by a history major in Wilmington, Massachusetts—one that now powers the robotics systems operating across all 42 Walmart distribution centers—to an idea conceived in a remote village in Nedre Vats, Norway; one that led to the birth of robots now used by companies such as Whole Foods and IKEA, these stories reveal deeper patterns. They illuminate the underlying forces embedded in the DNA of these founders and the ventures they built—patterns that transcend geography, background, skillset, and that ultimately lead to extraordinary outcomes.[3-6]

While many entrepreneurs in robotics encounter daunting hurdles along the way—from raising capital and securing early customers to making difficult strategic decisions and managing the long-term risks inherent in this path—these founders managed to confront those uncertainties and defy the odds, building companies that now serve an equally diverse set of industries, from dairy production facilities to nuclear waste repositories, and from logistics and energy to automotive and e-commerce. Their solutions span the full spectrum of modern robotics platforms—from autonomous indoor mobile robots and industrial robot arms to delivery robots, social robots, and even humanoids.

What I discovered through this research was an extraordinary diversity of founders and an equally inspiring spectrum of journeys. Each came from remarkably different backgrounds, navigating the complex intersection of engineering, capital, and market traction in their own way—hellbent on bringing their vision to life and making an impact.

Some of these founders grew up far from the Silicon Valleys of the world.

Far from Silicon Valley

One such founder was born in a remote region of China's Jiangxi province, a place known for its rugged mountains and long, harsh monsoon seasons that leave barely sixty clear days each year. Growing up in such unforgiving conditions, this young lad pursued his passion for robotics from an early age, went on to win the 2018 World Robotics Olympiad, and is now building one of the frontier humanoid robotics companies, AigBot, backed by leading investors including Tencent, LG, and Baidu. [7-8]

What makes his journey even more fascinating is the strategy he has adopted to build his company and bring his mission to life: launching early testbeds in Milan, Italy, and across Malaysia and Thailand, targeting the hospitality and tourism industries for humanoids, while simultaneously scaling production by establishing strategic manufacturing hubs in Kazakhstan and China. The approach adopted by this young founder to compete in a space flooded with entrepreneurs and investors armed with enormous capital and deep partnerships is truly awe-inspiring.

The journeys of other founders who came from very different backgrounds are equally remarkable.

From Farms, Labs, and Shenzhen Alleys to the Frontlines of Robotics

One grew up on farms in Illinois. Another worked in research laboratories at NASA. Today, both are building frontier humanoid robots designed to automate automotive production lines for companies such as BMW and Mercedes-Benz in the United States.

Another founder incubated his robotics venture in Shenzhen, the manufacturing epicenter of the world, and is now supplying humanoids to companies such as Airbus and Audi, targeting the aerospace and automotive industries.

Then there are founders who kickstarted their journeys in university laboratories—from Oregon State University to the University of Oslo, and from MIT to CMU's legged robotics laboratory—and went on to build companies supplying robots to enterprises ranging from Hyundai and Merck to ADT and DHL.[2-10]

One such founder who started his journey in the labs at Oregon State University has since built one of the largest robot manufacturing facilities, a 70,000-square-foot plant in Oregon that now supplies hundreds of humanoids to companies like Amazon, GXO, and Ford. Another founder, whose roots trace back to the University of Oslo, was among the first in the humanoid space to sell nearly 140 robots to the American security giant ADT and is now building what could become one of the most consequential consumer-centric humanoids of our era—designed specifically for our homes and families.[11-12]

Yet another founder who started his journey at ETH Zurich's robotics laboratory is now bringing ultra-dexterous quadruped robots capable of operating 24/7 in hazardous Zone 1 (explosion-prone) environments to oil refineries operated by companies such as Shell, Vale, and Petronas across Europe and Asia, enabling safer inspection workflows and minimizing workers' exposure to toxic gases at these mission-critical sites.[13]

Each of these founders is scaling their startups at remarkable speed and with a level of resource efficiency that is truly striking. Some have already achieved extraordinary exits, while others are already on their way.

But not every story in robotics ends with a successful exit.

What These Journeys Reveal

Some of the most ambitious consumer robotics companies of the past decade built extraordinary products and attracted significant capital, only to face challenges that ultimately led to their graceful closure. Companies like Anki, Jibo, and Mayfield Robotics' Kuri represent journeys that are just as important to understand as the successes.

Their stories are not simply about success or failure. They reveal patterns that often determine the fate of even the most promising robotics startups in this arena. Each of these trajectories represents a bold attempt by founders to bring robots into our society—and the immense challenges involved in marrying complex hardware and software systems at scale, along with the difficult trade-offs they must navigate along the way across both engineering and business frontiers to succeed in one of the most unforgiving arenas of the startup cosmos.

Together, these journeys reveal how robotics startups are born, the risks they face, how they attempt to reach global scale, and sometimes how they fail. They illuminate the critical decisions founders must make around technology, capital, partnerships, and market timing—decisions that offer valuable lessons for the next generation of entrepreneurs and researchers, people who are brave enough to build robots in the real world.

This book examines the trajectories of these startups and their founders—the risks they took, the decisions they made, and the outcomes that ultimately defined their exits. Some of these stories are deeply inspiring. Others are heart-piercing, given the caliber of the products and the enormous amount of effort and sleepless nights that went into building them, only for the journey to end in an unexpected exit.

At its core, this book examines those paths—decoding the risks, decisions, and strategies that ultimately shape outcomes in the robotics industry.

Now we turn to the first startup and the founder who dared to build it.

Chapter 02: Boston Dynamics — DARPA Origins to Hyundai Scale

Funded by DARPA in the early 1990s, it started as a bold experiment with the sole purpose of building robots that could navigate like animals across all terrains, including hills and mountains, while retaining high speed, balance, and autonomy. Over time, these lab experiments led to the launch of the world's first quadruped robots deployed for mission-critical applications, and later, in 2013, one of the first humanoid robots on planet Earth.[2,14-15]

Along the way, the company was nourished by some of the most powerful hands in technology, first acquired by Google in an attempt to marry world-class robotics hardware with cutting-edge software, then sold to SoftBank in 2016, and eventually landing with Hyundai Motor Group. It was these transitions that proved pivotal, helping these robots make the leap from the lab to the real world.[16]

Today, these robots power mission-critical operations around the world—from Hyundai's automotive assembly lines in Korea to DHL's logistics warehouses across Europe and remote inspection missions for NASA, BMW, and Michelin.[17]

This is the story of a company that has endured for a staggering 26 years in robotics and, equally importantly, has managed to secure strong capital lifelines along the way while building a robust product portfolio that spans nearly every major class of robot, from the Spot quadruped and the Stretch logistics robot arm to the Atlas humanoid, each emerging as a strong revenue stream for the company. What started as a research project in the laboratory of the Massachusetts Institute of Technology eventually transpired into one of the most iconic robotics companies in the world: Boston Dynamics.

Given this trajectory and the decisions made by its founders to catapult the company from the lab into industry while building sustainable capital and revenue streams and bringing frontier robotics into society, it merits closer examination. To better understand this trajectory, it helps to decode the origins of both the company and its founder.

Boston Dynamics was founded in 1992 by Marc Raibert, a professor at the Massachusetts Institute of Technology whose research work focused on legged robots. At MIT's Leg Laboratory, Raibert had spent more than a decade studying the biological locomotion systems that power animals. In simple terms, he was interested in understanding how animals are able to navigate a wide range of terrain profiles while retaining balance and speed. The goal was to study these locomotion systems and apply the underlying principles to develop outdoor legged robots with similar navigational capabilities.[2,15]

That raises an important question: why was Raibert focused on locomotion and not perception and spatial manipulation systems? Aren't these core components of all modern robots?

Yes, perception systems and spatial manipulators play a pivotal role in all modern robotics applications. However, if you trace back the history of robots, most systems at the time were static and designed specifically to automate industrial tasks. A useful reference point is Unimate, the industrial robot arm developed to automate car factories in the United States in the 1950s.[18]

Unimate was a powerful robotic arm, but at the end of the day, it was extremely bulky and inherently stationary. There was no easy way to transport it from one part of the factory to another without the help of heavy-duty forklifts. As a result, while industrial robot arms like Unimate gained widespread adoption in the 1950s, there were virtually no robots capable of autonomously transporting pallets filled with goods across factory floors. Thus, the real bottleneck in robotics at the time was not spatial manipulation, but locomotion. It was the inability of robots to navigate diverse terrain profiles and industrial environments that limited their adoption.

This was the problem that Raibert wanted to address, and that thesis ultimately led to the launch of Boston Dynamics.

This also raises another important question: where did the company get the core capital required to fund its early experiments?

The DARPA Years (1990s–2010s)

Given that robots had still not crossed the chasm of need in consumer and commercial markets, defense was the only door the company felt was worth knocking on—and that decision paid off over the next two decades.

During its first two decades, Boston Dynamics was funded primarily through government contracts, most notably from the U.S. Department of Defense and the Defense Advanced Research Projects Agency (DARPA). It operated as a research-driven engineering organisation rather than a conventional startup. This path is similar to the one followed by companies such as OpenAI and iRobot.

This funding model, and the research-focused approach it adopted, shaped both the company's operational velocity and its priorities. With its first target customer segment—defense—solidified, the team at Boston Dynamics became focused on solving fundamental engineering problems in robotics, among which the most important at the time was enabling autonomous locomotion, i.e., mobility. To address this challenge, this team of researchers started studying the biological locomotion systems of wild animals capable of navigating harsh terrain profiles with speed and reliability, such as cheetahs and lions, with the end goal of building mechanical equivalents of these systems in the robotics domain. The goal was to develop robots capable of navigating a diverse set of terrains on four mechanical legs. Among the most prominent outcomes of this period were the BigDog quadruped robot and its successor, the Legged Squad Support System (LS3).[15]

BigDog and Proof of Capability (2000–2010)

BigDog, developed in the mid-2000s around 2005, was designed to navigate harsh terrain profiles such as valleys, hills, and mountains while retaining balance and speed. Its success led to the launch of the Legged Squad Support System (LS3) in 2010, which extended BigDog's capabilities by enabling it to carry heavy payloads with extremely high reliability.[15] With this addition, this robot evolved into a go-to quadruped platform for defense applications.

By the end of 2010, Boston Dynamics had developed unmatched expertise in enabling advanced legged locomotion systems at scale. In simple terms, it was able to translate the locomotion capabilities observed in wild animals into the robotics domain, meaning robots could now traverse terrains and environments that were previously inaccessible. That said, the business was still fundamentally structured as a research organisation. There was no meaningful commercial revenue at this stage.

So what did this team do next?

Launch of Atlas — Humanoid Robot (2010–2013)

Boston Dynamics decided to expand its capabilities beyond quadrupeds. It became increasingly clear that if robots were to serve humans and operate around humans, they would need to be designed more like us. That realization naturally directed the company towards humanoid robots.

Between 2010 and 2013, the company pivoted much of its research focus toward building what can be considered some of the earliest humanoid robots in history. This body of work culminated in the development of Atlas in 2013—a humanoid robot built for the DARPA Robotics Challenge—establishing Boston Dynamics as a global leader in humanoids.[19]

Atlas was one of the first humanoid robots capable of navigating rough terrain and demonstrated advanced spatial manipulation capabilities. While its appearance often drew comparisons to the Terminator, it was designed as a disaster-response and rescue robot under the DARPA Robotics Challenge. It was intended to operate in environments that were either unsafe or inaccessible to humans.

This raises a critical question: how advanced was the robot from a technological standpoint, and more importantly, how did Boston Dynamics build it?

Atlas — Engineering Design and Capabilities

From a technical standpoint, it was one of the most advanced humanoids ever constructed at the time. The robot featured near-human anthropometry. In simple words, its design, performance, and capabilities were comparable to those of humans. It was powered by hydraulic actuators and featured a staggering 28 degrees of freedom. Some of you may ask: what about sensors and the perception system?

It was equipped with an equally sophisticated sensor package developed by Carnegie Robotics. This sensor stack was deployed in its head and primarily included LiDAR, stereo cameras, dedicated electronics, and onboard computers running perception algorithms. In addition to these sensors, Atlas was configured with two different hand designs—one developed by iRobot and another by Sandia National Laboratories—one optimised for agility and the other for strength. That raises another question: what about its body structure? How was it engineered?

Structurally, the robot was built from two core materials, aircraft-grade aluminium and titanium, with protective coverings designed to absorb shocks resulting from high-impact accidents. From a software standpoint, the system was written primarily in C++, with ROS serving as the foundational middleware. Atlas weighed roughly 150 kilograms and reportedly cost on the order of $2 million per unit. The launch of Atlas extended Boston Dynamics' footprint beyond the defense and research community, bringing the company under the radar of Silicon Valley for the first time in its history.[20]

Acquisition by Google (2013–2016)

In 2013, Boston Dynamics was acquired by Google, which at the time was assembling a portfolio of advanced robotics companies under its Google X research division. This naturally raises another question: why was Google interested in acquiring Boston Dynamics?

The rationale was strategic. Google was making long-term bets on advanced technologies spanning robotics and the broader autonomous mobility spectrum, technologies that had the potential to transform a wide range of industrial sectors, from manufacturing to transportation and beyond. It was on a mission to lead the emerging technology frontier with a clear application-centric focus on high-impact industrial verticals.[21] Executing on this vision made deep capabilities in both hardware and software non-negotiable. While Google had built a formidable software base to support this ambition, its hardware capabilities were not equally mature. Boston Dynamics brought deep expertise in hardware and control systems, particularly in legged robotics, expertise that complemented Google's strengths in software and artificial intelligence.

On paper, this marriage made sense. In practice, commercialising a product that was still at an extremely early stage quickly became a significant bottleneck for the company.

Why was commercialisation so important for Google?

As a public company, commercialisation was particularly important for Google because it had to justify its investment by demonstrating tangible returns in its quarterly briefings to its global stakeholders. This justification, however, was not possible, as Boston Dynamics' product portfolio remained largely in the research phase and had not yet matured enough for commercialisation.

If that were not enough, Boston Dynamics was a hardware-centric organisation characterised by long development cycles, high unit costs, and limited opportunities for large-scale deployment of its robot-centric products. Google's core businesses, by contrast, were software-driven, data-centric, and optimised for exponential growth. While research at Boston Dynamics continued largely uninterrupted, this acquisition did not result in a clear path towards commercialisation.

By 2016, Google started unwinding its robotics portfolio, signalling a pullback from the advanced robotics domain. That raises another question: if not a software giant, then who was the right long-term owner for a company like Boston Dynamics?

Commercialisation Under SoftBank (2017–2019)

That question was answered in 2017, when Boston Dynamics was approached by SoftBank and eventually acquired by it. That raises another question: why did SoftBank acquire Boston Dynamics?

SoftBank's investment thesis differed from Google's. At Google, the commercialisation of Boston Dynamics' robots had become a huge bottleneck—one the organisation was neither structured nor incentivised to resolve. SoftBank, by contrast, viewed this constraint as an opportunity.[22]

From a macro perspective, SoftBank saw robotics as a long-term and structured response to global societal challenges such as labour shortages, ageing populations, and the growing demand for technologies enabling industrial automation. In simple words, it deeply believed that robots had the power to address these challenges in society. Guided by this belief, and under its strategic guidance, Boston Dynamics made its most significant pivot yet—transitioning from a research-first organisation to one focused on building commercially deployable robots.

This transition marked the company's most decisive shift toward commercialisation and culminated in the launch of Spot in 2019—Boston Dynamics' first industrial-grade quadruped robot.

Launch of Spot — The Remote-Inspection Robot (2019)

Spot was conceived from the outset as a commercial product rather than a research platform. It was significantly lighter, smaller in size, easier to operate, and more portable than its predecessor, BigDog. The robot was positioned primarily as a mobile inspection and remote-operations platform, targeting a wide range of industry verticals, from energy and infrastructure to manufacturing and research.

With the launch of Spot, for the first time in its history, Boston Dynamics had a product generating sustained revenue. While this trajectory looked good on paper, and it seemed like the company was about to hit escape velocity, it immediately encountered its next major bottleneck: scale of adoption.

Spot solved extremely surgical and high-value problems for a specific set of industrial customers, but scaling this revenue stream proved much harder in reality than on paper. In simple terms, the market was not large enough for Spot to generate significant revenue for the company at the time. Robotics was still perceived as a "nice-to-have" rather than a "must-have" technology, and most enterprise organisations had yet to develop a strong appetite for large-scale adoption.

While Spot was adopted by organisations such as NASA, Michelin, and BMW, overall uptake fell short of the company's initial expectations.[17] Boston Dynamics quickly realised that although Spot was generating sustained revenue, it had unintentionally landed itself in a single-product "death zone"—a position in which one moderately adopted product becomes a single point of failure and is incapable of securing a company's long-term financial runway.

So what did it do next?

Launch of Handle: Entry into Warehouse Logistics (2021)

To scale its target pool of customers, the company turned its attention to the logistics and warehouse automation industry. Why turn to the warehouse and logistics industry?

The logic was straightforward. There are roughly 180,000 warehouses around the world, which meant the target market size was large enough for Boston Dynamics to solve a problem that could yield returns at scale. Warehousing also offered clearer return-on-investment models and significantly higher adoption potential than the remote inspection market addressed by Spot alone.

But did Boston Dynamics have any background in this vertical?

Not exactly—but the early foundations of this work had already been laid within the company's research labs.

In 2017, during the Google–SoftBank transition period, the company's research team had already started steering development efforts towards wheeled robots optimized for the warehousing and logistics industry. This line of research led to the development of a mobile robot arm called Handle. Handle was quietly launched in 2017, but at the time it was still perceived more as an internal R&D project than a commercial product.[15] Why was that the case?

Handle was equipped with a high-performance spatial manipulator, that is, a robotic arm mounted on a wheeled mobile base, but it was missing a critical component: a perception system.

If Handle was to succeed in the warehouse vertical, it had to be equipped with equally strong "eyes" to ensure it was able to navigate tight spaces with ease and reliability. Why was that the case? Warehouse environments are highly cost-sensitive and optimised for throughput. What matters most in these settings is a robot's ability to efficiently transport goods from point A to point B using the minimum amount of resources, namely time and capital.

Integrating Boston Dynamics' existing robots, such as Spot, Handle, and Atlas, into these environments meant the company had to equip them with a strong set of "eyes," that is, robust perception systems. This would allow the robots to better understand their surroundings and execute tasks with a high degree of accuracy.

So how did Boston Dynamics address this challenge?

Acquisition.

Addressing the Perception Gap: The Kinema Acquisition (2019)

In 2019, Boston Dynamics acquired Kinema Systems, a company specialising in developing vision-centric deep-learning models to power spatial manipulators designed exclusively for warehouse environments. These models enabled industrial robotic arms to locate, pick, and transfer boxes across a wide range of pallet configurations inherent in the warehouse industry, with minimal setup and retraining requirements.[23]

At its core, Kinema solved a critical bottleneck for Boston Dynamics: perception. Warehouse environments are constantly evolving. If you take a look at the floor traffic in the 180,000 warehouses around the world, you will notice that each of these facilities is loaded with thousands of boxes that come in different sizes and shapes. In addition to this warehouse-centric variable, these boxes are transferred across the warehouse on pallets on a near-weekly basis. As a result, the floor footprint undergoes significant change every week.

For spatial manipulators like Handle, developed by Boston Dynamics, to operate in such environments, they had to be equipped with a strong set of eyes that would help them deeply understand these evolving footprints and navigate them safely in each of these facilities. Kinema's deep-learning models were designed specifically to address this challenge. The acquisition filled an important gap in Boston Dynamics' stack. Kinema added the missing layer: a cutting-edge perception system, bringing Boston Dynamics one step closer to enabling the large-scale rollout of Handle across 180,000 warehouses worldwide.

While the acquisition clearly signalled intent to dominate the logistics and warehousing industry, it also introduced another bottleneck. Boston Dynamics now had to generate revenue in parallel while turning Handle into a commercially viable product. That required capital. The company needed to extend its runway. This need was clearly understood by one of the leading automakers of our time.

Long-Term Capital: Hyundai Motor Group (2021–Present)

In 2021, to secure long-term capital and extend its runway, the company turned to Hyundai Motor Group. The underlying thesis was that automotive factories were increasingly impacted by labour shortages, and that the robots developed by Boston Dynamics, particularly its robotic arms, quadruped, and humanoid, were strategically positioned to address these challenges in a time- and capital-efficient manner.

Given this hypothesis, in 2021, following a series of negotiation sessions, Boston Dynamics was acquired by Hyundai Motor Group.

Under Hyundai's ownership, Boston Dynamics gained financial stability and an industrial parent with an ultra-long investment horizon. More importantly, Hyundai approached this acquisition as an opportunity to automate its 12 major manufacturing facilities around the world and accelerate the production of 2.4 million vehicles each year with the highest levels of capital efficiency.[24-25]

So what did Boston Dynamics do with this newly extended financial runway?

Yet another commercial product launch. This time for the logistics and warehousing industry.

Warehouse Automation: Stretch and DHL (2020–2022)

The company poured that capital into marrying Handle—its robotic arm—with Kinema's advanced perception system. The result was the launch of Stretch, its advanced robot arm equipped with a high-definition perception system.

In simple terms, it successfully transformed Handle, which had largely been perceived as an internal R&D project constrained by perception bottlenecks, into a commercially viable product called Stretch. Stretch was designed as a commercial mobile robotic arm from day one, built specifically to automate warehouse and logistics operations. It was equipped with an omnidirectional wheeled base, a purpose-built robotic arm with seven degrees of freedom, and a robust vision-based perception system. This trio was designed and integrated with surgical precision to ensure it could not only fit into existing warehouse infrastructure, but also transport pallets filled with boxes at scale.

Stretch quickly marched beyond pilot deployments. DHL emerged as an early adopter, signing a memorandum of understanding to deploy up to 1,000 units across its warehouse network. Each system was designed to handle a staggering 700 boxes per hour.[26] This agreement represented Boston Dynamics' first serious attempt to scale a commercial robot within the logistics industry. So where does the company stand today?

Where the Company Stands

Today, Boston Dynamics remains a global leader in robotics. Specifically, it has developed three commercial products: Spot, the quadruped robot for remote inspection; Stretch, the mobile robotic arm for warehouse automation; and Atlas, the humanoid robot designed to automate car factories. With more than 1,500 Spot robots, over 1,000 Stretch units, and dozens of Atlas systems deployed across pilot and commercial settings by a diverse spectrum of customers ranging from NASA and Michelin to BMW, DHL, and Gap, the company has clearly addressed some of the most painful needs of its users. More importantly, it has done so with remarkable endurance and agility.

From early on, Boston Dynamics understood that capital was its lifeline. To secure that lifeline from Day 01, it ensured that its cutting-edge research translated into frontier robotic platforms capable of attracting industry giants with deep pockets, including Google, SoftBank, and Hyundai. That awareness reflects endurance on the capital front.

The next principle it demonstrated is agility. It understood early on that, to succeed in the long run, it needed a commercially viable product with a significantly large addressable market segment. So even though Spot alone did not offer that scale, its ability to expand beyond the remote inspection market into the warehousing industry with the launch of Handle and Stretch demonstrates the sheer level of agility embedded in its culture to accomplish its mission.

It is this combination of endurance and agility that enabled the company to overcome the many bottlenecks along its path. In short, Boston Dynamics remained true to its vision and is now widely regarded as best-in-class in robotics. So what are some of the lessons we can learn from this journey?

Lessons for Robotics Founders & Engineers

Lesson 01: Build Endurance by Surgically Defining Your Funding Strategy

As a robotics company, you need to adopt a surgical approach to secure your financial runway, given how capital- and resource-intensive this journey is. While Boston Dynamics was able to secure a series of long-term institutional investors, that approach is not always feasible for early-stage robotics startups. If you are at the early-to-mid stage with none to a handful of customers, there are two core pathways you can explore to secure your financial runway.

First is the hybrid capital approach, which consists of accelerators, venture capitalists (VC), and community funding rounds. Your goal here is to optimize for early-stage velocity through accelerator channels, mid-stage reach through retail, that is, crowdfunding channels, and late-stage scale and speed through the VC channel. This also helps you keep your team and operations lean in the early-to-mid stage, allowing you to focus surgically on your product and the problem you are trying to solve, thereby minimising the burn that comes with scale and speed.

Second is the research-oriented approach. If you are starting as a research organisation, you are better off building applications for the defense and academia sectors, then transitioning to the commercial segment by building world-class products that can later be funded through investments led by industry giants. Boston

Dynamics took the second route, while startups like Miso, Jibo, and Anki took the first. No matter which route you take, define it with surgical precision from Day 01 and secure your runway with extreme clarity.

Lesson 02: Market Strategy – Win Small Markets Before You Chase Big Ones

Do your homework and surgically study the potential and trade-offs of your target market segment. If you notice, in the case of Boston Dynamics with Spot, it was able to penetrate a reasonably niche market by targeting remote industrial inspection use cases with quadruped robots—but the trade-off was the size of that market and the urgency of the need. Many of these industrial operators didn't trust quadrupeds and were not willing to adopt them at scale. While a small market segment like this makes it easier to penetrate the market as an early-stage robotics startup, it will also eventually become your major revenue bottleneck. So how should robotics founders and engineers approach this challenge?

It is important to clearly define your target market segments for each stage of your startup. In the early stage, you want to remain laser-focused on small markets. Smaller markets provide significantly lower resistance for newcomers, which works in your favor early on.

Once you have refined your product and ensured it addresses the most pressing needs of customers in that market, you should adapt it to serve adjacent target markets—just as Boston Dynamics did by strategically extending the scope of its products from Spot to Stretch, pivoting from inspection-only workflows to enabling full-scale automation for warehouse operators. These larger markets provide the financial runway you ultimately need to sustain your company—so know when to pivot.

Lesson 03: Acquisitions – Acquire to Address Technical Bottlenecks, Not to Rent Users

The majority of acquisitions, at the end of the day, are pursued by startups with the sole goal of renting a user base. That's a trap. A rented user base is rented—not earned. Trust cannot be rented; it has to be earned. And you earn trust by delivering real value, which means building great products that address the core pain points of your customers. Period.

So if you're thinking about acquisitions, pursue them with the primary goal of unblocking technical bottlenecks. Boston Dynamics' acquisition of Kinema Systems is a great example of what acquisitions should truly serve: addressing critical technical constraints. The same goes for Amazon's acquisition of Kiva Systems and Canvas Technology. One provided the mobile robotic hardware Amazon needed; the other strengthened its software capabilities.

Together, these acquisitions helped Amazon overcome key bottlenecks and build a strong robotics foundation that ultimately enabled the deployment of a fleet of over one million robots operating across more than 1,200 logistics sites—quadrupling its throughput worldwide.

This approach to acquisitions helps you save time and commercialize your products with exceptional speed and efficiency. These are the core lessons for robotics founders and engineers from this journey.

Now that we have distilled the core lessons embedded in Boston Dynamics' trajectory, from its origins through its evolution and the pivotal decisions the team made along the way, it is time to decode the underlying business model.

In the final section of this chapter, we will break down its business model and illustrate how the product portfolio maps to select primary use cases.

Boston Dynamics: Product Snapshot I

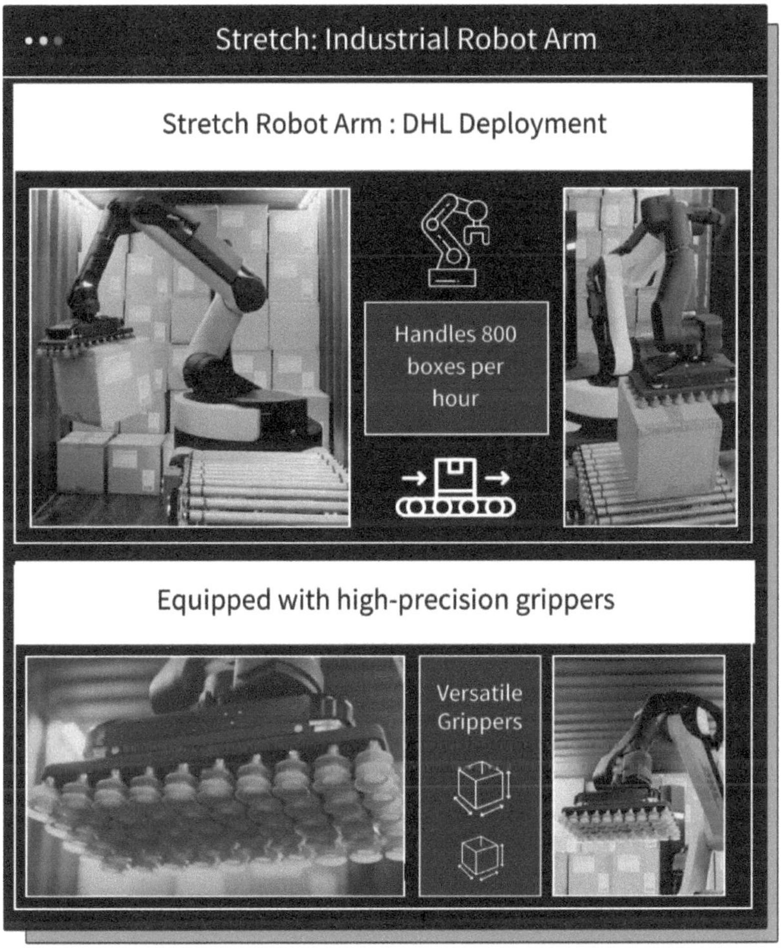

Figure 2.1 — Stretch: Industrial Robot Arm

Illustration by the author. Stretch image from DHL Supply Chain press release (2023).

Boston Dynamics: Product Snapshot II

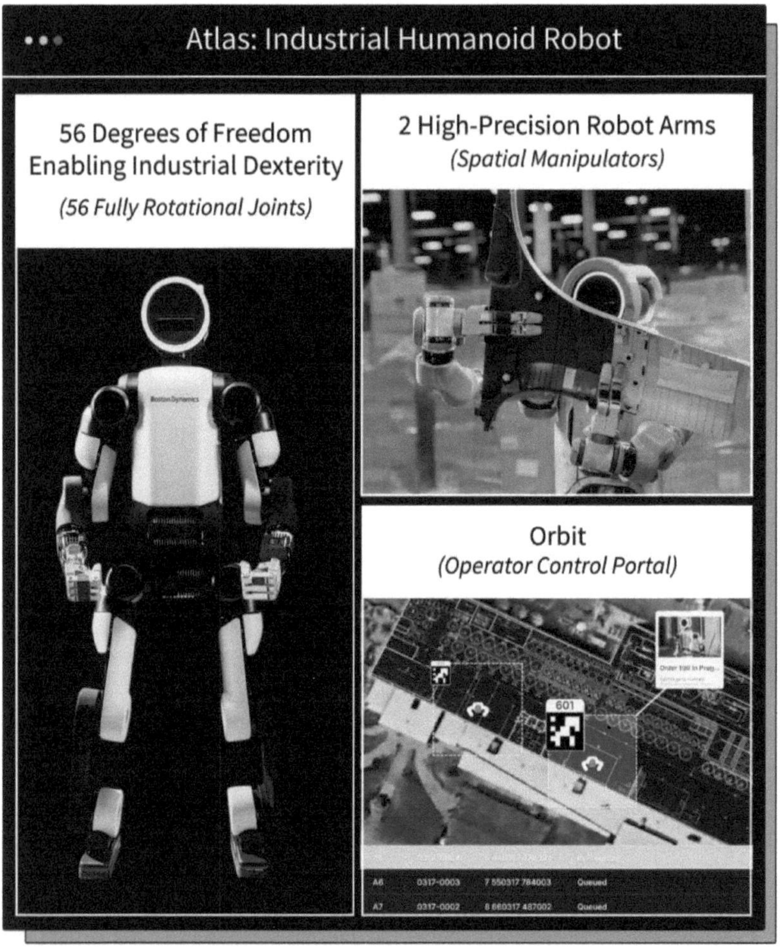

Figure 2.2 — Atlas: Industrial Humanoid Robot

Illustration by the author. Atlas image from Boston Dynamics product specification page.

Boston Dynamics: Product Snapshot III

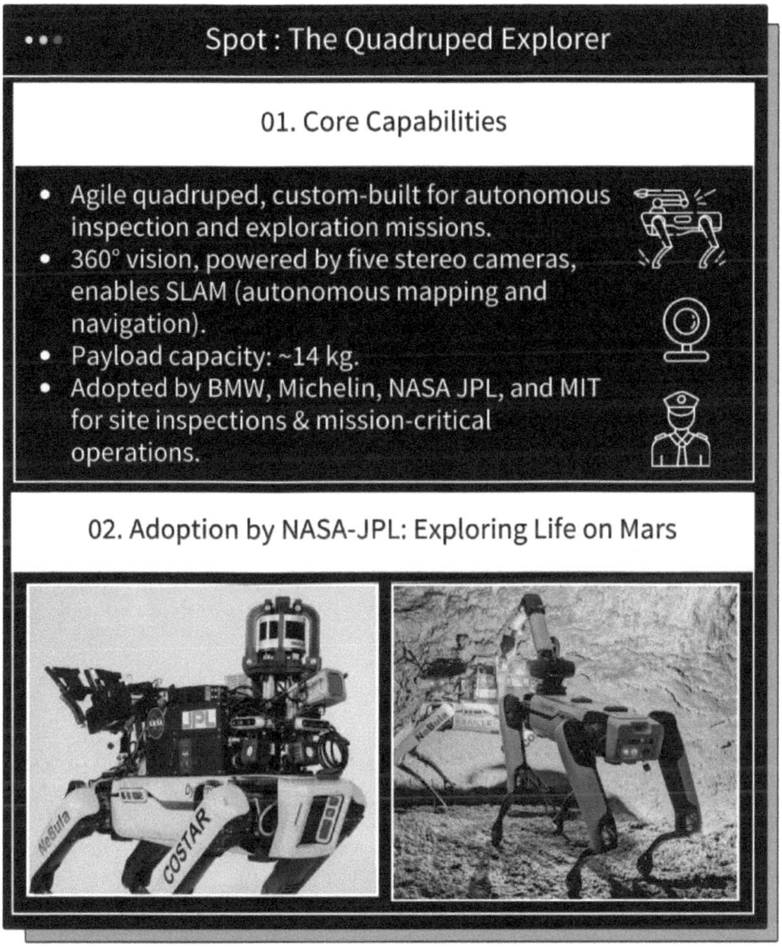

Figure 2.3 — Spot: The Quadruped Explorer

Illustration by the author. Atlas image from Boston Dynamics product specification page.

Boston Dynamics' Business Model — Part 1

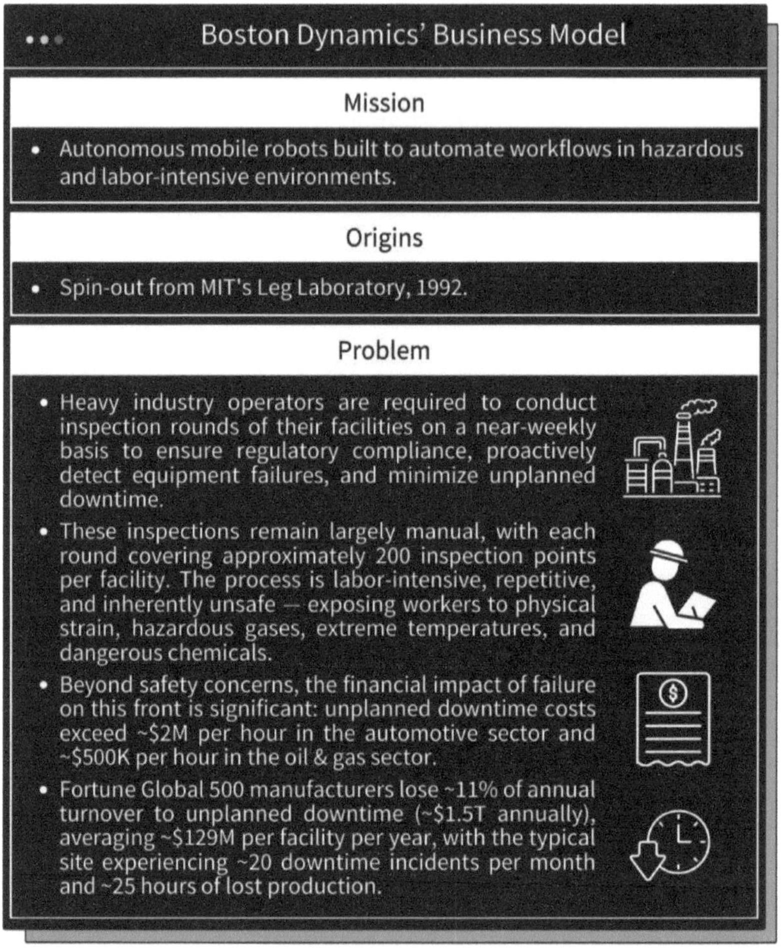

Figure 2.4 — Business Dynamics' Business Model Part 1

Illustration by the author.

Boston Dynamics' Business Model — Part 2

Figure 2.5 — Business Dynamics' Business Model Part 2

Illustration by the author.

Boston Dynamics' Business Model — Part 3

Figure 2.6 — Business Dynamics' Business Model Part 3

Illustration by the author.

Boston Dynamics' Business Model — Part 4

Figure 2.7 — Business Dynamics' Business Model Part 4

Illustration by the author.

Chapter 03: Amazon Robotics — Kiva Acquisition to 1M+ Robots

6 billion orders fulfilled across more than 25 countries by 1 million robots working in tandem with 1.5 million employees. These numbers reflect the sheer intensity with which the Amazon Robotics team has led this frontier over the past two decades.[27-29] From warehouse robots to last-mile delivery humanoids, robotics has become a core operational fabric for Amazon. That leads us to our first question: When and why did Amazon decide to explore robots as a potential solution to scale its operations? What was the underlying thesis?

Before we try to answer that question, it helps to understand the history of Amazon. The company was founded in 1994, and its founder, Jeff Bezos, initially named it Cadabra Inc., derived from the famous phrase *abracadabra*, which implies magical creation. In this context, the idea was simple: a book would be delivered to your doorstep almost magically after you typed its name on the company's website. While the idea of an online bookstore seemed logical to Bezos at the time, the name didn't make sense to his lawyer.

When Jeff tried to convey the name of the company to his lawyer, it was reportedly misheard as *cadaver*. To avoid confusion, Bezos decided to rename the company in 1995 as Amazon, after the world's largest river, reflecting his ambition to build the largest company of its kind.[30] Thus, Amazon started as an online bookseller offering a wider range of books than any physical store could carry at the time. As online adoption accelerated in the late 1990s and early 2000s, Amazon steadily expanded its offerings beyond books into music, electronics, household goods, and eventually nearly every retail category. By the early 2010s, this strategy had succeeded: demand for goods sold on Amazon was surging at a rapid pace, and the company had established itself as one of the most successful e-commerce giants of that time.

By 2012, while the e-commerce business was booming, the company encountered an expected bottleneck on its path: the classic demand-supply gap. In this case, it showed up in its fulfillment network. Demand for goods sold on Amazon was surging at a rapid pace, but its ability to serve that demand was not, and the underlying bottleneck was its fulfillment network. This quickly became the weakest link in its business.

So how did Amazon address this bottleneck? For the first few years, it tried to scale the number of human workers across its 200+ core warehouses worldwide to accelerate the pace at which it fulfilled orders. But that approach was not a long-term solution. Why? Amazon operated on thin margins, which meant it had to keep operational expenses low while maintaining an extremely high fulfillment speed for each order. A human workforce is a good solution in many cases, but here it was not going to resolve Amazon's fulfillment bottlenecks.

So where did it go from here?

Robots. How? The Kiva acquisition.

Acquisition of Kiva Systems (2012)

Back in 2012, a fast-growing robotics start-up called Kiva was on a mission to automate warehouses with robots capable of carrying pods of goods from one point to another in an autonomous manner. It had already launched with top-tier customers like Staples, Walgreens, Gap, Zappos, and Crate & Barrel. Kiva's robots were operating in their warehouses, and this adoption curve proved the value the company had to offer to its customers.

Amazon quickly recognized the implications. To address its own fulfillment bottlenecks, it needed to bring this capability in-house. At the same time, leaving Kiva as an independent operator in open markets meant competitors could gain access to the same productivity gains. Amazon responded decisively, making an acquisition offer that Kiva's founders and board could not ignore.

The $775 million all-cash deal, announced in March 2012, gave Amazon a clear operational advantage over its competitors.[1] While the acquisition attracted little attention outside logistics circles at the time, a decade later it is clear that it marked one of the most strategic bets a company has ever made in the history of the robotics industry.

That brings us to an important lesson on acquisitions: always pursue acquisitions to address your A+ technical bottlenecks, not to rent a user base. A rented user base rarely endures, because trust cannot be rented—it must be earned. Amazon's move here was strategic and was in alignment with this principle of acquisition.

The logic behind this acquisition of Kiva was simple: scaling its operations with a human-dominant workforce was not going to help Amazon meet emerging customer expectations around delivery speed, cost, and scale. Amazon had to drastically improve its speed of operations across its network of 200+ core warehouses and 1,200+ logistics sites, and that meant, in simple words, its ability to move carts full of goods from one point to another.[31-32] These carts were manual, and this was a major technical bottleneck. They had to be automated. Adding more human workers in these facilities was not the answer here.

Kiva's autonomous mobile robots, already proven in third-party warehouses, enabled Amazon to automate these carts—compressing fulfillment times while reducing labor costs. That was not the only problem it solved for Amazon. Even more importantly, this acquisition effectively withdrew Kiva from the open market, preventing rivals from accessing the same productivity gains. It was this acquisition that laid the foundation for its internal robotics arm, now known as Amazon Robotics.

Launch of Amazon Robotics (2015)

By 2015, Amazon formally rebranded Kiva as Amazon Robotics, signalling that robotics had become a core component of its operational infrastructure. It immediately integrated this fleet of Kiva robots across its core warehouses, and after seeing double-digit gains in operational efficiency, it became clear that if it was to continue fulfilling more than 6 billion orders each year—about a million each day—robots were the only way forward. Soon, it embarked on what one could describe as a warrior-style mission to develop the most advanced robotic capabilities to scale its operations engine, which serves over 310 million customers around the world.

As Amazon started deploying Kiva robots across its 200+ global core fulfillment centers, it quickly ran into its next major bottleneck: software. While Kiva provided the foundational version of warehouse robots, namely hardware-centric systems designed to transport goods from one area of the warehouse to another, at the scale at which Amazon was operating, this hardware alone was not enough. This was not a plug-and-play solution: Kiva's software stack was not designed to operate across a network of one-million-square-foot warehouses spanning 1,200+ sites at the pace Amazon required. As a result, the software system powering Kiva's robots became the next major constraint. So how did Amazon address this challenge?

AWS RoboMaker, an in-house developed software platform for building and operating robots, was its answer.

Building the Software Platform Layer (2018–2019)

In 2018, the Amazon Web Services team launched AWS RoboMaker, a cloud-based development, simulation, and deployment platform designed exclusively for robotics teams and built on ROS, the open-source software framework that powers most modern robots.[33] The underlying principle behind this strategy was the same one that guided the development of the AWS cloud platform: provide the infrastructure layer that others can build upon and, in doing so, entrench Amazon at the center of an emerging ecosystem. In this case, that emerging ecosystem was robotics.

To many, this software appeared to be built for external teams to develop and deploy robotics systems in the cloud. In reality, the core research behind it was driven by Amazon's own need to build a robust software foundation for its warehouse robots. In doing so, Amazon addressed a critical constraint within its own network while simultaneously building a platform it could offer to its customers.

As Amazon continued to build out its robotics software stack, it soon encountered its next major bottleneck. While Kiva robots provided the hardware foundation and basic navigational capabilities needed for its robots to traverse structured, pre-mapped warehouse environments, they were not designed to operate safely around dynamic obstacles such as pallets, forklifts, and people. Put simply, Kiva robots were built for highly structured, largely static warehouses. Objects in motion posed a fundamental challenge for these robots.

At Amazon's scale, however, warehouses are far from static; they are continuously evolving environments. The footprint of goods, people, and machines within these facilities changes on a near-daily basis. This meant Amazon had to upgrade the Kiva robots to handle the dynamic traffic patterns inherent in its warehouses. Doing so required adding a strong set of "eyes" and a perception system capable of navigating this constantly changing environment.

So how did Amazon solve this bottleneck? Yet another acquisition. This time, Canvas Technology.

Acquisition of Canvas Technology (2019)

In 2019, Amazon acquired Canvas Technology in a $100 million deal. Canvas was a startup building autonomous carts designed to operate safely around humans. While relatively less prominent than Kiva, Canvas addressed a persistent bottleneck for robots operating inside warehouse environments: the ability to maneuver safely around humans, carts, and forklifts—in other words, adaptive navigation.[34] In simple terms, Canvas Technology provided the vision-based, adaptive navigation system that Amazon's warehouse robots desperately needed to operate safely in dynamic and unstructured environments. Powered by 3D cameras and cutting-edge AI models, this system gave Amazon the "eyes" its robots desperately needed to operate safely in dynamic environments.

Once again, this acquisition signaled the same principle around addressing technical bottlenecks: acquisitions should be pursued with the sole goal of addressing technical bottlenecks, not renting user bases. While this acquisition strengthened Amazon's ability to automate its warehouses, it soon encountered the next bottleneck: last-mile delivery. As Amazon warehouses became increasingly automated, the outdoor delivery network remained largely manual. Humans still drove the delivery vehicles, and that meant last-mile delivery costs were rising rapidly. For reference, this segment alone can account for 30–50% of total shipping costs. At the same time, same-day and next-day delivery were becoming default customer expectations.

The team at Amazon had to answer this question: *Can we automate the final handoff to the customer? Can we bring the cost of delivery down?* If Amazon wanted to truly address the needs of its customers, it had to automate the outdoor delivery network as well.

So what did it do next?

Launch of Scout Delivery Robot (2019)

Scout was Amazon's answer.

Scout was designed as an outdoor delivery robot with the sole goal of automating last-mile deliveries and slashing delivery costs and time to meet customer expectations around same-day and next-day delivery. The underlying principle here was that if Scout could be placed into delivery vehicles and used to automate the last-mile delivery portion, then the amount of time required per delivery would drastically go down. With this thesis, in 2019, Amazon unveiled Scout—a small sidewalk delivery robot piloted in select U.S. neighborhoods.[35]

While Scout seemed like a promising last-mile delivery robot, it also revealed the next major operational bottleneck in Amazon's logistics network. Any guesses? Delivery trucks. The movement of goods between warehouses and major urban hubs remained largely dependent on trucks driven by humans. This meant that even if Scout could automate the final handoff, Amazon still had to absorb the significant delivery costs associated with trucks that were primarily human-operated.

So what did Amazon do next? Yet another acquisition.

Bet on Zoox (2020–2021)

In June 2020, Amazon made one of its most ambitious bets in the robotics world by acquiring Zoox for approximately $1.2 billion.[36] At first glance, one might ask: what does Zoox have to do with Amazon's logistics strategy? Zoox is not a warehouse technology provider; it is a full-stack autonomous vehicle company. That observation is correct—but only in part.

In reality, Amazon perceived Zoox as a company with a mature autonomous-vehicle (AV) stack. If Amazon was serious about automating its delivery fleet, it would need a robust AV stack. Zoox represented a long-term solution to that challenge.

But this acquisition raised another question: what about the trucks themselves? Were they AV-ready? Not exactly. That is where Rivian entered the picture. Amazon partnered with Rivian in September 2019 to modernize and electrify its delivery fleet. With Rivian providing the hardware platform, electric delivery vehicles, and Zoox supplying the autonomous-driving software stack, Amazon started assembling the core building blocks needed to automate large portions of its delivery network.

Now, with all segments of its fulfillment network potentially automated, Amazon realized that to better understand its customers—their behavior and choices—it needed more data. That meant entering their homes, quite literally. So what did it do next? The answer was Astro.

Launch of Astro (2021)

In 2021, Amazon introduced Astro, a consumer robot positioned as an extension of its Alexa-powered smart-home ecosystem.[37]

The goal was to place a robot inside the homes of its customers. Why was Amazon interested in putting a robot inside homes? On one hand, the company wanted to help safeguard customers' loved ones and assets, and on the other, to collect data that could enable Amazon to better serve those customers. Put simply, the experiment was intended to test whether a consumer robot deployed in millions of homes could create a mutually beneficial outcome for the company and its customers.

While the idea was strategic and logical on paper, reality told a different story. There was no demand for this kind of robot. Customers who needed home surveillance systems already relied on cameras, while those uncomfortable with a digital surveillance system often preferred alternatives such as pets. In effect, Amazon was attempting to solve a problem that was not sufficiently painful for most of its customers.

The market responded accordingly. Astro failed to gain traction. There is an important lesson to learn here: consumer robotics is an extremely hard game, and unless you have done the homework to truly understand the most painful needs of your customer, you are better off not entering these waters. Companies like Anki (Vector, Cozmo), Mayfield Robotics (Kuri), and Jibo tell the same story.

Here, Amazon had an almost unlimited supply of capital to make Astro successful, along with a well-established brand name. But those two ingredients alone will not serve the purpose if the very problem you are trying to solve is not the most painful problem for the majority of your customers. With that lesson firmly ingrained in its capital chest, Amazon started shifting its focus back to its core mission—automating its logistics and warehouse network through robotics to maximize fulfillment speed per order.

Launch of Proteus, Sparrow, and Cardinal (2022)

By 2022, Amazon's robotics strategy sharpened decisively around accelerating warehouse operations. The company announced a $1 billion Industrial Innovation Fund, designed to invest in technologies that had the potential to eliminate core bottlenecks in the warehouse, supply chain, and logistics industry. One major caveat Amazon made was that acquisition was not a requirement for a company to be eligible for this fund. In simple terms, accepting capital from Amazon did not obligate a company to be acquired by Amazon in the long run. This ensured that principles of fair competition were still upheld across the industry.

At the same time, the robotics team operating inside Amazon made some significant moves. That year, they launched three high-powered robotic systems to automate core segments of its fulfillment network—namely Proteus, Cardinal, and Sparrow.[38] Together, these systems targeted some of the most labor-intensive and injury-prone tasks in warehouse operations.

Proteus is Amazon's first fully autonomous mobile robot capable of operating safely alongside human workers and transporting pods of goods across the warehouse. Cardinal is a robotic arm built to lift and sort heavy packages, reducing physical strain on workers. Sparrow is a robotic arm equipped with an AI-driven perception system, designed to identify, pick, and sort millions of unique items that come in different shapes, sizes, and packaging.

To summarize, Proteus is all about autonomous mobility, Cardinal offers strength to lift heavy objects, and Sparrow offers a high-definition perception system to sort objects. One provides speed, another strength, and the third vision. Together, these systems reflected a clear shift in priorities. The emphasis was no longer on experimentation, but on delivering clear operational value—throughput, safety, and unit economics at scale.

In contrast, Amazon ended field trials of Scout in late 2022. The decision served as a reminder of the company's operational discipline: projects that failed to meet economic thresholds were shut down immediately. This renewed focus on automating its warehouse operations with specialized robots also marked Amazon's entry into the world of humanoids.

Humanoid Robot Trials with Digit (2023)

In 2023, Amazon kickstarted pilot trials of humanoid robots developed by Agility Robotics, specifically its Digit humanoid. These trials were focused on evaluating whether general-purpose, humanoid robots could perform repetitive and physically demanding tasks such as tote handling, package sorting, and more inside fulfillment and logistics environments.

At a strategic level, the team was attempting to answer a core question: instead of continuing to build highly specialized robots like Cardinal and Proteus, could Amazon deploy a general-purpose humanoid platform and train it to handle a wide range of warehouse tasks?

This initiative was explicitly experimental. The goal was not immediate workforce replacement, but evaluation: assessing whether a single, agile humanoid system could safely, reliably, and efficiently perform tasks that otherwise required multiple specialized robots working in tandem with human workers. Digit represented an early probe in that direction.[39-40] So what was the outcome of this trial?

While the trial confirmed that humanoid robots were capable of automating a limited set of warehouse tasks, it also revealed clear limitations. It became evident that current systems were not yet ready for full-scale warehouse deployments. They were not general-purpose in any meaningful sense. In simple terms, humanoid robots were not a silver bullet for Amazon's challenges, prompting the company to shift its focus back toward developing specialized robots to automate core segments of its warehouses.

Building on this momentum toward fully automated warehouses, Amazon introduced its next generation of fulfillment systems powered by a staggering 9 specialized robots. This next-gen warehouse system is referred to as Sequoia.

Launch of Sequoia: Robot-Native Warehouses (2023)

In 2023, Amazon introduced Sequoia, a redesigned fulfillment system with the sole aim of reducing order fulfillment time by integrating the end-to-end flow of goods within its warehouses, from the inventory storage area all the way to the delivery docks, using a coordinated set of nine specialized robots. We are talking about warehouses that span nearly one million square feet. In these facilities, goods often have to traverse diagonally opposite ends of these giant warehouses.

Sequoia is Amazon's answer to coordinating this flow of goods across these nine robots from one diagonal end of the warehouse to the other.[41] In doing so, Sequoia represented a shift away from optimizing individual robots toward rethinking the fulfillment system as a whole. That raises an important question: Why did Amazon build Sequoia? Why not keep building more specialized robots?

The answer here is scale. By this point, Amazon had deployed hundreds of thousands of robots across its network. Marginal improvements to individual robots were no longer sufficient for Amazon to meet customer demands around order fulfillment speed and cost. The next bottleneck was no longer automating individual segments of the warehouse, such as picking, sorting, or packaging, but rather enabling end-to-end coordination across all segments of the warehouse. This is where Sequoia came in.

Sequoia represented a shift towards redesigning the warehouse as a single, integrated system. Storage locations were optimized for robotic access, robot routes were dynamically planned based on real-time traffic patterns, and worker workflows were tuned to fulfill demand while minimizing congestion and downstream constraints. The objective was clear: minimize handoffs and maximize throughput—more orders processed per square foot, per hour, with greater predictability.

Did Sequoia deliver on this promise? It did.

Deployment of Sequoia resulted in 75% faster processing speeds for sorting and storing incoming inventory and up to a 25% reduction in order processing time for outgoing packages. In simple terms, Sequoia helps process and store incoming inventory items in the warehouse much faster than its predecessors and brings them to workers in the safest, shortest way possible. Thus, Sequoia is Amazon's system-level warehouse fulfillment architecture that integrates robots, software, and workflows to maximize throughput across each of its warehouses.

In essence, Sequoia marked a shift in Amazon's strategy. The company transitioned from building robots for warehouses to designing warehouses around robots. Naturally, with thousands of robots operating within each warehouse, traffic management became the next major constraint.

How do you keep 2,800+ robots navigating a 1 million sq. ft. warehouse floor from bumping into each other? Add to that variables like employee movements, malfunctioned robots, Wi-Fi outages, power failures, battery life, and more. Now add to that equation 200+ warehouses—each with its own layout, floor plan, and unique traffic patterns. Looking at this data, it became clear to the team at Amazon: a static robot-routing solution that was in use at the time wouldn't work in the long run because of the sheer number of variables involved in this equation. So how did they solve this problem?

That is where DeepFleet came in.

DeepFleet AI Models to Manage Robot-Led Traffic Jam (2025)

DeepFleet is Amazon's traffic router for its one million-plus robots operating across more than 200+ global warehouses.[42] The goal is to coordinate traffic among roughly 2,800+ robots operating inside each of these cosmic-grade fulfillment centers. But what exactly is DeepFleet? It is an AI model—or more precisely, a suite of AI models—designed to predict and prevent robot-led congestion in real time across Amazon's warehouse network. That raises an important question: how did Amazon build these models?

Instead of hand-coding routes for each robot or relying on compute-heavy simulations, the team building Amazon's robots developed this suite of AI models by training artificial neural networks on billions of hours of navigation data captured from nearly one million robots operating across its portfolio of 200+ warehouses worldwide. The goal was simple: predict congestion before it happens and reroute robots in real time.

Once these models were developed, Amazon evaluated them across its warehouse network. The company found that certain models were better suited to predicting the next move of individual robots rather than the behavior of the fleet as a whole, while others were more effective at holistically analyzing fleet-wide dynamics and forecasting congestion patterns. From a resource-efficiency perspective, the trade-offs were equally clear: some models were computationally intensive, while others delivered comparable insights at significantly lower cost.

After evaluating the trade-offs for each, Amazon decided to deploy the strongest-performing models across its warehouse network and relegated the remaining ones to research-only mode. While DeepFleet helped Amazon solve the warehouse traffic management problem for its one million robots, rising wages and persistent shortages of human workers once again became a major concern for the company on its path to automating last-mile deliveries.

This concern was clearly visible in its revenue profile for that quarter. Amazon had to make decisive moves to automate the last-mile delivery segment. While the original trials with the Scout robot had not resulted in the outcomes Amazon desired, the team approached humanoids as a potential solution this time. But this raised another challenge: how do you train a humanoid robot across millions of neighborhoods to enable last-mile deliveries?

Where do you get locomotion, navigation, and perception data for all of these neighborhoods? One might ask why we even need this data in the first place. To train a humanoid, Amazon needed to understand how delivery workers approach homes across different neighbourhoods, including how they walk, carry packages, move their limbs, and maintain balance in these diverse environments. This data is essential to teach a humanoid robot to perform the same tasks with ultra-high precision and dexterity. So how did Amazon solve this problem?

That's where the smart glasses came in.

Smart Glasses and Humanoid Delivery Systems (2025)

In 2025, Amazon equipped its delivery teams with smart glasses.[43] The goal was to capture real-world locomotion and navigation data from human workers by having delivery drivers wear these devices during their daily operations. The glasses captured perception data, giving Amazon a detailed view of each neighborhood and its physical constraints, as well as locomotion data, such as walking style, balance retention, and gait selection, and navigation data, including how drivers route themselves from the truck to the front door. Together, these three data streams could then be used to model real-world deliveries accurately and train humanoid robots to replicate them with extreme precision and reliability.

This raises an obvious next question: who is supplying the humanoid robots for Amazon?

An old friend: Digit is being trained to automate last-mile deliveries, and trials are underway at Amazon's San Francisco campus.

With this robot-led revolution in the e-commerce and logistics business, Amazon has developed a fortress of cutting-edge technologies needed to create an operational moat of a caliber that most competitors will find increasingly difficult to replicate.

So, what are the lessons next-generation robotics founders and engineers can learn from this journey?

Lessons for Robotics Founders & Engineers

Lesson 01: Use Acquisitions to Eliminate Technical Bottlenecks, Not to Rent Users

Amazon acquired Kiva to bring warehouse automation technology in-house, not to rent a user base. Similarly, it acquired Canvas Technology to add a strong set of eyes to its robots and unblock the perception bottleneck. In both cases, Amazon leveraged acquisitions to address a critical technical bottleneck; it did not attempt to rent a user base. A rented user base will not last long.

Trust cannot be rented; it can only be earned. Therefore, acquisitions should always be approached with this objective in mind.

Lesson 02: ROI Is the Ultimate Filter—No ROI, No Product

Projects such as Scout and Astro were shut down immediately once their unit economics failed to prove viable in the real world. You always want to focus on your A+ products—products that address real problems and that customers are willing to pay for. Projects that do not have a clear ROI are essentially a death trap. Scout and Astro fell into this category. The lesson is clear: decisively shut down projects that do not generate revenue and redirect your energy and resources towards those that do.

Lesson 03: Specialized Robots Outperform General-Purpose Robots

Proteus, Sparrow, and Cardinal solved A+ problems and delivered measurable throughput gains across Amazon's warehouse network. They addressed real operational bottlenecks, and the economic impact was clearly reflected in the company's operating expenditure each quarter. Amazon started seeing up to a 75% improvement in throughput for inventory sorting and storage workflows, and approximately 25 per cent in outbound package workflows. Multiplied across 1,200 sites, these gains translated into significant economic value for the company. Amazon was now able to process the 13 million orders it receives on average each day at twice the speed and at one-third of the cost. The lesson here is clear: build robots that deliver real value, robots that solve your most pressing A+ problems and generate returns at scale. Today, specialised robots consistently outperform general-purpose systems.

Lesson 04: Continuously Evaluate and Address Your A+ Bottlenecks

From addressing traffic congestion bottlenecks in its warehouses using DeepFleet AI models to acquiring Canvas and Kiva to solve perception and automation challenges, Amazon systematically tackled each constraint and remained true to its mission of ramping up warehouse throughput through robotics. In each phase of its journey since the acquisition of Kiva, it remained laser-focused on its A+ bottlenecks and addressed them sequentially. In summary, the lesson is clear: as a founder, you should identify your A+ bottlenecks and focus relentlessly on resolving them before attempting to address other challenges.

Lesson 05: Consumer Robotics Is a Trust Game; Industrial Robotics Is an Economics Game

Choose your battles wisely. Consumer robotics can take decades to succeed, even when the problem being solved is genuinely painful for users. The reason is simple: consumer robotics is ultimately about changing human habits. You are asking people to trust a robot to perform tasks they are accustomed to doing themselves. Habit change is extremely difficult for humans, far more so than for machines. As a result, adoption is slow, trust takes time to build, and scale is hard to achieve. Capital and brand recognition help, but they are not sufficient, as demonstrated by the trajectory of Astro and iRobot. In consumer robotics, the hardest problem is not building the robot; it is changing human behaviour.

Industrial robotics, by contrast, is fundamentally an economics problem. If a robot saves a company a dollar every minute, it creates immediate value, regardless of branding or emotional appeal. In simple terms, in industrial environments you are replacing machines with robots, whereas in homes you are replacing habits with robots. As a result, it is relatively easier to build robots for industrial environments than for homes. You can easily quantify and justify the return on investment in industrial robotics. Clear return on investment drives adoption. Choose your battles wisely.

Now that we have distilled the core lessons embedded in Amazon Robotics' trajectory, from its origins through its evolution and the pivotal decisions the team made along the way, it is time to decode the underlying business model.

In the final section of this chapter, we will break down its business model and illustrate how the product portfolio maps to select primary use cases.

Amazon Robotics: Product Snapshot I

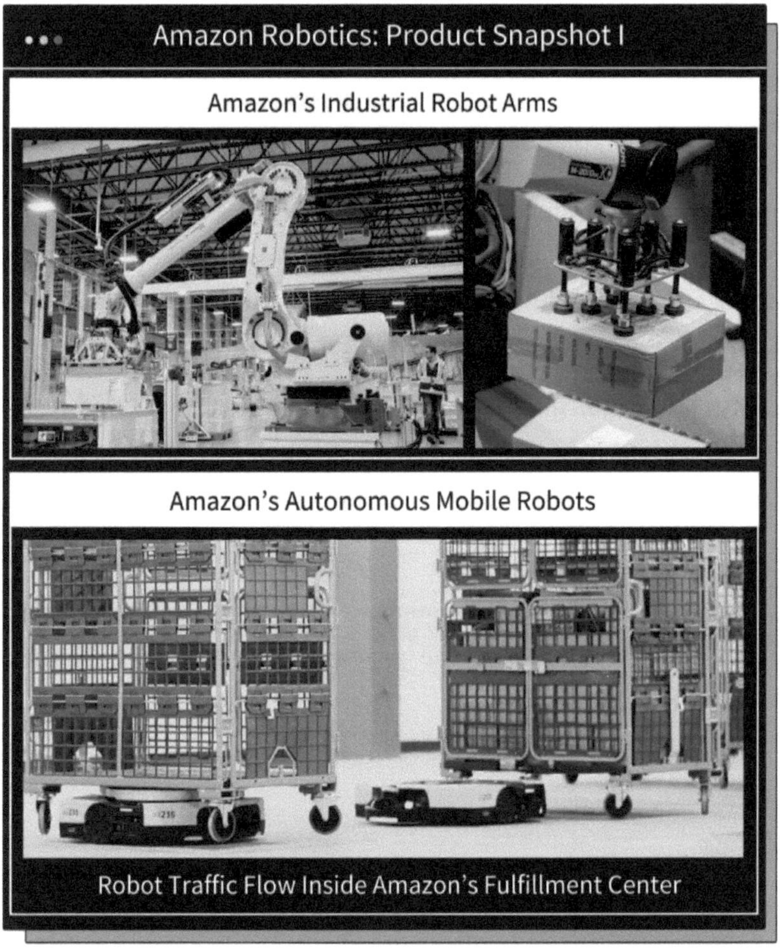

Figure 3.1 — Amazon's Industrial Robot Arms & Mobile Robots

Illustration by the author. Robot images from Amazon News article (2025).

Amazon Robotics: Product Snapshot II

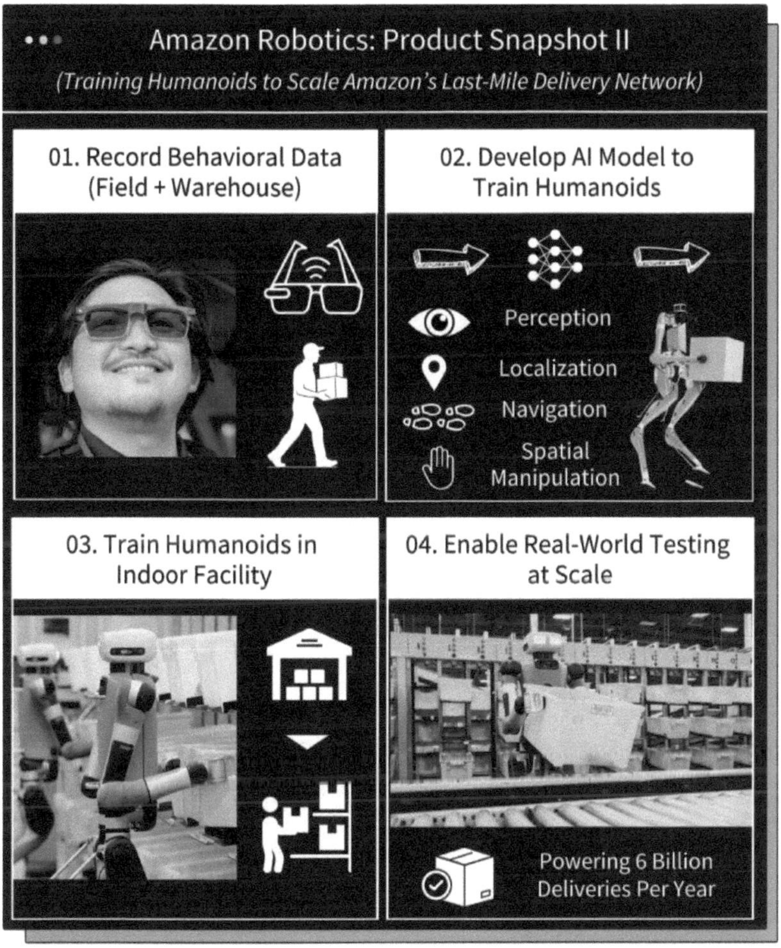

Figure 3.2 — Training Humanoids to Scale Amazon's Last-Mile Delivery Network

Illustration by the author. Digit robot image from About Amazon (2023).

Amazon Robotics' Business Model — Part 1

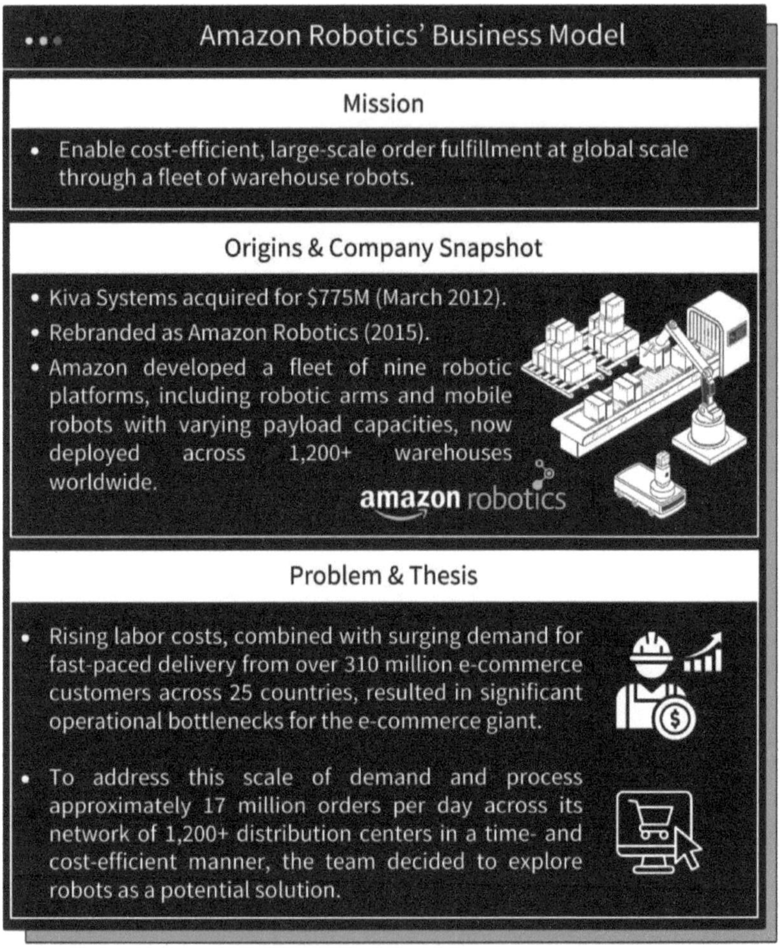

Figure 3.3 — Amazon Robotics' Business Model: Part 1

Illustration by the author.

Amazon Robotics' Business Model — Part 2

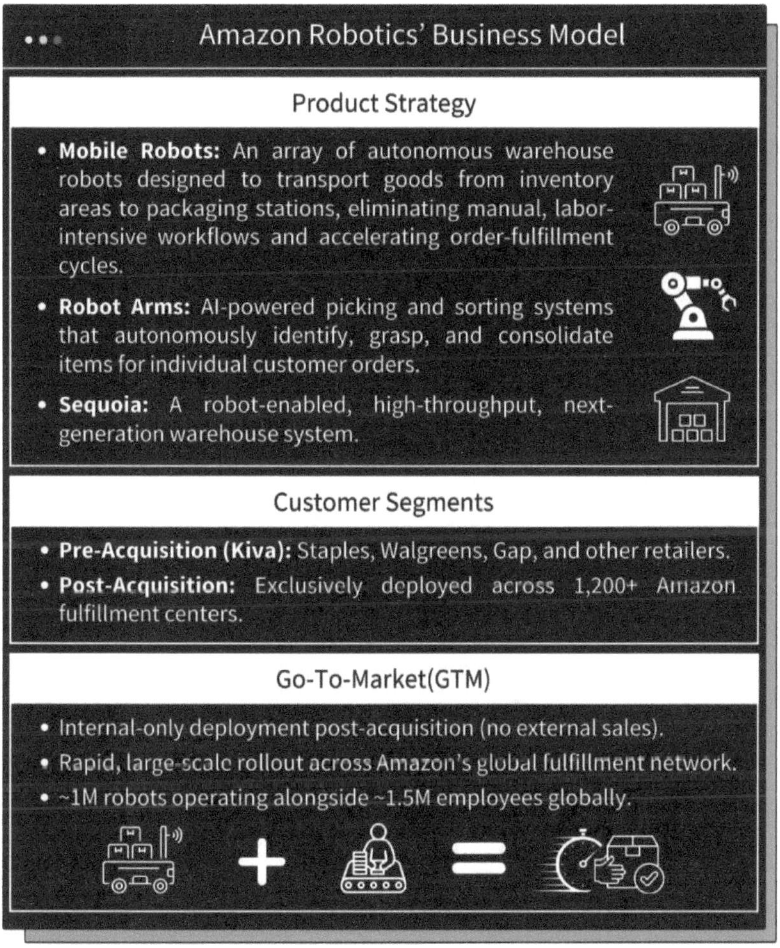

Figure 3.4 — Amazon Robotics' Business Model: Part 2

Illustration by the author.

Amazon Robotics' Business Model — Part 3

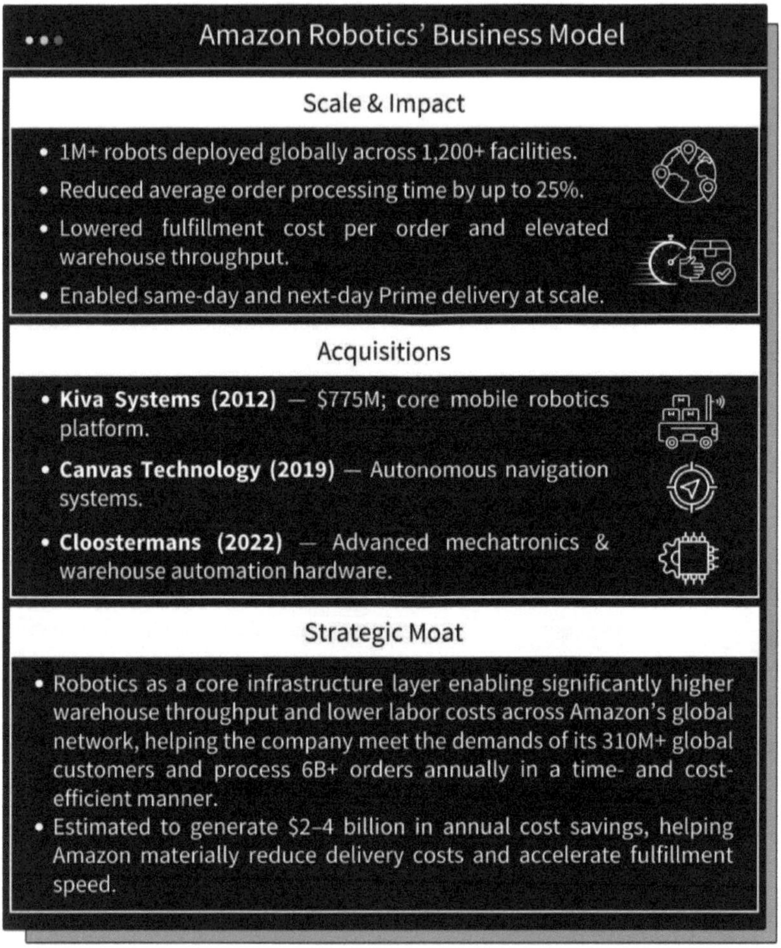

Figure 3.5 — Amazon Robotics' Business Model: Part 3

Illustration by the author.

Chapter 04: Kiva Systems — Webvan Roots to Amazon Exit

How do you go from shipping your first production-grade robot within just 20 months after incubation to landing more than 20 high-profile enterprise customers worldwide within the next few years, selling over 10,000 robots, becoming revenue-positive, and ultimately being acquired by Amazon for $775 million? All of this in less than a decade, in the robotics industry. That is equivalent to a one-year trajectory from launch to a successful exit in the software world.

This is the story of Kiva Systems.

Kiva is one of the rare gems in the robotics industry—an outlier whose trajectory has become a goldmine of lessons for next-generation robotics founders and engineers. In an industry where driving enterprise adoption is notoriously difficult, the speed and scale at which customers around the world embraced Kiva's warehouse robots was extraordinary.[44-45] After all, how do you land Staples, Walgreens, Zappos, Crate & Barrel, Office Depot, Saks Direct, The HON Company, Toys "R" Us, Drugstore.com, Timberland, Dansko, Von Maur, and Boston Scientific as core enterprise customers—and become revenue-positive in robotics within just ten years?

This trajectory reflects the calibre of the company, its team, and the surgical decision-making that went into building it from the outset. In this chapter, we will decode the journey of Kiva and extract some of the core lessons that you, as next-generation robotics founders, researchers, and engineers, can learn from it. To better understand Kiva and its journey, we must first understand the core problem that inspired its founders to launch the company in the first place.

So, who started Kiva, and what problem were they trying to solve?

To answer that question, we need to travel back to the year 1996.

Webvan: The Birthplace of Kiva (1996–2001)

The 1990s were infamous for the dot-com bubble. A lot of companies started during this era with the launch of the World Wide Web, which emerged as a commercial product out of its precursor, DARPA Net—i.e., the private version of the internet. With the World Wide Web being launched, websites became the go-to platform for startups, much like the AI models of today. During this time, a gentleman by the name of Louis Borders came up with the idea of building an online grocery delivery system, which he named Webvan. While most of you are now familiar with Instacart, DoorDash, Uber Eats, and other online delivery platforms, Louis was way ahead of the market. At the time, Webvan was the only startup working on this problem

With funding from venture capitalists and institutional investors, including Sequoia, SoftBank, Goldman Sachs, and Yahoo, Louis and his team raised about $800 million and started building state-of-the-art warehouses equipped with high-end conveyor-based systems to automate sorting and order-processing workflows for hundreds of grocery items at each of these facilities, enabling online deliveries. Each warehouse came at a cost of roughly $35 million.[36-38]

From San Francisco to Los Angeles, Chicago, and Seattle, the company launched a series of these tech-first warehouses to accelerate the order-fulfilment process. To enable last-mile deliveries at scale, they also acquired a fleet of delivery vans, similar to Amazon's. The mantra at the time was simple: build at scale, and customers will come.

Unfortunately, that was the same mantra that killed most dot-com-era companies, and Webvan was no exception. While the company managed to raise over $800 million within a span of just a few years to build infrastructure both inside the warehouse and outside to automate its grocery delivery system, the truth is that much of the operational workflows within its warehouses were still manual—driven by workers carrying pods of groceries from one end of the warehouse to another.

In addition, the company had deployed legacy, capital-intensive heavy machinery, i.e., static conveyor systems, to enable sorting workflows. This combination of traditional high-end conveyors used to sort items and workers manually transporting those sorted goods across the warehouse made the system operationally heavy and capital-inefficient. The same constraint was reflected in its outdoor delivery network.

With extremely high capex required to build warehouses, coupled with heavy opex to operate them and limited customer traction, Webvan became another casualty of the dot-com era. The company officially filed for bankruptcy and shut down operations in 2001, just a few years after its launch.

The Webvan Postmortem (2001)

While the company shut down, another gentleman named Mick Mountz, who was working at Webvan at the time, walked away reflecting deeply on this entire saga. Based on these reflections, Mick—an MIT engineering graduate and Harvard MBA—realised that while Webvan failed due to the extremely high operational cost of building and managing warehouses loaded with workers and heavy machinery, robots could have saved the company.

That realisation became the emerging thesis for the launch of Kiva: robots were the missing piece in eliminating Webvan's warehouse-level inefficiencies.

Distrobot: The Founding Years (2002–2003)

With the shutdown of Webvan in 2001, Mick contacted Professor Peter Wurman, another MIT engineer, with this thesis and a mission to automate warehouse distribution systems using robots. That raised an important question: what was Mick specifically trying to automate in warehouses, and why?

Mick had learned that the warehouse distribution system needed to be redesigned so that goods were transported to workers, not the other way around. Workers should be stationary, and goods should be autonomously transported to them using robots. In other words, this new model was strategically positioned around a "goods-to-workers, not workers-to-goods" philosophy. The underlying principle was efficiency: to reduce the time workers spent ferrying goods across the warehouse floor by transporting those goods autonomously to them using robots, thereby scaling sorting workflows and overall throughput while slashing operational costs for warehouse owners.

Convinced by Mick's thesis and its ability to solve the underlying problem, Peter and another professor, Raffaello D'Andrea, a professor of control systems from ETH Zurich, joined the founding team, and the trio launched what was originally named Distrobot in 2003.[45]

That raises another question: how did they secure the seed capital needed to launch Distrobot?

The company was initially funded by a private group of angel investors, who contributed about $1.6 million, followed by a seed round led by Bain Capital of around $5 million in 2004, and subsequent funding of approximately $15 million over the next few years.

Within just 20 months of its incubation, in 2005, the company renamed itself Kiva Systems and launched its first official production-grade robotic system, called the Kiva Mobile Fulfillment System (Kiva MFS).[49] In simple terms, this system can be described as a fleet of low-cost mobile robots designed to autonomously transport pods of goods to worker stations in warehouses. Think of Roomba vacuum-cleaning robots equipped with industrial-grade capacity to lift heavy goods—up to 3,000 pounds per robot.

This system became commercially available to enterprise customers in December 2005, offering both the robots and the underlying software required to orchestrate the fleet and automate warehouse operations.

That raises another critical question: how did they go about communicating this problem, and the value their solution would offer to warehouse customers? How did they even articulate the problem and their offering well enough to get in the books of customers such as Staples, Crate & Barrel, Gap, Walgreens, Office Depot, and others?

Kiva Mobile Fulfillment System: The Goods-to-Person Model (2005)

Mick and the team knew that warehouse operations were inherently capital-intensive, and one of the biggest contributors to this inefficiency was the way work was organized on the warehouse floor. At the time, warehouse workers themselves were responsible for transporting goods from one end of these massive facilities to the other. This labour-intensive setup meant workers spent nearly 70% of their time walking across warehouses rather than actually fulfilling orders. He arrived at this insight through his firsthand experience at Webvan and set out to solve this problem for customers operating large-scale warehouse facilities.

Kiva's Mobile Fulfillment System (MFS) was the answer to that problem. Mick's solution consisted of a fleet of robots designed to transport pods of goods directly to workers, orchestrated through a centralised software system. The goal was to keep workers largely stationary while mobile robots transferred goods from inventory areas to worker stations. As a result, workers spent less time walking across warehouses and more time fulfilling orders.

To make the model easier to communicate to customers, the company labelled it the "goods-to-worker" model. That was in the initial stages. In later stages, as luck would have it, since nearly every company that adopted Kiva's robots was eventually acquired by Amazon, Kiva updated its messaging to quip, "Buy our robots and get bought by Amazon." This was the reality that unfolded in the second half of the 2000s.[44]

It was this goods-to-worker model that created a true win–win scenario for companies, workers, and warehouse operators alike. More importantly, Mick had introduced a scalable model that made warehouse operations near-autonomous for enterprise organisations across verticals. Launched in 2005, the system quickly caught the attention of office-supply giant Staples.

First Pilot: Staples, Coppell, Texas, U.S. (2005)

In 2005, Staples built a pilot fulfillment center in Coppell, Texas, to test Kiva's system before committing to a full-scale rollout. The pilot was followed by the first full-scale deployment at Staples' 650,000-sq-ft distribution center in Chambersburg, Pennsylvania—marking a critical validation point for Kiva's goods-to-person model.[50] This rapid transition from pilot to full-scale deployment within just a matter of months with Staples raises an important question: how did a young robotics company manage to land a customer like Staples in the first place?

The answer boils down to the value it offered to customers. Kiva's messaging was never centered on robots as a product. Instead, it focused on outcomes—warehouse throughput and labor efficiency. Customers weren't buying "robots"; they were adopting a model designed to directly triple warehouse throughput. Kiva's marketing assets framed this payoff in the language warehouse operators cared about most: "triple the number of orders fulfilled per hour per person." By speaking directly to its customers' most acute pain points, Kiva positioned itself not as a robotics vendor, but as a productivity engine for warehouse operators. This messaging resonated with enterprise giants like Staples and soon became the organic secret sauce for translating pilots into production deployments at remarkable speed.

First Deployment: Staples, Chambersburg, Pennsylvania, U.S. (2006)

In the summer of 2006, Staples became the first customer to go live with Kiva at its 650,000-sq-ft distribution center in Chambersburg, Pennsylvania, starting cautiously with a 30-robot deployment inside a 50,000-sq-ft dedicated testing area carved out within the facility. The objective behind this deployment was pragmatic: to prove that the "goods-to-person" model could triple warehouse throughput without redesigning these giant facilities around robots.[45]

This mattered. If a robotic system demanded a ground-up rebuild of warehouses, adoption would remain extremely limited. No enterprise giant wants to redesign 100+ warehouses and invest billions of dollars simply to add robots. That kind of capital drain had already been witnessed in Webvan's case in the late 1990s. Kiva's model ensured that integrating 1,000+ robots under a single roof was possible within a week and did not require major changes to the warehouse itself.

The results spoke for themselves. Staples gradually expanded the deployment to approximately 500 robots spanning 140,000 sq ft of warehouse floor space. These robots went on to handle nearly 50% of the site's total throughput. Operator productivity reportedly reached 600–700 items per hour per worker, compared with roughly 200–400 items per hour in traditional pick-to-belt operations.[45] In simple terms, Kiva delivered on its promise: throughput inside Staples' warehouse effectively tripled. Workers were now processing three times more items per hour than under the conventional system. With these gains clear, Staples made a decisive move to scale Kiva's robots across its facilities. That raises an important question: where did it go next?

The company headed west—to Denver, Colorado.

Second Deployment: Staples — Denver, Colorado, U.S. (2007)

By mid-2007, Staples was no longer merely purchasing warehouse robots from Kiva in bulk. It had started designing its next generation of distribution centers around these robots. These robots had become the lifeline of Staples' warehouse operations and led to the launch of its first organically designed, robot-led 300,300-sq-ft distribution center in Denver, Colorado. The site was engineered from the ground up around Kiva's system to support Quill, Staples' dedicated B2B wholesale arm serving large enterprise customers.

Using about 150 robots across a 100,000-sq-ft operational footprint, the facility was shipping roughly 5,000 orders per day. Staples reported productivity levels nearly double those of comparable conveyor-based facilities, with "fulfillment quality" estimated to be about 30% higher than in conventional Staples warehouses.[51]

With this accelerated scale of adoption at Staples, the team at Kiva decided to expand its footprint beyond the orbit of the office-supplies giant. The question now was clear: where should they go next?

The answer was a pharmacy giant—Walgreens.

Walgreens: Mount Vernon Distribution Center, IL, U.S. (2007)

In November 2007, Walgreens invested in and deployed Kiva's robots at its 448,000-square-foot distribution center in Mount Vernon, Illinois, serving approximately 300 stores across the Midwest. Over the next few years, this deployment scaled rapidly—eventually reaching close to 1,000 robots operating under a single roof at each Walgreens facility.[52]

Walgreens quickly recognised the value Kiva's robots delivered: mission-critical speed and precision within its order-fulfilment operations. More importantly, the system once again effectively tripled warehouse throughput. At the time, this marked the largest single-warehouse robotic deployment anywhere in the world, naturally propelling Kiva to expand its footprint beyond the pharmacy giant.

The next enterprise to place its bet on Kiva? Zappos.

Zappos: Shepherdsville Distribution Center, KY, U.S. (2008)

Zappos started with a simple mission: make it easy for customers to order shoes online and have them delivered to their doorstep as quickly as possible. In many ways, the idea echoed Webvan—except this time, the product was shoes, not groceries. This naturally implied the company faced the same fundamental problem that had plagued Webvan. Zappos needed to transport goods through its warehouses at a galactic pace, and Kiva's robots offered a clear solution to that challenge.

By this point, Kiva's systems had already been validated through deployments at Staples and Walgreens. With that proof in hand, Zappos adopted Kiva's robots to scale its warehouse operations and keep pace with rapidly growing demand.[45] Needless to say, Kiva once again did its magic at Zappos and was able to deliver shoes to its customers at the same pace that groceries are delivered today by Instacart and DoorDash. While Zappos benefited from adopting Kiva's robots, it was Amazon that ultimately benefited even more. How?

Make no mistake—when Zappos was acquired by Amazon in 2010 for $1.2 billion, Amazon also inherited Zappos' robot-laden warehouses.[53-54] The acquisition gave Amazon a firsthand look at Kiva's impact inside a real, high-throughput fulfillment environment. These early-stage customer deployments became fertile ground that ultimately led to Kiva's acquisition by Amazon in 2012. But the story of Kiva doesn't end at Zappos.

With office supplies, medicines, and shoe orders now being processed by mobile robots, the company naturally started exploring new verticals. The question was obvious: where should the team go next?

Baby products turned out to be the answer. Why? Frequency. Customers buy diapers almost every month, worldwide. That meant order density was significantly higher—often many times greater than in the office-supply and pharmacy verticals. So how did Kiva enter the baby-products space?

It shipped diapers by putting robots inside Diapers. Literally.

Diapers.com: Multi-State U.S. Deployment (2008)

In October 2008, Kiva secured another high-velocity e-commerce validation when Diapers.com (later Quidsi) announced it was implementing the Kiva Systems Fulfillment System across all three of its distribution centers. These included Gouldsboro, Pennsylvania—a 1.25-million-sq-ft facility with roughly 260 robots operating within a 200,000-sq-ft zone—along with sites in Kansas City, Missouri (102,000 sq ft) and Tahoe-Reno, Nevada (566,820 sq ft).[52]

In total, the company had purchased more than 350 robots across these facilities, supporting a catalog of more than 10,000 baby products and serving over 200,000 customers across the nation. This deployment was a clear signal that Kiva was no longer a company selling robots to warehouse operators, but one providing the cutting-edge operating infrastructure needed to scale warehouse operations for fast-growing e-commerce businesses and retailers. It also made clear that Kiva's system was no longer confined to a single category, but ready to scale across entirely new industries.

Soon after, the team began expanding its footprint across apparel, healthcare, and retail.

Multi-Vertical Expansion: Retail, Apparel, Healthcare (2010)

Gap · Crate & Barrel · Von Maur · Boston Scientific

In March 2010, Gap Inc. went live with Kiva at its 1.2-million-square-foot fulfillment center in Groveport, Ohio, supporting e-commerce operations across its portfolio of brands—Gap, Old Navy, Banana Republic, Piperlime, and Athleta.[55] The deployment positioned Kiva as a credible alternative to traditional cross-belt sorting systems that dominated large-scale apparel distribution centers at the time.

Luxury and specialty retail followed. In January 2010, Saks Fifth Avenue, a high-end fashion retailer, kickstarted a multi-year deployment to implement Kiva at its 471,000-sq-ft facility in Aberdeen, Maryland. Later that year, Von Maur, a specialty department store, deployed 15 high-capacity Kiva robots and 937 custom-designed inventory pods, handling more than 1,000 orders per day at its 46,000-sq-ft e-commerce fulfillment center in Davenport, Iowa, effectively doubling its throughput in a matter of weeks.[52]

Large enterprise operators also validated the system at scale. Office Depot opened a 600,000-square-foot distribution center in Pennsylvania, equipped with one of Kiva's largest installations at the time—430 robots and 3,500 inventory pods capable of autonomously handling roughly 12,000 SKUs, or unique products. The company officially replaced its traditional conveyor-based system with mobile robots. This momentum continued. Crate & Barrel deployed 50 robots at its Tracy, California campus, dedicating a 400,000-square-foot facility to Kiva's robot-led model and shipping approximately 2,000 packages per day, effectively doubling its operational efficiency.[52]

Perhaps most telling was Kiva's expansion beyond retail altogether. In January 2010, Boston Scientific purchased Kiva Systems to automate its 800,000-square-foot global central distribution center in Massachusetts, as well as a 269,000-square-foot international hub in Kerkrade, the Netherlands.[52] Here, operational speed was not the primary driver—reliability was. The decision was driven by the need for absolute reliability, traceability, and rigorous inventory control. Failure in any of these areas was not an option for a company producing medical equipment designed to support the lives of millions of patients around the world.

By the end of 2010, Kiva was serving customers across industry verticals—footwear, apparel, office supplies, pharmacy, medical equipment, and baby products. Momentum was undeniable, and that naturally raised the question: where should the company go next?

Multi-Vertical Expansion Phase II (2011): Office Depot, Saks Direct, The HON Company, Toys "R" Us, Drugstore.com, Timberland

By this point, the team didn't have to go anywhere—customers were knocking at the door. Kiva was now in the driver's seat, advancing its mission of helping enterprise customers triple the number of orders they could fulfill per hour.

The first to knock was healthcare-focused e-commerce player Drugstore.com. The company adopted Kiva's robots at its 270,000-square-foot distribution center in Swedesboro, New Jersey, initially targeting roughly 10,000 of its 50,000 products to enable same-day shipping across its portfolio of health, beauty, and general merchandise.

Traditional retailers followed suit. In June 2011, Dillard's announced plans to implement Kiva in a new 850,000-square-foot e-commerce fulfillment center in Maumelle, Arkansas. That same month, Follett Higher Education Group revealed it would consolidate operations into a single 551,200-square-foot Kiva-enabled distribution center in Aurora, Illinois.

Kiva's appeal was not limited to large incumbents. Acumen Brands, an e-commerce apparel startup, deployed a Kiva system in just 14 weeks at its 60,000-square-foot facility in Fayetteville, Arkansas. The company reported that employee productivity more than tripled, while returns and cancellations dropped by nearly 50 percent.[52]

The toy industry provided another proof point. In September 2011, Toys "R" Us opened a 300,000-square-foot e-commerce distribution center in McCarran, Nevada. During its first holiday season, the facility shipped close to 100,000 units per day, with a significant portion of that volume handled by Kiva's robot fleet.[52]

This cosmic-grade adoption of Kiva's robots in North America naturally meant it was time to cross the Atlantic Ocean, quite literally. The next stop was Europe. In September 2011, Timberland announced plans to automate a newly constructed 430,500-square-foot distribution center in Almelo, the Netherlands, using the Kiva robots to fulfill consumer orders.[52]

By the end of 2011, Kiva's trajectory was clear. The system had proven itself across a wide range of industries, companies, and geographies. It had become a core operational fabric for startups and incumbents alike—serving both domestic and international enterprises facing seasonal demand spikes as well as perennial, year-round pressure to fulfill millions of customer orders.

Kiva's robots had, in many ways, become the Ferrari of the order-fulfillment racetrack. And just as Enzo Ferrari would leave no stone unturned to put the fastest possible car on the track, Jeff Bezos would leave no stone unturned to deliver customer orders at Formula One–level speed. Speed was not a feature. It was the race itself.

By then, Amazon—on its own path to becoming the Ferrari of the e-commerce world—had taken careful note of Kiva. It was already one of Kiva's largest indirect customers, operating dozens of robot-enabled fulfillment centers. Amazon had effectively acquired nearly half of Kiva's customer base, and with it, a significant portion of Kiva-powered warehouses. Now, it was time to do more than follow the race. It was time to lead it.

That meant acquisition.

Acquisition by Amazon (2012)

On March 19, 2012, Amazon announced it had reached an agreement to acquire Kiva Systems for approximately $775 million in cash.[1] The rationale was clear: bring a proven "goods-to-person" automation system in-house, deploy it across Amazon's rapidly expanding fulfillment network, and ship those 7 billion+ orders each year with twice the speed and half the cost.

This was not a speculative bet by Amazon. It was the absorption of the most powerful F1-grade warehouse operations engine on Earth—one it desperately needed to win the e-commerce race and one that had already been battle-tested across industry verticals: retail, apparel, healthcare, and e-commerce. For Kiva, this deal marked the finish line of a decade-long race that had started in the foothills of the dot-com era and ended as one of the most consequential exits in the history of robotics to date.

So, what are the lessons next-generation robotics founders and engineers can learn from this journey?

Lessons for Robotics Founders & Engineers

Lesson 01: Solve One A+ Problem That Truly Matters

In the case of Kiva, the company did not attempt to make the warehouse fully autonomous, i.e., it was not trying to automate every workflow in the facility with robots. A "lights-out" model was not its mission. That would have been an extremely difficult problem to solve because robots at the time were only capable of automating specific workflows and were not capable of replacing a trained human workforce in these facilities.

Instead, Kiva focused on one specific part of the problem: ensuring workers did not have to move around the warehouse transporting carts loaded with hundreds of goods. That was a strenuous workflow for workers and capital-intensive for operators, since it added up to nearly 70% in operating expenses per warehouse. That was the A+ problem Kiva was trying to address. The goal was to keep workers stationary while robots transported goods around the warehouse.

The lesson here is simple. In robotics, you must clearly define—and be surgical about—the problem you are solving. If you go too broad, you will burn a significant amount of capital just building the product itself, let alone driving adoption. That brings us to Lesson 02.

Lesson 02: History Is the Best Teacher.

In the case of Kiva, the founders had learned from history that there was little value in building custom robots to enable end-to-end automation of warehouses focused strictly on online grocery delivery or other niche problems. Not only was there limited value, but it was also an extremely hard problem to solve—automating the entire warehouse end to end at that time. However, there was immense value in automating specific warehouse workflows using robots in a way that was industry-agnostic.

Reflecting deeply on history helps you understand what the core problems really are, how others have attempted to solve them, and where those approaches failed. This perspective alone can save years of effort and help you launch production-grade robots with lean resources that deliver real value to your customers. So how do you sell these robots?

That brings us to the next lesson.

Lesson 03: Lead With Outcomes, Not Robots

Kiva never sold robots. It sold outcomes. The message was simple and quantified: triple warehouse throughput per worker. At scale, that outcome mattered deeply to customers.

In robotics, outcomes close deals. The lesson here is to define and quantify the results you are guaranteeing in clear, measurable terms—and commit to them. To deliver consistent outcomes, you need high-quality products. That brings us to our next lesson.

Lesson 04: One Great Product Beats Many Customized Ones

Kiva stayed true to its mission by continuously refining a single product: a high-powered robot designed to transport goods autonomously across warehouses. It did this for more than 12 years without fragmenting its focus. It didn't try to build custom warehouse robots for each vertical.

That discipline helped Kiva secure high-profile customers across multiple industry verticals. In summary, focus compounds; customisation efforts do not. That brings us to Lesson 05.

Lesson 05: Know When to Exit—and When to Endure

While a $775 million acquisition looks impressive on paper, make no mistake: Kiva could have been a trillion-dollar company had it remained independent. An acquisition within 12 years may seem like an excellent outcome for many robotics founders, and in many cases, it absolutely is. But if you have a product that consistently generates large-scale positive returns year after year, it is worth asking a harder question: why exit?

Why not double down? Why not scale your impact? Why not help more companies automate their workflows and become that F1 car that wins the race?

This lesson is partly philosophical and will vary based on personal circumstances. But when you have a product that truly solves customer pain points at scale and you are clearly winning the market, the greater opportunity may lie in scaling your footprint rather than exiting early.

Now that we have distilled the core lessons embedded in Kiva Systems' trajectory, from its origins through its evolution and the pivotal decisions the team made along the way, it is time to decode the underlying business model. In the final section of this chapter, we will break down its business model and illustrate how the product portfolio maps to select primary use cases.

Kiva Systems: Product Snapshot

Figure 4.1 — Kiva's Autonomous Mobile Warehouse Robots

Illustration by the author. Robot images from IEEE Spectrum article (2008).

Kiva Systems' Business Model — Part 1

Figure 4.2 — Kiva Systems' Business Model: Part 1

Illustration by the author.

Kiva Systems' Business Model — Part 2

Figure 4.3 — Kiva Systems' Business Model: Part 2

Illustration by the author.

Kiva Systems' Business Model — Part 3

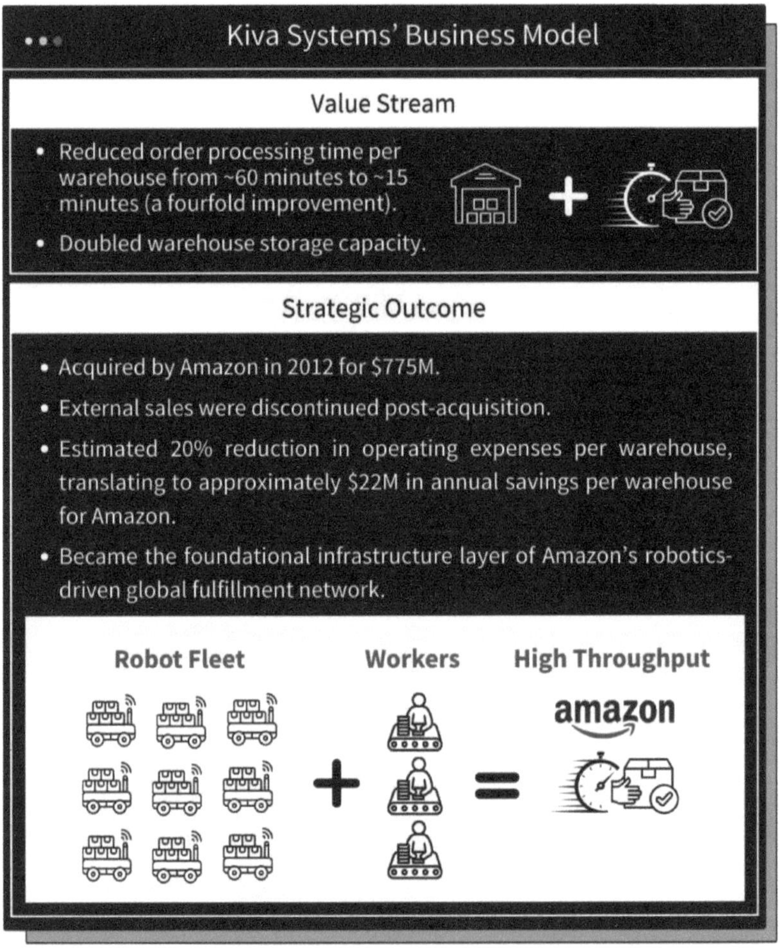

Figure 4.4 — Kiva Systems' Business Model: Part 3

Illustration by the author.

Chapter 05: Fetch Robotics — Willow Garage Lab to Zebra Exit

Seven years. Two production-grade robots designed exclusively to automate warehouse workflow. One cloud-based software platform to manage them. One strategic partnership with Honeywell. And half a dozen core customers operating across warehouses and logistics hubs in North America and the country known for its windmills and Gouda cheese—the Netherlands.

Seven years from the start line to a successful acquisition by an industry leader, Zebra Technologies, for a staggering $305 million.[56]

This is the story of Fetch Robotics.

For context, it took iRobot more than 30 years to come onto the radar of an industry titan, Amazon; Kiva Systems 10 years to get acquired by one; and Fetch Robotics did it in just seven years. While building a robotics company is itself an Ironman race, building one that actually solves real customer problems, creates value, turns revenue-positive, and exits at a fair price is at least 100× harder in robotics than in the software world.

From maintaining a relentless focus on solving A+ problems for customers, to marrying hardware and software the right way, prioritizing deployments at scale across the Atlantic, and, most importantly, choosing the right partners to scale with, the trajectory of Fetch Robotics serves as a goldmine of lessons for the next generation of robotics founders and engineers.

That leads us to our first question: Who came up with the idea of Fetch Robotics? More importantly, how did they come up with this idea, and what was the underlying mission?

To answer that question, we need to travel back in time to 2004.

Founder DNA (2004–2006)

In 2004, a fresh engineering graduate who had developed a deep passion for robotics at the age of eight completed dual bachelor's degrees in Mechanical Engineering and Physics at the University of Illinois. Two years later, she earned her MS in Mechanical Engineering.

Determined to pursue her passion for robotics, she interned at Alcoa, DaimlerChrysler, and Honeywell Aerospace, working at the intersection of systems and mechanical engineering.[57] In the years that followed, she helped build some of the most consequential warehouse robots of her generation and later contributed to the development of cutting-edge humanoid robotics systems powering large-scale industrial workflows at companies such as Amazon, including Digit at Agility Robotics.

That engineer was none other than Melonee Wise, one of the core founders of Fetch Robotics.

Which brings us to the next question: how did Wise evolve from a mechanical engineering student at the University of Illinois into one of the most respected contributors to modern robotics?

During her time at the University of Illinois, Derek, a friend and future co-founder of Fetch Robotics, approached her with a proposition.

His offer: "Hey, I'm thinking about building a robot. You want to build it with me?"

Wise accepted the offer.[58]

And so, the duo built their first robot, called Zippy. While some of you might think of Zippy as the equivalent of a ROS-based two-wheel-drive robot today, make no mistake: it was this very project that seeded a deep desire in them to build production-grade robots. With this first project successfully executed at the University of Illinois Urbana-Champaign, Wise headed to California in 2007 after graduating, to continue her work in robotics. She started working on the DARPA Urban Challenge while pursuing a PhD program.

Just as Wise had started her PhD program, one of the founders of a robotics company that is very well known today approached her with an offer to join the company if she would leave the program. Faced with the opportunity to work full-time in robotics and pay off her education debt, Wise chose to take that path.

This company was none other than Willow Garage, the birthplace of ROS and PR2, and the company known for building the platforms on which many modern robots are designed and launched today. Willow Garage soon became Wise's training ground.

That leads to another question: Who founded Willow Garage—and why?

The Training Ground: Willow Garage (2007–2013)

Willow Garage was founded in late 2006 in Menlo Park, California, by Scott Hassan, a Google engineer who had worked closely with Larry Page and Sergey Brin.[59] The company was primarily launched as a research lab with a clear mission to accelerate the development of robots around the world by providing next-generation builders a standard, application-agnostic base platform.

By eliminating the need to design a custom mobile base and spatial manipulator from the ground up for every new robotics project, and by offering a standardized base platform, Willow Garage enabled engineers to focus on the application layer of the stack rather than rebuilding hardware and middleware for each new robot. Here at Willow Garage, Wise served in engineering and leadership roles from 2007 to 2013. During this period, she managed robot development programs, contributed deeply to ROS (Robot Operating System), helped develop affordable robot platforms such as TurtleBot, and worked on advanced spatial manipulation platforms including the PR2.

While Willow Garage, like many robotics research labs, shut down its operations in early 2014 as the commercialization chasm proved far harder to cross than the team had initially thought, it served as the birthplace of yet another robotics startup. That company was Unbounded Robotics, a spin-out founded by Wise and her colleagues. It carried forward Willow Garage's platform-focused thesis.

Although short-lived, lasting only a few months, Unbounded Robotics served as a critical bridge in translating the platform-focused research work conducted at Willow Garage into production-grade systems. It ultimately became the direct precursor and birthplace of Fetch Robotics, founded in 2014.

Dawn of Fetch Robotics (2014)

Fetch Robotics was officially incubated in 2014 in San Jose, California, by Melonee Wise, Derek King, Eric Dyer, and Michael Ferguson, engineers and robotics veterans who had worked together at Willow Garage on robotics platforms such as ROS and PR2. This team wanted to translate the research conducted at Willow Garage into production-grade robots to solve an extremely painful problem for warehouse operators.[60]

At the time, warehouses around the world were primarily driven by workers manually sorting, picking, processing items, and transporting them from one end of the warehouse to another. This meant workers had to travel on foot, often hundreds of kilometres each week within a warehouse. This was a manual, tedious, straining, and highly inefficient approach to transporting goods that cost warehouse operators millions of dollars each year in operations.

Fetch wanted to solve two subsets of problems here. First, they wanted to automate the process of picking and sorting items located on pods in the warehouse. Second, they wanted to automate the process of transferring these pods from one area of the warehouse to another. This naturally meant the company came up with two products: one, a robot arm mounted on a mobile base to pick items, called Fetch; and second, a strong autonomous mobile base designed to transport not just individual items but entire pods of goods from the inventory area of the warehouse to dispatch stations or worker stations, called Freight.

With two prototypes developed within a matter of twelve months, the team now had to raise seed capital and secure its financial runway. How did they pull it off?

They turned to two Silicon Valley firms: one known for backing frontier technologies, i.e., O'Reilly AlphaTech Ventures; the other for investing in hardware-centric emerging technologies, i.e., Shasta Ventures, both with a proven track record of over a decade in the space, making them natural partners for Fetch.

Raising Seed Capital (2015)

In 2015, Fetch Robotics raised a $3 million seed round from O'Reilly AlphaTech Ventures and Shasta Ventures. From the outset, the company ensured its investors understood that Fetch was a robotics company, and that investment maturity cycles would therefore be significantly longer, nearly ten times those of an equivalent software startup.

If you are a founder or an aspiring founder, it is extremely important to be transparent and clear with your investors about these ultra-long investment maturity cycles inherent in robotics. You want to do so because in robotics, as most of you can probably relate, a large portion of early-stage capital goes into building reliable hardware and software systems that can operate at scale. Unlike software upgrade cycles, where you can fix a series of software bugs and upgrade the entire system overnight, robotics is essentially a marriage of hardware and software.

This marriage implies that the first few years in robotics translate into an extremely high-burn phase, because getting these hardware and software components to work as a unified system and communicate with each other in a reliable and secure manner across a diverse set of enterprise and global telecom networks, while adhering to regulatory frameworks such as GDPR and SOC, is an extremely challenging task. One that demands a significant amount of time, capital, and engineering resources.

With funding secured and terms communicated clearly to investors, the team then geared up for the launch of its first set of production-grade robots.

First Product Launch: Fetch & Freight (2015)

In 2015, Fetch publicly unveiled two robots: Fetch, a mobile robot equipped with a 7-degree-of-freedom (DoF) robotic arm capable of carrying payloads up to 6 kg, and Freight, an autonomous mobile base designed to follow workers in warehouses and transport carts filled with goods, thereby minimizing their workload and elevating warehouse throughput.[61] That raises an important question: why did the company build two separate robots instead of one unified system?

This decision was strategic. While spatial manipulators mounted on a mobile base are extremely valuable in the warehouse industry, at the time, getting them to work reliably across all warehouse profiles was an extremely hard goal to accomplish. Each of these warehouses handled a wide spectrum of goods, ranging from spare parts to baby products to books to electronics and more. Training spatial manipulators to operate effectively across this diverse range of object profiles was, at the time, a monumental task.

You may ask why that was the case. Artificial intelligence (AI) models at the time were not as capable as they are today, and more importantly, the data required to train them did not exist at the required scale. So while spatial manipulators were extremely valuable in theory, building ones that were truly reliable was an extremely difficult task.

Mobile robots, on the other hand, similar to high-powered vacuum robots, were relatively easier to build and deploy. Most importantly, they came with minimal training requirements, if any, thereby lowering barriers to adoption. ROS already provided the autonomy stack needed to navigate within these spaces. That meant they offered immediate productivity gains to warehouse operators. As a result, Fetch positioned this dual-robot product portfolio around "follow-me" and "goods-to-worker" workflows.

The follow-me mode ensured that if workers needed to carry large volumes of goods across the warehouse, they could deploy multiple Fetch mobile robots, load them with all the goods they needed to carry, and have this fleet follow them along their routes. This robot-enabled approach effectively scaled the capacity of each worker while minimizing physical strain. That translated into up to 10× greater efficiency and significantly reduced the distance each worker needed to travel per day within these warehouses.

The goods-to-worker mode, on the other hand, was more of an autonomous configuration. In this mode, the robot would independently, i.e., without having to follow or be guided by workers, transport goods directly to a worker or packaging station. This was enabled by a ROS-based site-mapping feature that allowed operators to create a detailed map of their facility, configure pick-up and drop-off points, and upload them to the robot so it could handle these tasks 24/7 in a fully autonomous manner.

Together, these two robots were designed to eliminate the daily marathons workers had to run across warehouse floors and to maximize overall productivity. In addition, the team positioned these two robots as a unified warehouse automation platform: mobile robots coupled with cloud-enabled software systems. One was capable of transporting goods, and the other capable of sorting them. This platform-centric approach differentiated Fetch from other providers of autonomous mobile robot systems.

With this platform and messaging finalized, it was now time to land their first customer. So where did the team go next?

41 miles north of their headquarters in San Jose, to Livermore, California.

First Production Deployment: RK Logistics (2016)

Around 2016, a logistics provider operating a series of warehouses located in Livermore, California, was in desperate need of reducing the time its warehouse workers were spending manually transporting goods—walking from one end of the warehouse to another—and instead focusing more on processing orders. Fetch was naturally destined to address this problem.

The company introduced its mobile robots, namely Fetch, to this operator, i.e., RK Logistics (RK). Over this initial six-month period, the robots operated 24/7 across three shifts to help workers transport carts full of goods throughout the warehouse. During these six months, the company reported 1,000 km of worker transport time saved by these robots, which successfully completed 4,500 deliveries on this warehouse floor in California.[62]

This naturally implies the robots saved these workers from running about 40 full-scale marathons, or about 80 per year, within these warehouses. Those 1,000 km of footsteps saved for workers also contributed significantly to productivity gains and labor cost savings for the warehouse operator.

RK later reported full ROI within months—a rare statement in industrial robotics. Over the next two years, the company doubled down on these mobile robots to support its 600+ employee base across 15 warehouses in the greater Bay Area region.

For Fetch, this deployment validated the founding thesis: reducing walking distance by offloading transport tasks to mobile robots minimized strain on workers, maximized their productivity, and significantly elevated warehouse throughput. While this validation was a strong inflection point in its trajectory, it also revealed its next major bottleneck: software. The software layer that powered these robots was built largely around ROS and was not designed to handle the scale at which Fetch was about to deploy these robots for its customers. It was designed for prototypes, not for production scale. So, what did the team do next?

Platform Infrastructure: FetchCore & Fleet Software (2016)

Fetch started developing a proprietary software stack tailored specifically to handle fleets of robots deployed across geographically dispersed warehouse sites. This effort culminated in the launch of FetchCore, a cloud-based software platform enabling centralized robot dispatch, fleet-level coordination, and system-wide health management at scale.

While ROS was retained as a valuable software layer for enabling inter-system messaging, tooling, and internal development, it was not designed to handle the operational demands of large-scale, multi-robot production systems. This meant it was not designed to control large fleets of robots, assign tasks to them in a centralized manner, and monitor their health. ROS was designed to operate individual robots, not production-scale fleets. This is where FetchCore came in. It ensured Fetch was ready for large-scale deployments.

While FetchCore helped solve the fleet management bottleneck for Fetch, naturally the next major bottleneck on its path to commercialization was solidifying its manufacturing and supply chain strategy.

Manufacturing Strategy and Supply-Chain Bottlenecks (2016)

As Fetch started deploying its robots for its customers, manufacturing and supply-chain bottlenecks surfaced in no time. To begin with, the base platform developed by Fetch was built by assembling about 300+ total parts and 100+ unique parts. Beyond the number of parts required to assemble these robots, the cost of these parts contributed heavily to the overall cost of building each robot. For example, the laser scanner was sourced from Germany from a company called SICK and was priced in the range of about $7,000 to $9,000 at the time. That sensor was reliable for warehouse autonomy, but it also contributed significantly to the capital expenditure for each of these robots. In addition to the cost, international sourcing meant it had a large influence on the manufacturing schedule of these robots. This same pattern around cost, sourcing, and scheduling bottleneck was reflected across other subsystems.

Fetch designed its own PCBs and had them fabricated in California, while sheet metal work, powder coating, and related fabrication tasks were handled through Bay Area suppliers, including vendors located very close to Fetch's office. At the same time, certain parts were sourced globally: the robot "skins" were injection molded in China, even though the designs and molds were developed in the United States. Similarly, the drivetrain for these robots, or in simple terms the motor–wheel assembly, was engineered in the United States but manufactured through a motor partner in China.[60]

In other words, during the early stages, Fetch adopted a two-pronged operations model. The supply chain was prong one and spanned the United States, Germany, and China, while assembly and testing were done in-house and served as prong two at the Fetch office. The goal was not to build a factory in-house and instead minimize operating costs by outsourcing certain processes to geographies that offered better pricing for these tasks. In simple words, Fetch developed the design for parts and engineered the underlying software. That software-and-design combination enabled scale, and this manufacturing and supply-chain model made that scale deliverable.

With software and manufacturing bottlenecks addressed, Fetch was finally ready for global deployments. The next proving ground for Fetch was across the Atlantic, in the Netherlands, the land of windmills and Gouda cheese. This country would soon become Fetch's next operational fortress.

International Expansion: Europe — Wärtsilä and DHL (2017)

In 2017, Fetch Robotics completed a production-grade pilot with Wärtsilä and DHL at one of their core distribution centers in Kampen, Netherlands.[63] The facility spans 37,000 m^2, manages over 80,000 unique spare parts, and ships more than 100,000 deliveries annually to over 180 countries. Warehouse operations at this site are managed by DHL, Wärtsilä's long-standing logistics partner.

The challenge this customer faced was that, as in many industrial logistics environments, a large portion of employee time was consumed in point-to-point material transport. This meant employees were running marathons on a pretty much weekly basis across these warehouses. This approach of transferring goods from one end to another required employees to walk on foot from one end to the other, which meant less time to process orders and more time spent moving goods. This was a big challenge, as this operational model was not scalable for this customer and was extremely cost inefficient.

The goal of this pilot was to reduce physical strain on warehouse workers, minimize their time spent on the floor, and maximize their time at order-processing workstations, thereby elevating their productivity levels. In practical terms, the pilot evaluated whether autonomous mobile robots developed by Fetch could be safely integrated into a production-grade European warehouse environment with stringent operational standards, while delivering measurable productivity gains.

To address this challenge, Fetch deployed its autonomous mobile robots (AMRs). Each robot supported a payload of up to 78 kg, operated at speeds of up to 2 m/s, and delivered up to nine hours of battery life. This system was deployed in a matter of weeks, and the outcomes spoke for themselves. During live operations, this fleet of robots absorbed a significant amount of distance—approximately 30 km, per reports—previously traveled by workers on foot across the warehouse floor each day. This manual workflow, spanning 30 km across the warehouse floor and consuming an enormous number of labor hours each day, costing millions each year in operating expenses, was now successfully offloaded to robots, significantly reducing physical strain on workers while elevating warehouse throughput.

Wärtsilä and DHL described this first deployment as successful and committed to continued scaling of the system across their warehouse operations. This deployment proved that the solution offered by Fetch was universally applicable and could scale across diverse global warehouse models. Thus, by reducing the time workers spent walking across the warehouse, which accounted for approximately 70% of their daily routine, the system significantly elevated warehouse throughput while ensuring compliance with

European operational and safety standards and integrating seamlessly into a mission-critical global logistics hub.

Now, while Fetch's first product, called Freight, provided customers with the mobile robot base they needed to move goods across these warehouses, and FetchCore provided the fleet-management software for these robots, there was one core bottleneck that was not yet addressed. That was the cart profile.

Warehouse operators around the world use carts that come in different profiles, i.e., sizes, shapes, and structures, to transport goods. The challenge with the current approach was that Fetch had built only one version of its mobile robot, which meant customers had to design their carts around this robot for those carts to fit on top of it. This was a major adoption bottleneck.

One approach to address this challenge was that Fetch could continue asking customers to redesign their carts or design a custom cart for them. Another approach was for Fetch to design robots around these existing carts. Fetch chose the latter and, in doing so, ended up landing CartConnect.

CartConnect & Product Line Expansion (2017–2018)

From 2017 to 2018, Fetch focused on addressing one of the biggest barriers to product adoption: carts. Warehouse operators around the world rely on carts that vary widely in size, structure, and payload capacity. Asking customers to redesign these carts around Fetch's robots created friction and slowed adoption. To address this challenge, Fetch designed and launched a series of cart-compatible mobile robots under the CartConnect product line.

These robots were purpose-built to transport existing carts from inventory areas to worker stations and didn't require operators to redesign their warehouse infrastructure. CartConnect robots were designed to handle diverse payloads, with models such as CartConnect 100 capable of carrying lighter loads (100 kg and below) and CartConnect 500 capable of handling up to 1,100 lbs (500 kg). In effect, CartConnect enabled Fetch's robots to integrate directly into existing, cart-based warehouse workflows.

At this point, Fetch's product stack was unified in the true sense and designed exclusively to elevate warehouse throughput. Freight provided the autonomous mobile base, FetchCore delivered the fleet-management software, and CartConnect enabled seamless adoption by ensuring Freight and FetchCore were compatible with the type of carts used by its customers around the world. Together, these three components helped Fetch address the core technical and workflow bottlenecks that had previously limited the adoption of its products.

Now, with the right robot and the right software layer in place, Fetch had to land more customers and secure its financial runway. This meant it had to scale the pace at which it onboarded customers. That was a big challenge. Why was that the case?

The GTM Scaling Constraint

Fetch's existing go-to-market approach was sequential and painfully slow. Fetch was building trust with one customer at a time and onboarding them sequentially. The process typically started with a single customer pilot, followed by a six-month wait to prove return on investment (ROI). This would then lead to a small-scale deployment, another year of operation, and yet another round of ROI validation before expansion. In total, it took roughly 18 months to transition a customer from the ROI validation phase to the commercialization phase required to make each deployment operationally profitable for Fetch. The team realized this approach would not allow them to scale their customer base—and, worse, could erode their financial runway.

While Fetch had already demonstrated the value it had to offer to customers such as DHL and RK Group, this model was not sustainable for global expansion. So how did Fetch handle this challenge?

To address this challenge, the company adopted a centralized, hub-and-spoke model. The idea was to partner with a trusted industry player that could accelerate adoption at scale. In Fetch's case, that strategic anchor was Honeywell—a partner with deep reach, credibility, and existing relationships across global warehouse and logistics operations.

GTM Shift: Systems Integrators & Honeywell Partnership (2018)

Systems integrators like Honeywell are essentially the sun of the warehouse galaxy. Everything revolves around them for a reason. They own the customer relationships, design the facilities, and control deployment timelines.

Partnering with Honeywell Intelligrated meant Fetch was now close to its sun.[64] It gained direct access to hundreds of potential customers around the globe. While this partnership unlocked scale for Fetch, the trade-offs were shared margins. That trade-off proved to be a win-win for both companies.

There is an important lesson here for those building industrial robotics startups: building robots is the easy part; getting distribution channels and margins right is far more difficult. One effective way to address this challenge is to conduct a thorough analysis of the integrator landscape in your domain and select a partner that can address your needs across the spectrum, including customer reach, margin structure, operational ease, and integration capabilities. Few companies exemplify this principle better than Fetch. This strategic shift in 2018 proved pivotal for the company.

On one hand, it provided access to hundreds of customers worldwide. On the other, it gave Fetch critical insights into what these customers truly needed beyond an autonomous mobile robot capable of moving carts across their warehouse. It turned out that a vast majority of these customers were in desperate need of automating the inventory management process—a process that was still largely manual.

Workers would walk aisle by aisle with RFID-enabled handheld devices, scanning inventory and registering it into a centralized inventory management system. To understand the scale of this problem, consider that each worker was spending roughly 8–10 hours per day scanning thousands of items with these handheld devices. Applied across an entire warehouse network, this translated into millions of items scanned manually each year. This was an extremely tedious and time-consuming process. Fetch learned this through its partnership with Honeywell and decided to address the challenge head-on for its customers.

TagSurveyor was its answer.

New Product Launch: TagSurveyor & Fleet Telemetry (2019)

In 2019, Fetch launched TagSurveyor, an RFID-enabled robot designed to automate the inventory management process. With this move, Fetch expanded its capabilities beyond developing robots that could transport goods across the warehouse and into understanding what those goods were and precisely where they were located. This level of goods-level intelligence became a critical factor in driving customer adoption.

At the same time, data generated by its fleet of robots—traffic patterns, congestion points, forklift proximity, and the overall warehouse throughput per day—was no longer viewed as a byproduct of operations. It was increasingly framed as a product itself. This combination of goods-level and fleet-level intelligence meant a shift in its positioning was on its way. At this point it evolved from a hardware company, to a software company, and ultimately into a data-centric organization providing the core fabric needed to scale warehouse operations. In simple words, it was no longer selling robots; it was selling the most powerful warehouse operating engine on Earth—one capable of doubling warehouse throughput for major brands around the world.

That positioning made collaboration with top-tier strategic partners possible. The 2019 collaboration with Ryder Systems on smart warehouse initiatives, followed by the 2020 partnership with Zebra Technologies, reflects this shift.

Partnership with Ryder Systems (2019)

In 2019, Fetch Robotics partnered with Ryder Systems, an American transportation and logistics company well known for hauling goods for major brands such as Walmart, Starbucks, Amazon, and FedEx across its warehouse network.[65] This partnership was established to support Ryder's *smart warehouse* initiative, which was incubated to automate warehouse operations using cutting-edge technologies spanning robotics, data intelligence, and more. Fetch's autonomous mobile robots were deployed within Ryder-managed facilities to enable autonomous movement of pods filled with goods across the warehouse, minimizing physical strain on workers while maximizing their productivity.

For Fetch, the partnership served as a strong validation point for the hub-and-spoke–led go-to-market model it adopted around partnerships in 2018. While it started with Honeywell, it did not stop there. The team realized that strategic partnerships had the potential to scale its customer base in a far more time- and resource-efficient manner than the linear approach it had relied on in its early days. In simple words, it was much easier to scale with integrators and strategic partners than to chase individual customers one by one.

This hub-and-spoke model, adopted to scale its customer base, was validated yet again when another top-tier integrator, Zebra Technologies, approached Fetch in 2020 to integrate its mobile robots with Zebra's warehouse stack.[66]

That raises a critical question: Why was the company interested in integrating its stack with Fetch's mobile robots? What was the underlying thesis?

Zebra Technologies: From Robots to Warehouse OS (2020)

At the time, Zebra's portfolio primarily consisted of legacy inventory management devices, such as RFID scanners and barcode readers, that had to be carried by workers to scan and track inventory across warehouses and logistics facilities. From a logistical standpoint, there are only so many devices a worker can carry. The first challenge was that the approach itself was physically burdensome. Second, it was inherently inefficient. While the devices themselves were powerful, at the end of the day workers still had to navigate these massive facilities on foot, aisle by aisle, and manually scan each and every item. As a result, the workflow was operationally inefficient, tedious, and therefore not scalable.

This was the problem Fetch had the potential to solve for Zebra.

Fetch offered both the mobile base needed to carry these inventory management devices autonomously across customer sites and a spatial manipulator capable of handling and operating them in an automated manner. Thus, this product configuration was capable of transforming previously manual workflows into fully autonomous processes for Zebra and its customers.

Recognizing this potential, Zebra Technologies approached Fetch Robotics to integrate its mobile robots, designed to carry carts filled with goods across warehouses, into Zebra's FulfillmentEdge software system.

Zebra's software was designed to assign tasks to individual workers and to track, manage, and coordinate end-to-end workflows across warehouse operations. With Fetch's robots now integrated into this software stack, Zebra could offer its customers across more than 175 countries the ability to not only track inventory and assets using handheld devices, but also offload a significant portion of that work to Fetch's mobile robots, including TagSurveyor and other robotic systems.[67] In simple words, customers could now add robots to their warehouses and double warehouse throughput without having to redesign their existing warehouse workflow management systems (WMS) or scale their workforce.

Zebra provided the software needed to coordinate tasks among workers, while Fetch provided the hardware and software required to coordinate tasks among robots. This unified workflow management system, designed exclusively for warehouse and logistics operators, positioned Fetch as the ultimate throughput engine for Zebra's customers. It did not take long for customers to adopt the solution, and the results spoke for themselves. The robots had done their work, and the integrator had now made an offer—not for robots, but for the company itself.

The Zebra Acquisition: Strategic Fit Over Scale (2021)

In 2021, after integrating, testing, and validating Fetch's robots with its warehouse workflow management systems, Zebra Technologies recognized the long-term value this mobile robotics company brought to its customers. Recognizing both this value and the competitive advantage it would gain in the industry, Zebra immediately made an offer to acquire Fetch Robotics for $305 million.[56]

The rationale was straightforward. Zebra already owned scanners, RFID, WMS, and IoT infrastructure—much of the digital backbone required to automate warehouses and industrial operations. While this stack had served Zebra well for a long time, it depended heavily on adding more devices onto the shoulders of frontline workers. In simple words, Zebra's existing model was no longer scalable and had reached its limits. Fetch Robotics helped address this constraint and significantly increase throughput for Zebra's customers. Together, the acquisition resulted in an end-to-end, industrial-grade warehouse automation platform—one capable of orchestrating workflows across both workers and robots, and nearly doubling warehouse throughput.

While Fetch may not have deployed hundreds of thousands of robots around the world, it successfully developed a platform that addressed a first-order problem in warehouses. More importantly, it managed to land strategic partners along the way—first with Honeywell to demonstrate real productivity gains for customers, then with Ryder Systems, and later with Zebra Technologies as a long-term platform partner and, eventually, a parent.

This journey of Fetch Robotics, which started in Silicon Valley in 2014 and culminated in a strategic acquisition within just seven years, likely represents one of the fastest and most decisive exits a robotics company has achieved to date—and offers a goldmine of hard-earned lessons for the next generation of robotics founders.

So, what can we learn from it?

Lessons for Robotics Founders & Engineers

Lesson 01: Be Explicit About the Problem You Are Solving

In Fetch's case, mission clarity was non-negotiable. The company not only steered away from consumer robotics, but even within the industrial robotics vertical it eliminated the ambition of "automating the entire warehouse," a trap that eludes many robotics founders. Instead, Fetch focused on one core problem: minimizing the time workers spend walking across the warehouse and maximizing warehouse throughput.

It was this level of clarity in its mission statement that shaped Fetch's product offering, roadmap, customer messaging, and deployment strategy from day one. As a founder, if you can define and consistently adhere to the exact problem you are trying to solve for your customer with extreme clarity, you stand a strong chance of leading the robotics race.

Lesson 02: Prove ROI First, Expand the Product Portfolio Later

Fetch devoted its first several years almost exclusively to two products: Fetch and Freight. It ensured there was no deviation from this path until it had clear, measurable ROI from early deployments and had established the hub-and-spoke go-to-market model needed to scale its customer base. Only after proving value in production environments with customers and integrators did Fetch expand its product offerings with TagSurveyor and similar products. The lesson here is clear: prove return on investment (ROI) first with a few A+ products before deciding to scale your portfolio.

Lesson 03: Demonstrate Quantified Outcomes Through Early Adopters

Fetch, through its first deployment with RK Logistics, demonstrated a strong, quantified outcome: nearly 1,000 kilometers saved in cumulative distance traveled on foot by workers within a single warehouse. This metric was not a marketing claim; it was validated and reported by the customer itself. Quantified, customer-verified outcomes like this built trust and became compelling proof points for future customers.

Lesson 04: Distribution, Not Autonomy, Determines Scale

It goes without saying the role DHL, Honeywell, Ryder Systems, and Zebra Technologies played in Fetch's journey. DHL enabled the evaluation and validation of Fetch's solution on international soil, demonstrating that it had the potential to make an impact on a global scale. Honeywell and Ryder helped validate its integrator- and strategic-partnership-led distribution thesis, while Zebra served as yet another strong proof point for that strategy.

This hub-and-spoke distribution model demonstrates the importance of evaluating and selecting the right distribution partners. If you do this right, it can significantly accelerate your journey toward the commercialization phase and ultimately help you achieve the scale velocity needed to escape the startup orbit. In doing so, you can meaningfully scale your revenue streams.

Lesson 05: Partnerships Compound Faster Than Capital

As a startup founder, capital is often the most sought-after resource at the early stage. While it is understandable that you need capital to fund your team and product, it is often overvalued and has the potential to blur your real needs. Capital is effective at unlocking technological and talent-related bottlenecks, but its power often fades when it comes to your ability to land early-stage adopters.

In the early stages of your journey, what you truly need is a strong and trusted helping hand that can open more doors for you and help you demonstrate the value your product has the potential to offer to your target customer base. In the case of Fetch, Honeywell and Ryder Systems became that trusted helping hand and opened doors to hundreds of customers, while Zebra Technologies ensured Fetch had the warehouse-compatible technology stack needed to close those deals.

Now that we have distilled the core lessons embedded in Fetch Robotics' trajectory, from its origins through its evolution and the pivotal decisions the team made along the way, it is time to decode the underlying business model.

In the final section of this chapter, we will break down its business model and illustrate how the product portfolio maps to select primary use cases.

Fetch Robotics: Product Snapshot

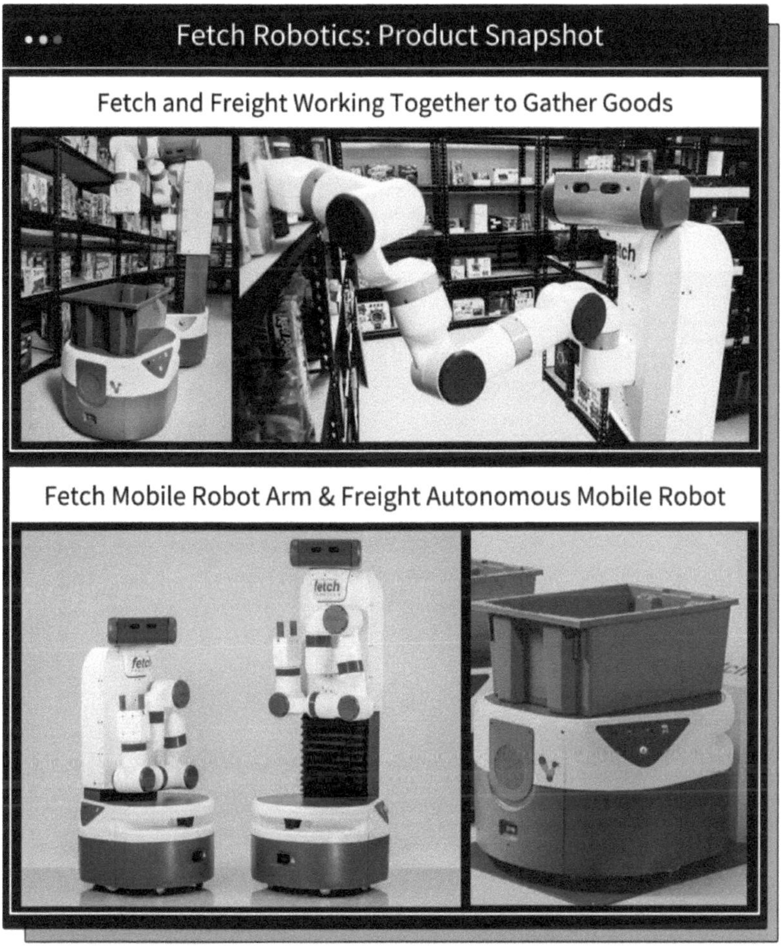

Figure 5.1 — Fetch & Freight Robots

Illustration by the author. Robot images from IEEE Spectrum article (2015).

Fetch Robotics' Business Model — Part 1

Figure 5.2 — Fetch Robotics' Business Model: Part 1

Illustration by the author.

Fetch Robotics' Business Model — Part 2

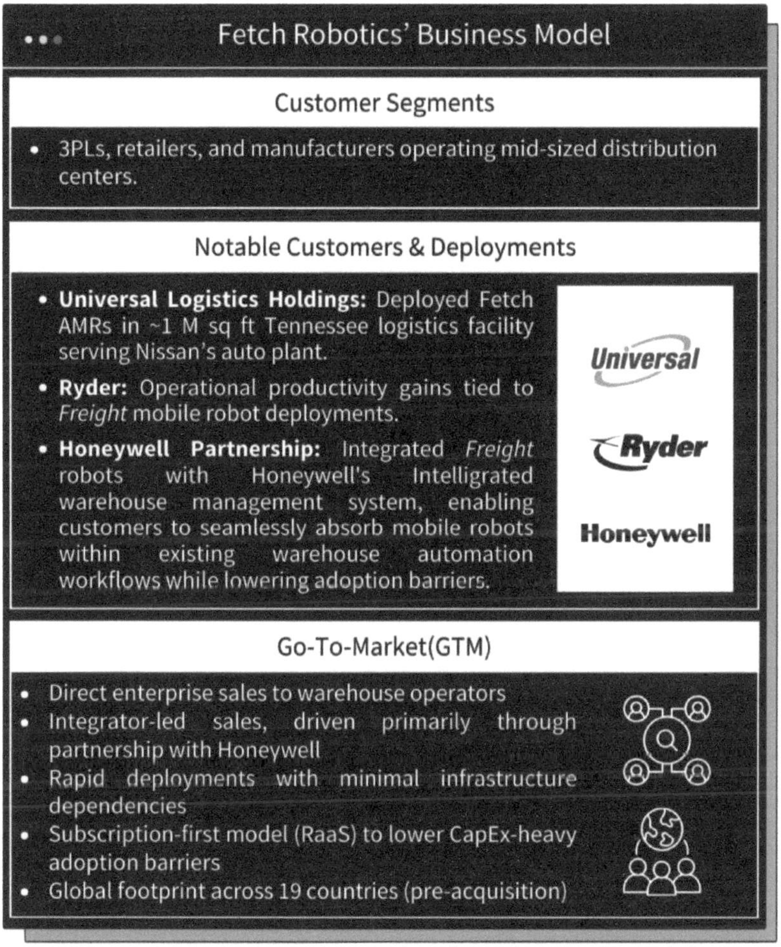

Figure 5.3 — Fetch Robotics' Business Model: Part 2

Illustration by the author.

Fetch Robotics' Business Model — Part 3

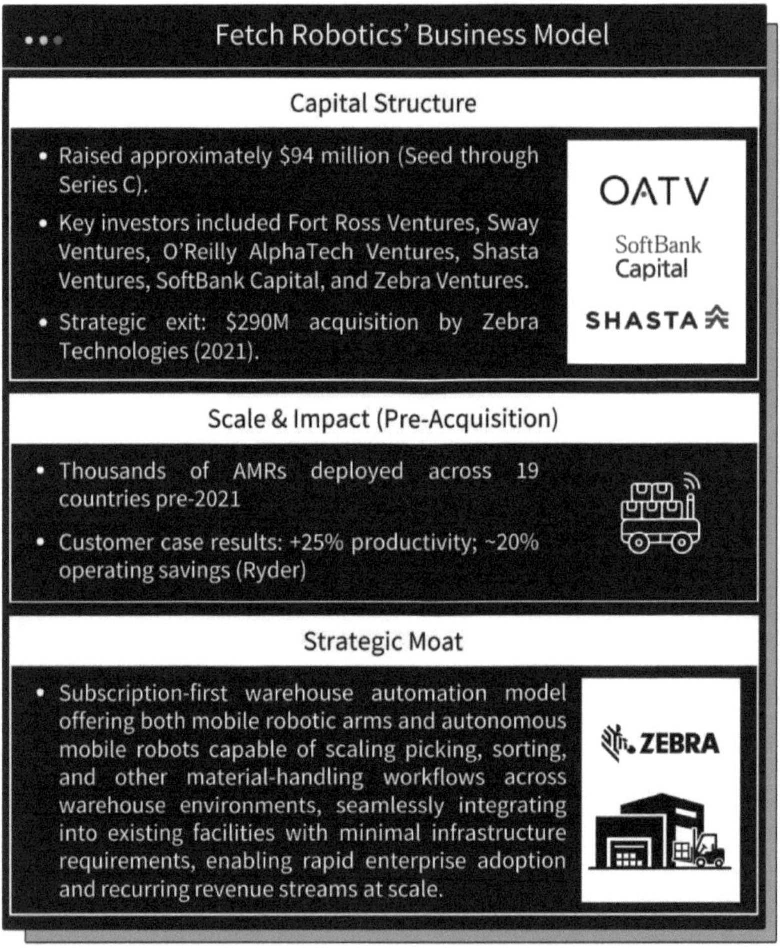

Figure 5.4 — Fetch Robotics' Business Model: Part 3

Illustration by the author.

Chapter 06: 6 River Systems — Six Years to Shopify Exit

Launched in 2015 in Waltham, Massachusetts, by the alumni of Kiva Systems, the warehouse automation pioneer that powers Amazon, this robotics company was acquired by Shopify in 2019 for approximately $450 million.[68] That is five years from incubation to an exit. In robotics, that is almost unheard of.

That company was none other than 6 River Systems.

While Kiva took nine years to be acquired by Amazon, founded in 2003 and acquired in 2012, and Fetch Robotics took seven years to be acquired by Zebra Technologies, founded in 2014 and acquired in 2021, it only took 6 River Systems five years to reach that point. While the context for each of these acquisitions is different, and the aftermath of each is equally unique, the path this company and its founders took, the decisions they made, and the corresponding lessons they offer are universally applicable in this domain and deserve a deeper look.

After all, how do you go from the start line to a successful and, most importantly, a strategic exit in robotics in just a matter of five years? How do you land enterprise customers like DHL, XPO, and Office Depot with a product built around an emerging technology for which customers have yet to develop an appetite, all within 60 months of receiving seed funding, and position yourself squarely on Shopify's radar? Where should we even begin this research?

That was the question on my mind as I started researching about these founders and their journey.

Well, to answer those questions, we need to study the DNA of these founders. To do that, we have to travel all the way back to 2001.

Founder DNA (2001)

In 2001, a young enterprise architect kickstarts his career in the software space, working on the system architecture of a warehouse management system developed by Yantra Corporation.[69] He spends the next seven years perfecting this craft of iteratively understanding the needs of his customers in this space and optimizing the system design to better serve those needs.

Armed with this experience, he joins Kiva Systems, then the most prominent name in the warehouse automation space, where he leads commercial operations and manages key customer accounts. During his tenure at Kiva, the company is acquired by Amazon, and fast forward to 2015, he lands on the business development side at Amazon. Within two years of this acquisition, now a seasoned veteran of the warehouse solutions space with a spectrum of expertise under his belt ranging from software architecture all the way to commercial strategy, he decides to launch his own venture.

This individual is none other than one of the founders of 6 River Systems, Jerome Dubois.

The story does not end here. Another gentleman by the name of Rylan Hamilton, who had also worked at Kiva, encounters Dubois again following the acquisition, with both eventually landing at Amazon. While Dubois remains at Amazon for a few more years after the acquisition, Hamilton moves to Shopify, where he serves as Vice President of Shopify Logistics, the company's logistics arm, for four years.[70]

By 2015, the two reunite with a shared conviction: to serve the warehouse industry with autonomous mobile robots designed to automate material transport workflows, not restricted to a select few customers but made available to all. This is important to note because, after its acquisition by Amazon, Kiva's robots were officially not available to customers outside of Amazon.

Together, they decide to launch 6 River Systems.

They were joined by Chris Cacioppo, an engineer with a computer science degree from MIT and deep experience spanning hardware engineering, software engineering, and computer networking.[71] With this addition, the founding team was established. Dubois brought deep pre-sales and post-sales expertise from the warehouse automation space, Hamilton brought post-sales expertise, and Cacioppo built the company's technological base.

That leads to a critical question: What was the exact problem this team wanted to solve, and how did they fund this venture?

To better answer that question, let's first try to understand the core problem that exists in the warehouse industry.

The Early Days (2015–2016)

In the warehouse industry, labour accounts for roughly 70% of operating costs per warehouse. Roughly 70% of that labour cost is time spent by workers walking across the warehouse on foot to transport goods from one end to the other.[68] To better understand the severity of this problem, it helps to visualize the flow of goods inside a warehouse.

Operationally, a warehouse can be dissected into three critical zones. The first is the inventory zone, which serves as the entry point for goods into the warehouse. The second is the processing zone, where goods are picked, sorted, and packed. The third is the shipping zone, where goods are transported close to the loading dock so they can be transferred into trucks and shipped to their destination. In effect, the warehouse functions as a routing system for physical goods, with each item making three primary "hops" from entry to exit.

Historically, these hops were executed by human workers manually pushing carts loaded with goods across large warehouse floors. This was an extremely manual and time-consuming process, one that accounted for a significant share of warehouse operating costs and cut its throughput in half.

While Kiva Systems had managed to solve this problem, it was later acquired by Amazon, which meant its robots were no longer available to other customers. 6 River Systems wanted to address this challenge.

The company's objective was twofold. First, to automate internal goods movement for warehouses that did not have access to proprietary robotics platforms. Second, to do so in a way that was warehouse-agnostic—requiring no structural redesign, new infrastructure, or custom layouts. The goal was to develop a plug-and-play system using robots to enable warehouse operators to double their throughput.

In addition to these robots, the company developed a backend software ecosystem to manage robot-led workflows. This software integrated seamlessly with the traditional warehouse management systems (WMS) that were already in place in the majority of these warehouses. During initial trials, the company was able to prove that these robots more than doubled warehouse throughput with the same number of workers by enabling workers to transport roughly twice as many goods from one point to another, directly offloading this task from workers and, in effect, making the workers the robots' guides.

The workers became guides for the robots, and the robots carried the goods.

This initial thesis and the corresponding operating model led to the launch of the company's first production-grade robot, Chuck.

Launch of "Chuck" Collaborative Mobile Robots (2017)

Chuck was launched in March 2017 as a collaborative autonomous mobile robot designed to work directly alongside warehouse workers and carry goods across the warehouse floor.[72] The funding required to build Chuck and assemble the corresponding engineering team came through a seed round raised in March

2016, which enabled the company to scale the production of Chuck.

Physically, Chuck was approximately three feet long, two feet wide, and stood up to four feet tall, with a shelf positioned at about 3.5 feet. What differentiated Chuck from its predecessor, Kiva's robots, was its human-centric design. The height of the robot could be adjusted on the fly to match the needs of workers and warehouse configurations prominent across the globe.

Chuck was also equipped with an 11-inch display that guided workers through their task flows. The display showed images of the items a worker was about to pick, the required quantity, and numeric identifiers such as SKUs or barcodes. It also let workers know which direction they should head next inside these massive warehouses. Chuck could lead workers to the precise shelf location, communicate pick instructions, and carry up to 160 pounds of merchandise.

Beyond guiding workers and transporting goods, Chuck made real-time decisions about the route that both the robot and the worker would take through the warehouse, ensuring efficient path planning and almost doubling order fulfillment velocity for the crew. This capability was powered by a robust vision system the company developed by integrating feeds from SICK LiDAR sensors and two Intel RealSense cameras, processing them through a software stack built on ROS, the open-source Robot Operating System, to generate traffic-aware navigation plans.[73] As Chuck navigated the warehouse, it continuously collected data, which was transmitted to operators in real time to help them track overall throughput and diagnose fulfillment bottlenecks across facilities.

One important thing to note here is that Chuck operated autonomously using onboard sensors and eliminated the need for customers to redesign or customize their warehouse infrastructure. It was designed from the outset to fit directly inside existing warehouse footprints and workflows, which gave the company the acceleration it desperately needed to transition from the pilot phase to production-grade deployments at near-galactic speed.

From Demos to Deployments (2017–2018)

By late 2017, the company was already running successful pilot programs across multiple sites in North America. One of the core drivers cited by early adopters was the company's ability to set up its robots in virtually any warehouse around the world within weeks, not months, and secondly, without requiring any infrastructure redesign. This meant that by 2017, within just two years of launch, the company had already secured its early-stage adopters. One of them was a company called Barrett Distribution Centers.[74]

In 2017, 6 River Systems actively deployed its collaborative 'Chuck' robots in the warehouses of Barrett Distribution Centers. These early-stage deployments validated its product and service offering in a production-grade setting, and now the company had to secure the next leg of its financial runway. It was time to scale its customer base and build its financial muscle.

While the $6 million raised in seed funding had supported the company's first two years of development and early deployments, 6 River Systems needed more capital to scale its customer base, its team, and execute emerging deployments. It was time for Series B.

Series B Financing & Revenue Model (2018)

In April 2018, 6 River Systems raised a $25 million Series B, led by Menlo Ventures.[74] The company had three core goals in mind: scale enterprise adoption, upgrade its product to enable multi-site rollouts, and ramp up its engineering team to support this enterprise scale. That raised an important question: what was the revenue model the company adopted to sustain its runway?

Per reports, the company offered two primary pathways for customers to adopt its system. The first was a capital purchase model, where customers bought the robots for a fixed price and paid an annual fee for software and support related services. The second option was a rental model, allowing customers to lease robots for a one- to two-year term.

For 6 River Systems, small-scale customers adopted the rental model, while enterprise-grade customers adopted the hybrid model. These enterprise customers would 'buy for average and rent for peak,' purchasing enough robots to support day-to-day operations while renting additional units during periods of peak demand, such as the holiday season. This model proved to be capital efficient for enterprise customers while providing them with the operational agility they needed to handle peak-season demands. In return, it provided 6 River Systems with a sustained and recurring revenue stream that scaled equally well across small-scale and enterprise customers.

With Series B capital secured and a revenue model finalised, the company was ready to onboard enterprise customers. So where did the team go next?

It was time to expand across North America and, more importantly, across the Atlantic into Europe.

Enterprise Expansion – DHL Multi-Site 3PL Deployments (2018)

In November 2018, at DHL's annual robotics event called *Robotics Day – Sense the Future*, 6 River Systems, along with 28 other robotics companies, demonstrated its robot Chuck and its capabilities to the executive teams at the company's Innovation Center in Germany.[75] This demo was extremely vital for the company because it led to a series of deployments with DHL, a top-tier third-party logistics provider (3PL), across its heterogeneous warehouse environments around the world in the years that followed.

Over the next few years, DHL deployed more than 30 robots across its facilities in the Czech Republic. In a publicly available report, the company cited two core drivers behind the adoption: minimal setup and training requirements, and the tangible operational value the robots delivered. Employees could be trained to use the system in under an hour.

More importantly, during peak seasons these robots could be easily transported from one warehouse to another across the country, providing critical operational agility. Overall, with just these 30 robots assisting a workforce of 1,200 employees, DHL reported 30% productivity gains at that facility alone. This initial deployment paved the way for multi-site rollouts with DHL across North America and Europe. By 2025, the company stated that it had more than 750 pick-assist robots operating across its global facilities, supplied by two of its strategic partners, Locus Robotics and 6 River Systems.[75-77]

By this stage, the robots developed by 6 River Systems were fulfilling millions of units per week across customer sites, demonstrating sustained performance at enterprise scale. With DHL, the company had tapped into the honeycomb of the supply chain and logistics world in Europe. It was now time to tap its sister network in the United States.

This time, the company partnered with XPO, a logistics giant of North America.

Enterprise Expansion – XPO's Collaborative Robotics Program (2018)

For context, XPO is one of the largest logistics operators in the region, serving 55,000 customers with 38,000 employees across North America and Europe. It is indeed the honeycomb of the logistics world in the U.S., transporting about 17 billion pounds of freight per year across the country.

If you are in the U.S. and have ever purchased a refrigerator or patio equipment online, chances are it was delivered to your home by a truck owned by XPO. Given this scale, the company leaves no stone unturned when it comes to adopting cutting-edge technologies and maintains an innovation budget that should put it in the permanent customer registries of Silicon Valley start-ups. The company spends about half a billion dollars, or $500 million per year, to augment cutting-edge capabilities across its warehouse and logistics operations.

Given this budget, the company made a strategic decision in 2017 to adopt approximately 5,000 collaborative robots to augment its operations. This initiative was enabled through partnerships with companies such as Locus Robotics and 6 River Systems. While the exact allocation per vendor was not disclosed, the company did state that it had deployed more than 100 Chuck robots from 6 River Systems alone that year across sites in North America and the UK. In one documented deployment, XPO rolled out 40 robots supporting 25 warehouse workers tasked with identifying and picking items per customer order from inventory, effectively doubling the throughput of this workflow from roughly 60 units per hour per worker to 120 units per hour.[73,78-81]

During this same period, the company also secured deployments with Office Depot, supporting its network of retail distribution centers, which grappled with highly asymmetric seasonal demand curves for office supplies and had limited tolerance for operational disruption.

Companies like Office Depot, which typically operate a network of more than 100 large-scale warehouses averaging roughly 500,000 square feet each, benefit from Chuck's guided workflows, dynamic route optimization system, and the ability to scale robot fleets during peak periods—eliminating the need to expand the labor workforce or reconfigure warehouse infrastructure to manage seasonal demand.[81-82]

In summary, these deployments demonstrated that 6 River Systems could serve third-party logistics providers, large-scale logistics operators, and retail distribution networks using a standard and scalable product architecture capable of generating sustained throughput for its customers. These early deployments with DHL, XPO, and Office Depot gave the company the insights it needed to better understand these customers and expand its product portfolio.

This time, the company went beyond automating material transport inside these warehouses with autonomous robots and set its sights on the next major bottleneck: scaling picking and sorting workflows across these facilities.

To enable that scale, the company launched its next product: Mobile Sort.

Launch of "Mobile Sort" AMR-Enabled Sortation System (2019)

Before we try to understand the need behind launching Mobile Sort, it is important to zoom out a bit and visualise the end-to-end flow of goods in these warehouses. Goods are delivered to the warehouse and staged in the inventory area. That is phase one, also referred to as Inbound Transfer. Then there is phase two, also known as the Order Retrieval phase, where goods that belong to each customer order are picked from the inventory area and transferred to packaging or the consolidation stations. Then starts phase three, where these goods that belong to each customer order are then consolidated and packaged. That is the order consolidation or the sorting phase.

This sorting process was performed both at the customer order level and at the geographic level. All items per customer were sorted into a single tote or package, and then all orders headed to a given geographic region and allocated to a single truck were again consolidated and sorted, and routed to the appropriate conveyor belt that took the orders close to the outbound dock. This routine process was entirely manual and tedious at the time.

6 River Systems wanted to address this bottleneck and make this process semi-autonomous. The answer? Mobile Sort.

Mobile Sort was launched in 2019 to work hand in hand with its Chuck mobile robot and enable customers to automate the order fulfillment workflow in the warehouse.[83] From transferring goods from one end to another to sorting them per customer order and per geographic region, customers were now able to quadruple the throughput of their warehouse. This was a massive leg up for those warehouse operators who were struggling to meet the demands of their customers in a capital-efficient manner, given that 70% of operating costs were tied to the labour workforce. The old model that relied solely on workers was neither scalable for operators nor ergonomic and sustainable for employees.

Chuck and Mobile Sort, in combination, addressed both of these bottlenecks and made these workflows almost semi-autonomous, thereby providing these warehouse operators with the cutting-edge operating system they desperately needed to ramp up their warehouse operations.

6 River Systems, now equipped with two surgically engineered robots dedicated to automating two of the most critical workflows in the warehouse world and offering its customers quadruple productivity gains overnight, started appearing on the radar of Silicon Valley in the US and its sister ecosystem in Canada.

Acquisition by Shopify (2019)

On September 9, 2019, Shopify, the e-commerce platform giant headquartered in Ontario, Canada, announced the acquisition of 6 River Systems for approximately $450 million. The deal was formally sealed on October 17, 2019.[68-69] That raises a critical question: Why did Shopify acquire 6 River Systems? What has Shopify got to do with warehousing?

To answer that question, it helps to understand Shopify's business model. Shopify enables merchants to sell goods online. That means it provides merchants with the ability to launch a website for their brand and make their products available to consumers across geographies almost overnight. The challenge here was that, as some of these merchants started gaining traction in the industry, demand for their products had skyrocketed to an extent where they started leasing massive warehouses to manage the order fulfilment process.

For many of them, this proved to be an extremely tedious undertaking, especially given that most of these merchants had no background in warehousing operations. Shopify wanted to help them address this bottleneck.

To do so, it started building the Shopify Fulfillment Network, essentially a network of Shopify-managed warehouses paired with end-to-end order fulfillment services that it could offer to its merchants as a unified solution—allowing them to stay focused on building their brands while Shopify enabled them to process orders online and ship those orders at scale.

Over time, Shopify came to view logistics capability as a core competitive requirement. Rather than relying on third-party warehouses and technologies, Shopify chose to build these capabilities in-house for its merchants. This meant integrating warehouse robotics platforms directly into its stack and offering customers the ability to ramp up their warehouse operations using a tech-first robotics approach.

In doing so, Shopify was positioning itself as a logistics enabler for merchants competing against Amazon's vertically integrated model. 6 River Systems' globally validated, robot-led warehouse automation platform was a natural fit for this mission. As a result, the company decided to acquire 6 River Systems' portfolio of robots.

That leads us to a critical question: What happened to 6 River Systems' robots post-acquisition?

Operations Under Shopify (2020–2022)

Following the acquisition, 6 River Systems continued to operate as an independent business unit within Shopify. Chuck-based mobile robot systems remained commercially available, and deployments scaled beyond Shopify-owned facilities.

During this period, the company scaled its talent base from approximately 120 employees at the time of acquisition to over 420 employees and expanded its footprint across more than 110 sites enabling 50+ logistics customers, including industry leaders such as DHL and Office Depot.

The deployment and pricing model was also clearly defined: a minimum order of eight robots, available for rent or purchase.

Renting eight Chucks was priced at approximately $200,000 per year, while purchasing eight cost approximately $400,000, plus an ongoing annual fee of about 20 per cent for software licensing and maintenance services.[84]

During this period, 6 River Systems' trajectory became increasingly intertwined with Shopify's logistics ambitions. Shopify invested heavily in building an end-to-end fulfillment network using a mix of owned facilities and third-party operators, with 6 River's technology serving as the backbone of warehouse operations.

Importantly, Shopify did not restrict 6 River Systems' robots to internal use, eliminating the closed-ecosystem approach Amazon adopted after acquiring Kiva. Instead, 6 River Systems continued to offer its solution to all potential customers, reinforcing its position as an infrastructure-agnostic and open-market-first provider of robotics systems.

While, on paper, this marriage made perfect sense, in reality the e-commerce giant probably underestimated the effort it would take to make this service available to all its customers and, more importantly, to make it profitable. Over the course of the next few years, Shopify spent millions of dollars building the large-scale fulfillment operations systems needed to offer this service to its global array of merchants. As it did so, the sheer amount of cost and resources it had to dedicate to build even the first 20% of this system meant Shopify had started losing its edge in the e-commerce business. The very business that was sustaining it had now started losing its edge.

To minimise the impact on its primary business and maximise returns for its investors, and to continue its core mission, it decided to part ways with 6 River Systems and began searching for a safe home for 6 River Systems. So where did 6 River Systems ultimately land?

This time, yet another leap, but now west across the Atlantic—to the UK.

Divestiture to Ocado Group (2023)

On May 4, 2023, Shopify announced the sale of 6 River Systems to Ocado Group.[85] This divestiture enabled Shopify to pull back from a capital-intensive logistics-focussed business and refocus on its core software-led business aimed at providing the right tools for its merchants to enable online sales of their goods. Put simply, it decided to double down on software and divest from hardware and operations-heavy business verticals.

Under Ocado, 6 River Systems was integrated into Ocado Intelligent Automation, the group's warehouse automation and robotics division that builds end-to-end order fulfillment solutions for retail giants and logistics operators. This unit primarily develops software systems, robots, and conventional material-handling systems to automate warehouse workflows.

Within this portfolio, Chuck was positioned as part of Ocado's mobile robotics offering, often referred to as Ocado Mobile Robot Systems (OMRS).[86] The acquisition aligned 6 River Systems with a parent company whose core business was anchored in the warehouse automation orbit. In effect, Ocado secured a proven, commercially deployed mobile robotics platform, while 6 River Systems found a strategic home to continue its long-term mission.

While 6 River Systems ultimately found a strategic home for its robots, the path it took, the decisions it made, the partnerships it established, and the business models it developed to enable enterprise-scale adoption around the world within a matter of just a couple of years after its incubation offer an extremely important set of lessons for the next generation of robotics founders and engineers. We will now cover each of these one by one.

Lessons for Robotics Founders & Engineers

Lesson 01: Founder DNA Matters More Than Capital

In the case of 6 River Systems, the founders brought deep expertise in warehouse automation and robotics and had previously worked at Kiva, the warehouse automation pioneer that was later acquired by Amazon. Even more importantly, one of the founders had previously worked at Shopify, leading its logistics business segment. In summary, this trio had the perfect DNA for this company to succeed. One brought deep expertise in warehouse automation and robotics, another in technology, and the third a deep bond with Shopify. These three capabilities were critical to its success. Hence, founder DNA always outweighs capital, and 6 River Systems exemplifies this principle like no other. Once you have the right DNA in place, the next logical step is to ensure you have a proven customer insight in place, and that leads us to our next lesson.

Lesson 02: Deep Customer Insight Trumps Technological Excellence

The founders of 6 River Systems identified two core problems in the warehouse industry. First, labour accounted for roughly 70 per cent of operating costs for warehouse operators. Second, the technology needed to automate these warehouses was no longer widely available after Kiva's acquisition. The marriage of these two insights is what led to the launch of 6 River Systems and, ultimately, its success.

To summarise, in robotics it is critical to deeply understand customer needs and, more importantly, to ensure those needs are not already being served at scale by other providers. A deep pain point that customers face at scale, combined with few or no competitors addressing that need, creates a blue-ocean opportunity for launching a new product. Basically, in such markets, there is very minimal competition for you as a startup, and that means a big opportunity for you to scale with fewer resources and make an impact.

Now, with the right DNA and deep customer insight, you need to translate the two into the right product, and that leads us to our next lesson.

Lesson 03: Product Discipline: A Lean Two-Product Portfolio

Robotics being such a capital-intensive field, one of the core factors that separates successful companies from others is product discipline. 6 River Systems retained a strong product focus and did not venture beyond two core products until its exit. Kiva did the same with its autonomous mobile robots, and Fetch followed a similar two-product strategy: a mobile robot and a robot arm.

In summary, it is extremely important to craft and stick to a lean product portfolio that ideally should not go beyond two products. Each new product that you add cuts your resources for sales, engineering, development, and operations in half. As a start-up in robotics, the odds are stacked far more heavily against you than in software. That means you have to minimise your resource burn, and one way to do that is to keep your product portfolio lean.

Once you have the right product, the right founder DNA, and deep customer insight, the next logical step is to land the right customers. How do you do that? That is where the next lesson comes in.

Lesson 04: Hub-and-Spoke GTM as a Scaling Engine

There are two different approaches companies take to land customers. One is the sequential model, where you build trust with one customer at a time and then grow in a linear manner. The second is a hub-and-spoke model that was adopted by 6 River Systems, partnering with the honeybees located at the centre of the honeycomb, namely Honeywell, XPO, and DHL.

While the GTM model you adopt will ultimately depend on the vertical you are serving and your product maturity profile, in most cases you are better off adopting the hub-and-spoke model. The returns here are non-linear, to say the least. Even if you spend time and resources working with five centralised authorities in your domain, like Honeywell and XPO in the case of 6 River, and land even two of those, that is much better than landing five customers with a sequential approach, unless you are able to land those with scale.

So there is a trade-off. However, the hub-and-spoke model in most cases will perform better simply because enterprise sales cycles take years to mature into deployments, and as a start-up you do not have that time or those resources. This model speeds up that process for you, and the only downside is shared margins, which is fair given the speed and long-term value it offers to you, your employees, and your investors.

Once you have the right strategic partners and you are able to scale your customer base, the next important question is: what do you do when an industry giant makes an offer for acquisition? How do you evaluate the offer?

Lesson 05: A Mission-Aligned Exit Matters in Robotics

While there are many ways to answer this question and evaluate an offer made by a company to you as a start-up, there is one principle that should serve as a good north star to drive these decisions. Beyond capital, legal terms, and post-acquisition protocols, as a founder one of the key areas to look for in this offer is whether it will allow the company to continue its mission post-acquisition. In other words, will the mission that you, as a founder, started with this company be sustained beyond the acquisition?

If there is a definite yes here and if you have a commitment from the acquiring company, then that is a good starting point to evaluate the offer. If there is no definite commitment here, then no matter the financial or the materialistic incentives associated with that offer, it will not do justice to the people, time, and effort that went into building this start-up and advancing its mission.

As a founder, you have a responsibility towards the underlying mission that goes beyond the lifecycle of your role in the company. If you do choose to exit, ensure you find the right home for your startup. In the case of 6 River Systems, the founders did their part to ensure Shopify was aligned with the mission. While what followed may not have been fully in their control, Shopify honoured that responsibility by finding an appropriate long-term home for the company. A mission-aligned exit matters more than capital terms. Honour the mission.

Now that we have distilled the core lessons embedded in 6 River Systems' trajectory, from its origins through its evolution and the pivotal decisions the team made along the way, it is time to decode the underlying business model.

In the final section of this chapter, we will break down the business model and illustrate how the product portfolio maps to select primary use cases.

6 River Systems: Product Snapshot

Figure 6.1 — Chuck Mobile Warehouse Robot

Illustration by the author. Robot images from TechCrunch article (2017)

6 River Systems' Business Model — Part 1

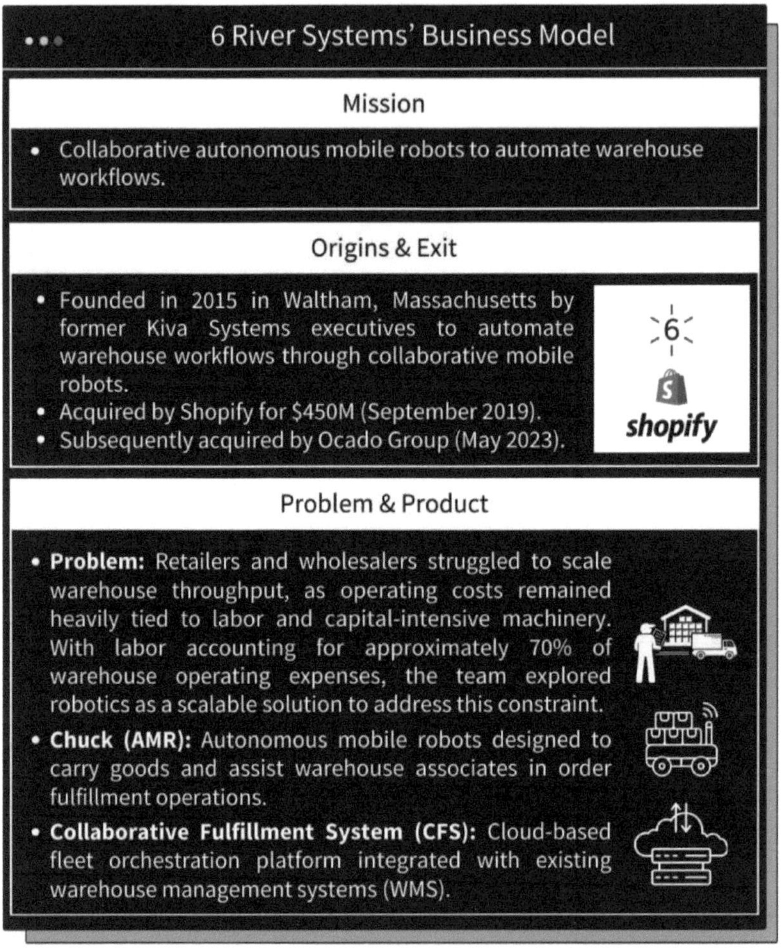

Figure 6.2 — 6 River Systems' Business Model: Part 1

Illustration by the author.

6 River Systems' Business Model — Part 2

Figure 6.3 — 6 River Systems' Business Model: Part 2

Illustration by the author.

6 River Systems' Business Model — Part 3

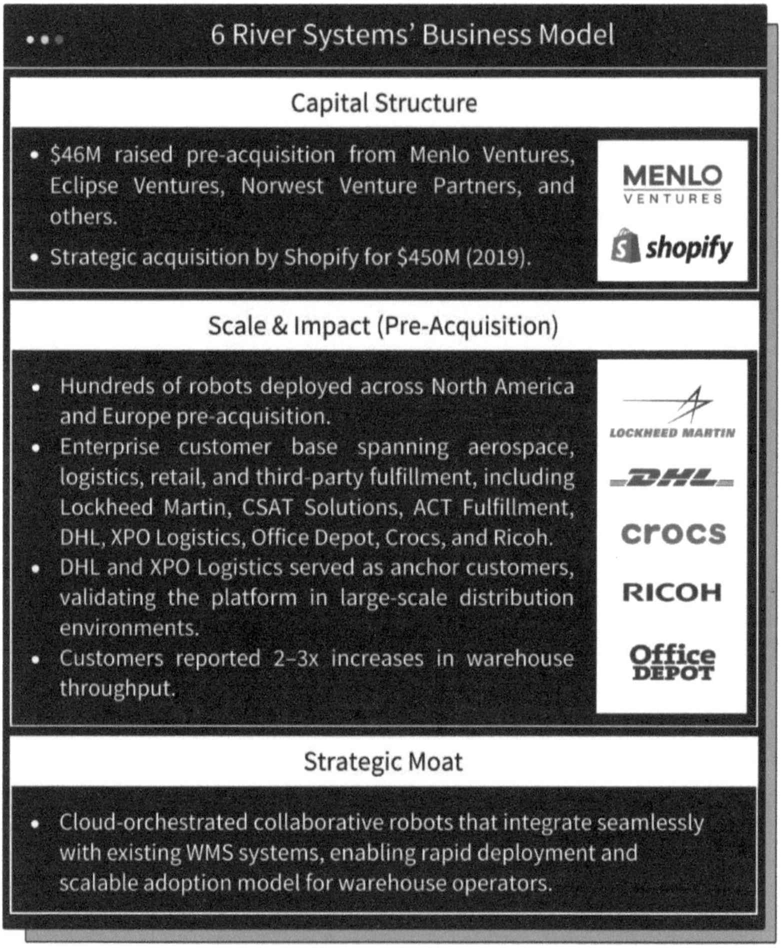

Figure 6.4 — 6 River Systems' Business Model: Part 3

Illustration by the author.

Chapter 07: Symbotic — The F1 Engine Powering Walmart

A robotics company incubated in 2007 by a history major with a deep-rooted passion for technology and funded by a wholesale grocer went on to serve the nation's largest big-box retailers, including but not limited to Walmart, Target, Albertsons, and Giant Tiger, across North America. It ultimately became the sole robotics enabler and acquired the entire robotics division of one of them.

This might sound like a Silicon Valley startup story. But it is not.

The seeds of this company were planted in Wilmington, Massachusetts, by a grocer-innovator duo: one brought firsthand insight into the operational bottlenecks that plagued grocery distributors nationwide, while the other brought a lifelong passion for applied innovation. Together, they launched what is now known as Symbotic.

Today, Symbotic serves several of the world's largest big-box retailers and has secured a long-term contract to power all 42 distribution centers of Walmart, the seventh-largest retailer in the United States. Given its unconventional origins, remarkable velocity, and extraordinary scale, this robotics powerhouse transforming grocery distribution centers across the nation warrants deeper examination. To truly understand its trajectory and the core lessons it offers, we must first decode the DNA of its founders.

To do that, we must travel back in time, all the way to the year 1974.

Founder DNA — C&S Wholesale Grocers (2000s)

In 1974, a fresh graduate of the Wharton Business School at the University of Pennsylvania joins his family-owned business, C&S Wholesale Grocers (C&S), which primarily serves as a small-scale grocery distribution center for local grocers in the Worcester, Massachusetts region. Over the next fifteen years, he works in multiple internal roles before becoming CEO of this company in 1989.[5]

By this time, he develops a clear understanding of the structural bottlenecks embedded in traditional warehouse operations: nearly 70 percent of costs were labor-related, and more than 70 percent of workflows were manual. Throughput was constrained. Productivity stagnated. Efficiency was underwater. This became a major problem for warehouse operators, and the corresponding consequences rippled across the industry in the form of soaring operating expenses, contributing to the demise of many grocers nationwide.

Determined to address this challenge, this young executive launches a series of warehouse modernization efforts aimed at automating workflows across the company's distribution centers. Specifically, he seeks to improve throughput, productivity, and space utilization rates across C&S's growing national distribution network.

As part of this push, in 1989 he rolls out his first major technological innovation: a wireless warehouse management system that enables forklift operators to communicate in real time with other workers on the warehouse floor. At the time, it was a breakthrough and effectively the first wireless warehouse management system in the world.[5]

Building on this momentum, in the late 1990s he spearheads the construction of C&S's first automated warehouse in York, Pennsylvania, using off-the-shelf components and technology stack. The 1.4 million-square-foot facility takes years to complete and, while it does improve operational efficiency, ultimately proves too capital-intensive to justify the returns. In the process, it teaches him one crucial lesson: what a warehouse automation system should not look like.

That hard-earned lesson serves as a launchpad for his entry into the robotics cosmos.

This CEO of a grocery distribution company based in New Hampshire would go on to become the founder of what we now know as Symbotic. This was none other than Rick Cohen, the architect behind the grocery giant.[5]

That raises an important question: how did a grocer leading the nation's largest grocery distribution company meet his counterpart, the history major turned innovator?

Launch of CasePick Solutions (2007)

In 2007, Cohen crossed paths with John Lert, a history major turned inventor and self-taught automation specialist who now runs a robotics company called Alert Innovation. Lert had been working on a concept-driven solution to automate retail distribution centers and was seeking funding to bring it to life. He approached two major grocery wholesalers, Supervalu and C&S. Supervalu declined, but C&S invited him to Keene, New Hampshire, for a meeting with its head of supply chain management.[3-5]

At the time, C&S was transporting enormous volumes of cases across its portfolio of warehouses, making the cost of case handling a strategic priority for the company. Given the potential of Lert's solution to address warehouse throughput challenges the company had wrestled with for decades, Cohen agreed to invest a small sum in the company, then known as CasePick Systems, in 2007. The duo worked together on this system and optimized it for warehouse distribution centers, designing robotic carts, which they called, that could navigate around a warehouse, picking up and dropping off cases filled with goods using a robotic arm.

This partnership lasted until 2011, when Lert transitioned out of the company and Cohen decided to continue its mission by enabling the rollout of their first warehouse-grade robots designed exclusively for C&S Wholesale Grocers' distribution centers across the nation.

First Warehouse Prototype — C&S Wholesale Grocers (NY, 2011)

During this period, the company built and tested eight different robots, all aimed at automating the flow of goods within these distribution centers and elevating their throughput. After these capital-intensive iterations, it finalized one design for production, which was deployed at a C&S warehouse in Newburgh, New York. There, the facility assembled cases full of goods on wooden pallets that were then trucked to nearby stores across the state. This operation was automated by roughly 168 production-grade autonomous robots engineered in-house and deployed at the facility, capable of assembling pallets and transporting boxes at speeds of up to 25 miles per hour.[87] Alongside the underlying hardware that served as the foundation for these robots, the company also engineered their soul: the software required to orchestrate a fleet of them at scale in production-grade warehouses. These efforts paid off big time in the years that followed.

First Enterprise Retail Customer — Target (2014)

At the time, Target was a customer of C&S for frozen goods, and one of its supply-chain executives had gone on a tour of this Newburgh-based robot-automated warehouse and saw the prototypes. While on this tour, when asked what problem Target was trying to solve, the executive explained that the retailer needed more space to keep up with rising sales in the Southwest, but was constrained in how much square footage it could add to its existing facilities—especially at one of its large distribution centers in Woodland, California.[5]

Symbotic's robot-enabled solution was a perfect fit, since it allowed retailers to scale the capacity of their warehouse without scaling its real estate footprint. How was that possible?

The answer boils down to the engineering behind the robots. They were designed to assemble pallets, flat platforms used to stack and prepare goods for outbound shipments, with significantly greater precision and efficiency than conventional labor-intensive methods. This meant Symbotic's robots could handle a disproportionate amount of goods per pallet. In effect, they served as mobile storage units while also automating the task of pallet assembly and transport for these warehouse operators. This was a major advantage and signaled significant gains for retailers like Target on the warehouse throughput front. Why was that the case?

A typical manually operated grocery distribution warehouse might hold 10,000 to 12,000 individual items, while one using Symbotic's technology could accommodate up to 300,000 items within two-thirds of the space.[5] Because these robots were capable of building more efficient pallets than humans, warehouse operators spent less time and effort assembling them. In addition to this core benefit, the robots also functioned as mobile storage units, effectively adding capacity within existing warehouses.

Given these benefits—higher capacity, lower labor requirements, and ultimately greater throughput—Symbotic signed Target, the seventh-largest retailer in the U.S., as its first external customer in 2014.

Following this success, Albertsons came on board in 2017. Walmart came next.

Commercial Validation — Walmart & Albertsons (2017–2018)

C&S had handled distribution for some of Walmart's Supercenters in the 1990s, and they remembered the founder. This relationship enabled Symbotic to secure a meeting with Walmart's executives. Given that Walmart is a big-box retailer operating at national scale, and Symbotic was a startup providing robot-led warehouse automation solutions still in its early stages, the retail giant honored that trust code but demanded a pilot demonstration of Symbotic's system at Walmart's Brooksville, Florida, distribution center, about an hour north of Tampa, in 2017.[5]

During the pilot, while the robot-led automation system performed as intended and demonstrated to Walmart executives that it was able to automate certain segments of its warehouse workflows and raise throughput, the results were not impressive. A mere 30% productivity gain was not enough to land a deal with the nation's largest big-box retailer.

This meant Symbotic and its team had to head back to their labs and improve their product. It was not yet ready for a nationwide rollout. That raises a critical question: what was the bottleneck? Why weren't the robots able to deliver on their promise? More importantly, what exactly was the throughput gain that Walmart expected from these robots?

The challenge was that, during the pilot, the inbound robots, which unpack pallets as they come into the warehouse, were able to provide a throughput of only about 900 cases per hour, which did not justify the cost of a nationwide rollout. In order to secure the contract with Walmart, the founders had to double that number. That meant, from December of that year through the next 12 months, product iteration at a massive scale. The results paid off. In just one year, the system was optimized to handle 1,650 cases per hour, a rate unheard of in the distribution center world.[5]

That number was enough not only to land a contract with Walmart, but also with Albertsons, the second-largest supermarket chain in North America after Kroger.

Albertsons, having tested the fully automated, state-of-the-art Symbotic system at its distribution center in Tolleson, Arizona, in 2018, decided to expand its adoption footprint and launch the system in a second distribution center in Melrose, Illinois. This marked a major win for the robotics startup. Albertsons is one of the largest food and drug retailers in the United States, with more than 2,300 locations across 35 states. The Boise, Idaho–based company operates stores under more than 20 well-known banners aside from Albertsons, including Safeway, Vons, Jewel-Osco, Shaw's, Acme, Tom Thumb, Randalls, United Supermarkets, Pavilions, Star Market, Haggen, Carrs, Amigos, and a meal kit company called Plated.[88]

Contracts with Albertsons and Walmart put the company on the radar of big-box retailers in Canada. That meant it was time to head north.

First International Customer — Giant Tiger, Canada (2018)

Giant Tiger, a Canadian big-box retailer known for operating over 250 stores across the country and serving a customer base of over 40 million, was struggling to scale its warehouse infrastructure and supply chain channels to keep pace with surging demand. The company was in desperate need to double the rate at which it transported goods from its network of warehouses to its 250+ stores.

To address this bottleneck, Giant Tiger, having learned about the Symbotic system that had already been adopted by Albertsons and Walmart, its peers in the U.S., decided to put the technology to the test. In response, it opened a high-tech distribution center in Johnstown, Ontario, featuring the first Symbotic system outside the United States. This 600,000-square-foot facility was equipped with 210 autonomous robots that worked hand in hand with 300+ full-time workers and automated the process of palletizing, meaning assembling pallets full of goods based on the orders received per store at the warehouse.

The system not only delivered a precision rate of 99.99%, but also ensured the warehouse was able to process 100+ large-scale orders received from its network of 250+ stores each day, at roughly twice the speed and efficiency of the conventional systems operating across the retailer's other warehouses.[89]

Given this success, the team at Walmart that had been testing Symbotic's solution at its Florida facility since 2017 was now convinced that these robots were not only capable of delivering the throughput it had originally envisioned, but were also ready for nationwide deployment. This meant scale was now the answer.

The following year, Symbotic was able to land a national-scale deal with the big-box retailer.

National-Scale Expansion, Walmart & UNFI (2021–2022)

On July 14, 2021, Walmart announced plans to deploy the Symbotic system across 25 additional regional distribution centers in the United States, following a successful testing and validation phase at its Brooksville facility. Less than a year later, on May 23, 2022, Walmart, convinced by the throughput gains delivered by the system and its potential to double the speed at which it could move goods through its distribution network to more than 10,000 stores serving over 260 million customers, revised its commitment to support not just 25 but all 42 of its national-scale distribution centers.[91-94]

That deal, already worth billions to Symbotic, put the company on the radar of another distribution giant within just three months: United Natural Foods, Inc. (UNFI).

UNFI is North America's premier grocery wholesaler, delivering a wide variety of fresh products to more than 30,000 stores across the continent. In September of that year, the two companies announced a commercial agreement to implement Symbotic's AI-powered and robotics-enabled warehouse automation system in UNFI's five major distribution centers over the following four years. Under the agreement, UNFI also retained the option to expand Symbotic's systems to additional facilities.[90]

With Target, Walmart, UNFI, and Giant Tiger on board, the company was now ready to kickstart its IPO engine.

NASDAQ IPO & SoftBank Joint Venture (2021–2023)

To accelerate its path to an IPO and secure strong long-term business and financial partners, the company struck a $200 million deal with SoftBank and, through a SPAC-led IPO, landed on NASDAQ at a whopping valuation of $5.5 billion. For reference, iRobot's 2005 IPO valued the company at around $559 million, while Norway's robotics giant AutoStore went public in 2021 at a valuation of $1.8 billion in the country's largest IPO.[5]

Following the listing, Symbotic established a joint venture with SoftBank Group, GreenBox Systems LLC, headquartered in Japan, to deliver warehouse automation as a service to big-box retailers around the world.[95] While Japan opened the door for the company's expansion into Asia, its crown customer Walmart, having rolled out the system across the U.S., was now ready for international expansion as well.

The question was: where should it expand next?

The retail giant decided to head south. Mexico was the next stop.

Latin American Customer — Walmart Mexico, Walmex (2024)

In 2024, Walmart de México y Centroamérica (Walmex) signed commercial agreements to deploy Symbotic systems at two newly constructed sites near Mexico City, marking the company's expansion into Latin America. Under the agreement, Symbotic would first deploy its fleet of robots at a site located in Mexico's Bajío region.[96] This facility will be one of the largest in Walmex's distribution network and would represent the largest system Symbotic has ever deployed in a single rollout in its history.

With thousands of robots now deployed across the retailer's distribution network spanning the U.S. and Mexico, Symbotic was no longer merely a robotics provider to Walmart. It had become a strategic partner, and that meant acquisition was on the horizon. More specifically, a reverse acquisition.

Strategic Platform Expansion — Walmart Advanced Systems & Robotics (2025)

In 2025, Symbotic announced an agreement to acquire Walmart's Advanced Systems and Robotics business for $200 million in cash, alongside a $520 million funding commitment from Walmart to develop store pickup and delivery automation systems for its network of more than 4,000 stores in the U.S.

Symbotic's Next Frontier

Today, Symbotic continues to operate as a publicly traded warehouse automation company headquartered in Wilmington, Massachusetts, with active deployments across North America. While the company has successfully automated warehouses for retail giants across the country, it is now bracing for its next major bottleneck: bringing automation into the coldest corners of the warehouse world.

While the current system supports palletization and transport of dry goods within these facilities, it must now handle wet goods, which demand sustained operation in sub-zero conditions. Getting robots and electronics to function reliably in the arctic conditions inherent in these freezer-dominant warehouse environments presents a new class of engineering challenges.

The system must now become truly mission-critical. From incorporating military-grade computing components to deploying automated lubrication systems and providing sustained maintenance services, operational bottlenecks are emerging at scale. As the company prepares for its next era on the arctic frontier of the warehouse world, the lessons distilled from its current trajectory are astronomically important and should not go unnoticed.

Lessons for Robotics Founders & Engineers

Lesson 01: Founder DNA Makes or Breaks the System

Whether you are a solo founder or part of a founding team, it is important to remember that the DNA of the founders determines the DNA of the system you build. To succeed in the robotics arena, this DNA must include a strong blend of industry expertise and technical capability. In this case, while neither of the two founders had an engineering degree, both brought deep and complementary expertise: one from the grocery distribution center world, with a strong understanding of customers, and the other focused on using technology to address those problems. This marriage led to the launch of CasePick and, eventually, Symbotic.

If you are a solo founder, you must be acutely aware of where your founder DNA is underdeveloped. While you may have deep insights into the industry based on the problem you are trying to solve, make no mistake: you cannot compete with those who have 40+ years of experience in that domain. The breadth and depth of industry knowledge they bring will often outweigh any high-end robotic system a technologist could develop by relying solely on information available online.

Hence, it is important that you carefully evaluate and choose a co-founder with the right DNA. This evaluation process is not optional; it is, in fact, foundational.

Lesson 02: Production-Grade Pilots Beat Big Marketing Budgets

In the case of Symbotic, the company first deployed its own pilot system at scale, with 168 robots operating in a production-grade facility at the C&S grocery distribution center, and demonstrated the resulting throughput gains to its own team. This naturally became fertile ground for its sales efforts. The executive from Target touring this facility experienced the value the system had to offer in a live production environment with his own eyes. As a result, the company was able to land a deal with one of the nation's largest retailers.

The lesson here is simple: you need to land at least one early adopter for your system who is willing to offer their production-grade facility so you can both automate their operation, elevate their throughput, and in return bring other prospects to this site to demonstrate the value your system has to offer. This approach creates a win-win for you, your early adopter, and your prospects. You have to prove your system works in production environments. That is the fastest way to build trust with your customers. This strategy outweighs any marketing strategy you may employ at this stage.

Lesson 03: Chase Few High-Quality Customers That Offer Scale

Whether you are targeting small businesses or enterprise customers, you are better off chasing a few high-quality customers. Quality here matters far more than quantity. These are customers that either have the potential to grow at scale or are already operating at scale. In the case of Symbotic, all four of its early customers—Target, Walmart, Giant Tiger, and Albertsons—were already operating at scale. This implies that even though it might take a six-month-long pilot with each to secure a small-scale deal in the early stages, it is worth it because, from a long-term perspective, it unlocks scale and operational efficiency for a startup trying to build its financial runway.

In the case of Walmart, the pilot system deployed at the Florida facility, even though it did not immediately result in a contract for Symbotic, allowed the team to iteratively improve the fleet of robots operating at the site over a period of three years. This was critical because, after this phase, Walmart returned and decided to roll out the robots across all 42 distribution centers nationwide. That is revenue for a lifetime. Hence, it is important to chase a few high-quality customers that offer scale.

Lesson 04: Industrial Robotics Is a Throughput Arena

In the case of industrial robotics, you are, in most instances, replacing machines with robots. In the case of Symbotic, conveyors, forklifts, and other manual systems were being replaced by robots. That implies your ability to convert a prospect into a customer ultimately boils down to your ability to demonstrate that your system delivers at least a 50% improvement in throughput. If you can prove that in a real warehouse setup, the case is made. The deal is sealed.

In the case of Symbotic, the company first proved this at the C&S grocery distribution center with a fleet of a staggering 168 robots, which led to the Target deal, and then at Walmart's Florida facility, which took three years to materialize but ultimately landed the company a deal of a lifetime with the retailer.

On the other hand, consumer robotics is about habit change. You are asking individual customers to change their behavior, and that is a much harder game. Habit change requires time. Hence, if you are in industrial robotics, find a way to demonstrate the throughput gains your system brings in a production-grade environment, and that will help you close more deals than any sales demo ever will.

Now that we have distilled the core lessons embedded in Symbotic's trajectory, from its origins through its evolution and the pivotal decisions the team made along the way, it is time to decode the underlying business model.

In the final section of this chapter, we will break down the business model and illustrate how the product portfolio maps to select primary use cases.

Symbotic: Product Snapshot

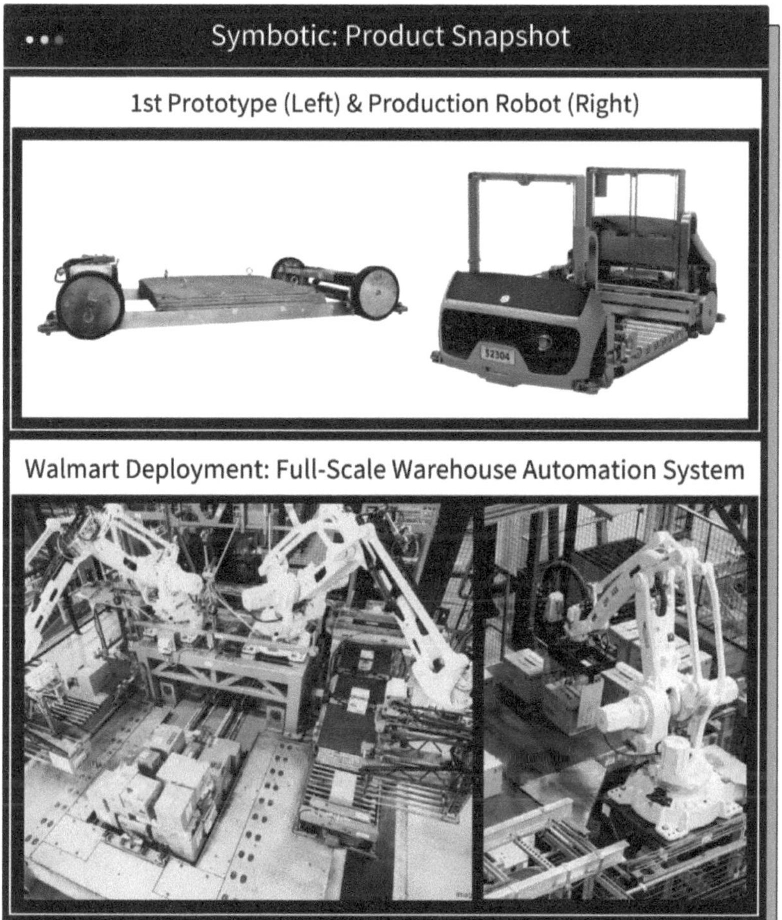

Figure 7.1 — Symbotic Prototype & Production Robot

Illustration by the author. Robot images from Symbotic "Our Story" page.

Symbotic's Business Model — Part 1

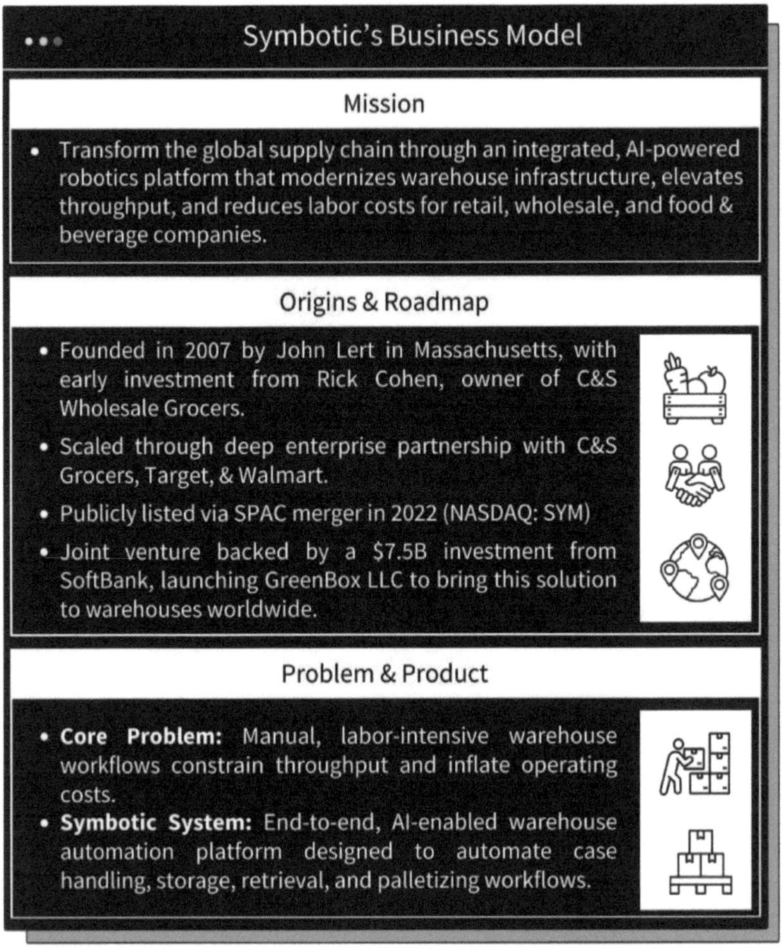

Figure 7.2 — Symbotic's Business Model: Part 1

Illustration by the author.

Symbotic's Business Model — Part 2

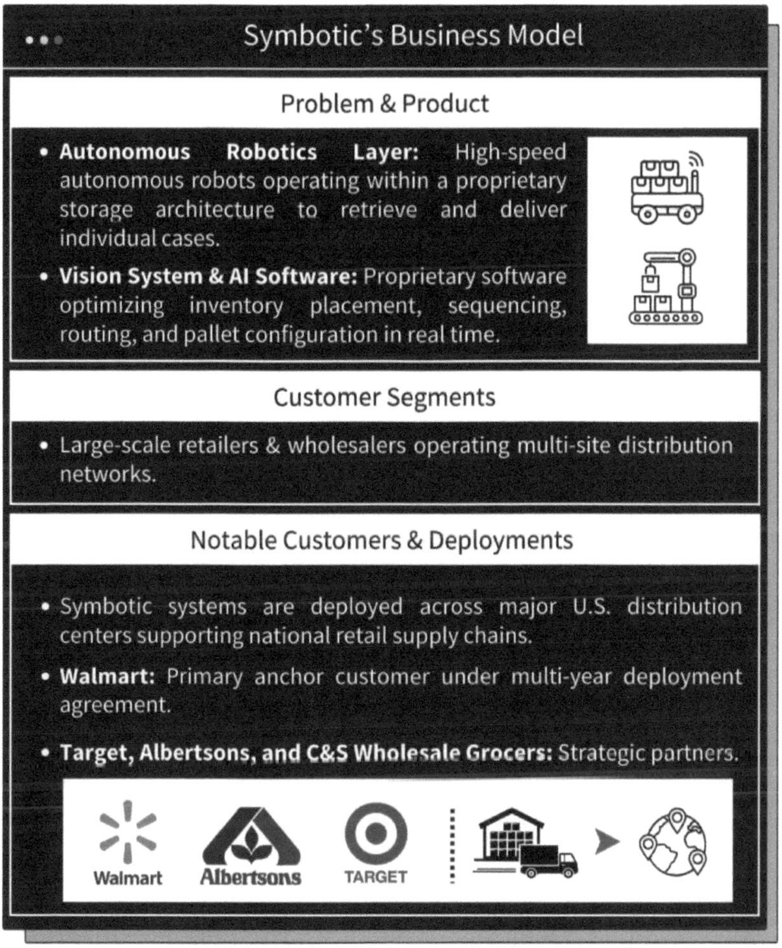

Figure 7.3 — Symbotic's Business Model: Part 2

Illustration by the author.

Symbotic's Business Model — Part 3

Figure 7.4 — Symbotic's Business Model: Part 3

Illustration by the author.

Symbotic's Business Model — Part 4

Figure 7.5 — Symbotic's Business Model: Part 4

Illustration by the author.

Chapter 08: Clearpath — University of Waterloo Lab to Rockwell Exit

1999 was one of the most important years in the history of Canada. The country officially gave birth to one of the most beloved consumer devices in the world — BlackBerry. Born and forged in the tech nestbed of the Silicon Valley of Canada, i.e., Waterloo, it seeded the entrepreneurial spirit in the soul of this country. That soul was revived a decade later in 2009, when a group of engineers at the nearby University of Waterloo decided to carry this spirit forward and launch what is now referred to as the BlackBerry of the robotics world in Canada, i.e., Clearpath Robotics.

Incubated in the basement of one of these engineers in Waterloo, the company was founded with a singular mission: build the best mobile robots in the world exclusively to accelerate academic research in robotics. With this mission finalized, the founders secured early adoption at leading institutions around the world including but not limited to MIT, UC Berkeley, CMU, Princeton, University of Virginia, and Portugal's historic University of Coimbra, founded in 1290. Over time, these early deployments helped the founders nurture their product through the feedback received from academia and mature it into an industrial-grade platform that now powers global industrial giants such as Toyota, GE, Bosch, and even the U.S. Army.[97]

A company that was incubated in 2009, established a global footprint across more than 35 countries, built strong adoption roots spanning academia, agriculture, defense, manufacturing, and the logistics vertical, and was acquired by Rockwell Automation for $600M in 2023 within a span of a mere 14 robot years, which in the robotics startup world is roughly equivalent to five years in the software startup world, warrants deeper examination.[98-99]

The lessons embedded in the journey of these founders, who were fresh university graduates at the time, and the path they took to fund their dream, pivot their product strategy, and scale their customer footprint across academia and industry, offer insights worth studying. To better understand these lessons, we must travel back in time to the year 2009.

Founder DNA — University of Waterloo Engineering (2009)

Back in 2009, four fresh graduates from the University of Waterloo's Mechatronics Engineering program, who were at the time involved in the university's robotics team, decided to carry their final-year project work in robotics beyond academia after graduation. The underlying thesis that had led to this project was that there are many jobs today that are strenuous, repetitive, and, most importantly, dangerous. Think about the workers who labor in tunnels 300 meters below ground to extract the minerals we need to build modern products. Think about those who work on oil rigs to extract the fuel that powers our cities, and those in chemical factories producing the lubricants and colors we rely on every day. These founders wanted to address those problems, and the solution they devised was a mobile robot that these professionals could operate remotely, thereby enabling them to work in hazardous environments without putting their lives at risk.

With this thesis in mind, the founders decided to launch their first prototype, a four-wheel-drive mobile robot, in the basement of one of the founders in Waterloo. Ferrying between this basement and a rather cozier space provided by the university startup accelerator, the Accelerator Centre at the University of Waterloo Research and Technology Park, the team started building this first prototype. These efforts culminated in the launch of what the company now refers to as its outdoor autonomous robotic platform, Husky. Those four engineers were Matthew Rendall, Ryan Gariepy, Patrick Martinson, and Bryan Webb, the four co-founders of Clearpath Robotics.[100-101]

Matthew Rendall (Matt) became the CEO. Ryan, with deep expertise in mechanical engineering, assumed the position of CTO. Patrick formally joined as co-founder, while Bryan took charge of angel fundraising efforts and the company's financial strategy.

That raises an important question: how did they fund this prototype?

With angel funding and grants from government agencies such as the Ontario Centres of Excellence and the Natural Sciences and Engineering Research Council of Canada, the team managed to secure enough funds to build this prototype, namely the Husky Unmanned Ground Vehicle (UGV). With funding secured and the first prototype ready, it was time for these founders to secure their early adopters. Given their extensive background and interest in nurturing research projects in academia, they decided to focus on local universities in Canada, and first among them was the University of Toronto.

First Field Deployment — Husky UGV, University of Toronto (2010)

The Institute for Aerospace Studies at the University of Toronto, in 2010, had developed an algorithm for the Canadian Space Agency to enable detailed mapping of the lunar surface. To test this algorithm at its local Mars Dome facility, the university was exploring a research platform to mount its heavy-duty laser sensor. Given the characteristics of this payload, the university needed a rugged mobile robotic platform.[102]

The research team evaluated multiple robotic platforms available in the market but ultimately selected the Husky UGV by Clearpath, as it was one of the few robots that could carry its laser scanner and yet fit inside the testing facility—thus serving as a compact yet mission-critical solution for this test.

This level of product–market fit, established at such an early stage in the academic domain, i.e., its target domain for this phase, meant the company was now on the radar of academic institutions across the U.S. and Canada.

Early Academic Adoption — Universities & Funding Constraints (2010)

Over the next few months, the company inked deals with more than half a dozen universities spanning North America, including the University of California, Berkeley, one of the leading engineering schools in the region, to drive adoption of its robots in the research labs of these institutions. Specifically, around this time, the company also managed to land a fairly large-scale deal with the University of Calgary.[100] This deal provided the company with enough capital to stay afloat and sustain its operations for the next three months.

Given how capital-intensive it is to run a robotics startup, the founders knew this deal was a sword hanging over their necks. Once the fresh supply of capital infused through this deal was exhausted, their operations would come to a standstill. They had to extend their runway, and that meant they needed to figure out ways to drive adoption for their product.

The question was: should they continue pitching this outdoor robotic platform to universities to drive adoption, or was there an alternative unmet need in the robotics research community that they should give a second thought?

Given this dilemma, the founders reflected on the numerous conversations they had had within academia with researchers. They took a step back to assess the series of unmet needs that were still not being addressed by their outdoor robotic platform in the research community.

At the time, they realized there were a growing number of research projects centered on maritime robots, designed exclusively for deployment in open-water environments. In addition to this multi-terrain requirement, robots capable of carrying heavy payloads were also in high demand.

Armed with these direct insights gained from academia, and the demand they had validated through ongoing conversations, they decided to expand their product portfolio and launch two new products.

Platform Expansion — Kingfisher M200, Maritime Applications, and Grizzly, Heavy-Duty Operations (2012–2013)

In November 2012, the company launched its first amphibious robotic platform, called the Kingfisher M200, to enable maritime research applications in academia. In addition to the M200, the company also launched its heavy-duty robot, called Grizzly, an ATV-sized mobile robot capable of carrying payloads of up to 600 kg for research and mission-critical applications. It demonstrated these products at local trade shows and to nearby universities during that year. Now, the company officially had robots that could traverse land and water, and more importantly, its potential customers were well informed about these products.

Given this launch in the middle of peak Arctic season that year, one of the founders, Matt, who was driving on a snowy winter day in Waterloo, received a call from a gentleman named Alejandro Ramirez-Serrano. This wasn't just any other call, but an open inquiry from a University of Calgary engineering professor about

Clearpath's robotic platform and, in particular, the need for an amphibious robotic platform for his research project.[100] The professor had been searching for a robotic platform for his research project centered on harbour security, but all he could find were expensive robots made by suppliers serving the defense industry. He had seen Clearpath's product that year at a local trade show and wanted to inquire about it.

After a conversation that lasted for less than half an hour, in the middle of a frosty winter, Matt heard those words that every robotics founder pines for: "Let's do it." Matt could hardly believe that the professor was interested in actually buying his robots. This call, which came on a snowy winter day, resulted in an immediate deal with the university for the M200 robots.[100]

Later that year, in November, Clearpath Robotics signed yet another deal with members of the Canadian Innovation Commercialization Program to provide M200 robots to accelerate maritime research at the Water Survey of Canada, an organization focused on mapping water bodies across the country. These robots enabled researchers and workers to autonomously map local rivers, streams, and pond floors, particularly in shallow, inaccessible, dangerous, or remote areas, thereby helping advance environmental research programs while minimizing risks to their lives.

With this growing adoption footprint in academia across North America, the company had officially cemented its name on the global academic canvas. Universities across North America and Europe had started exploring mobile research platforms for a range of applications, and the company's name had made its way to Portugal.

High-Risk Field Research — Landmine Detection, University of Coimbra (2014)

In 2014, researchers at the University of Coimbra in Portugal, one of the most historic universities in the world, officially chartered in 1290, were exploring a mobile platform to build an autonomous robot for landmine detection, otherwise referred to as minesweeping in the field. Their underlying thesis was that minesweeping is an extremely dangerous process that is still largely carried out by human operators.

This process involves detecting, locating, neutralizing, and extracting explosive mines in order to make corresponding waterways or land areas safe. About 20,000 people die each year trying to manually detect mines in these hazardous environments. With 110 million active landmines in the world, only about 100,000 are cleared each year using existing processes. Their thinking was that, at this pace, it would take 1,100 years for humans to clear all active mines on the planet.

To address this challenge, Dr. Lino Marques, a faculty member at the university, launched a research program to evaluate robotic platforms that could automate this process and minimize the risks faced by human operators. Based on their research, the team decided to reach out to Clearpath Robotics, given that its platform was readily available for academic applications, rugged, offered the payload capacity needed to mount the mine-detection equipment, and was equipped with autonomous navigation capabilities.[103]

With a 30-minute setup time out of the box and support for one of the most widely adopted open-source robotics operating systems called ROS, the team not only selected this platform to accelerate their research but, within six months, integrated their payload, and deployed these robots in the field to detect live mines.

That six-month runway, from customer adoption to deployment in the field, meant Clearpath now had sufficient traction in academia and enough capital to scale its product portfolio. The question was: how should they scale this portfolio? After having already launched robotic platforms for mission-critical applications and maritime research, where should the company head next?

To answer that question, the founders reflected on the feedback they had received during this phase as they expanded their footprint across Europe and North America. The team realized that one of the major barriers to adoption of their outdoor mobile robot, Husky, was cost, and the second was its size and profile.

While most researchers in academia needed rugged platforms for mission-critical and maritime applications, many were also looking for low-cost systems that were easy to program, easy to procure, offered a lighter profile, and could operate at higher speeds. Husky was not designed to address this need.

In addition to this realization, the founders, around this time, having conducted a deployment in Europe for researchers in the defense sector to enable autonomous landmine detection, received a prophetic request from the U.S. Army Research Laboratory (ARL) to develop an affordable, portable research platform for mission-critical applications.

Now, with this incoming request from a potential customer in the defense industry, along with continued feedback from academic researchers, the company decided to address this demand head-on with the launch of its next product, the Jackal Unmanned Ground Vehicle (UGV).

Product Line Expansion — Jackal UGV Launch, IROS Chicago (2014)

At the International Conference on Intelligent Robots and Systems (IROS) in Chicago in 2014, the company launched its portable, weatherproof, low-cost, rugged mobile robot platform, called the Jackal UGV.[104] In addition to being portable and durable, the robot was equipped with pre-coded programs that allowed universities and defense operators to focus more on developing applications on top of the platform, and less on navigation- and perception-related bottlenecks.

The platform was ROS-compatible and came equipped with all the standard sensors needed to enable safe and reliable autonomous navigation in outdoor environments, including, but not limited to, GPS, IMU, LiDAR scanners, and more.

This launch significantly raised demand for the product, which meant the company now had to secure funds to deliver on incoming contracts. It needed capital to manufacture these products and execute these contracts.

That meant it was time to extend its runway, and time for Series A.

Industrial Pivot — Series A Funding & OTTO Motors Launch (2015)

In 2015, the company raised a Series A round of US$11M led by iNovia Capital, with participation from Caterpillar Ventures, RRE Ventures, and GE Ventures. With this funding secured, Clearpath positioned its robots as autonomous mobile platforms capable of automating tasks across domains ranging from academic and environmental research to industrial automation and defense applications.

With this positioning, and fresh capital in place, in addition to scaling its production, the company decided to launch its indoor robotics division to serve warehouses, called OTTO Motors.[105-107]

The OTTO Motors division was incubated to develop self-driving vehicles exclusively to automate material transport inside manufacturing facilities, factories, and warehouses across the country. The underlying thesis was that these sites operate like small indoor cities, complete with roads, traffic, intersections, and pedestrians. Unlike city streets, however, a factory floor is a reasonably controlled environment, making it an ideal place to introduce self-driving vehicles at scale.

At the time, companies like Google and Uber were already testing autonomous vehicles on public roads. Clearpath wanted to offer similar capabilities for indoor environments, which led to the launch of OTTO Motors, autonomous mobile robots designed specifically for warehouses and factories.

With the launch complete and funding in place, the Clearpath team spent the next year refining the platform and started pitching it to industrial customers. These efforts resulted in multiple contracts within the industrial sector, and that meant the company had to ramp up its production. To ramp up production, it needed capital. It was time to raise funds. It was time for Series B.

Series B Funding & Amphibious Warthog UGV Launch (2016)

In 2016, Clearpath raised a $30M Series B round led by iNovia Capital, with participation from Caterpillar Ventures, GE Ventures, Eclipse Ventures, RRE Ventures, and Silicon Valley Bank. The goal was clear: scale the production of its four robotic platforms—the Husky and Jackal UGVs for outdoor environments, OTTO warehouse AMRs for indoor logistics, and the Kingfisher M200 for maritime applications.[108]

With this fresh supply of oxygen, the company upgraded its Kingfisher M200 platform and subsequently launched a mission-critical variant called the Warthog UGV, capable of traversing both land and water. With all four platforms ready for scaled deployments, it was now time to take them to market. It was time to ink deals in academia.

Academic Scale Adoption — MIT, University of Waterloo, Queen's University (2017–2018)

In 2017, the company launched a samurai-styled, highly disciplined sales effort to scale adoption of its four-line robotic platform portfolio within academia, inking deals with institutions such as MIT, the University of Waterloo, Queen's University, and others. The first such deployment was with the University of Waterloo.

University of Waterloo: Robot-Enabled Bridge Stabilization System

In 2017, researchers at the University of Waterloo's Structural Dynamics Lab had developed a novel and cost-effective mobile damping solution to stabilize lightweight bridges that are often deployed in disaster-relief scenarios by first responders and military professionals. The research team deployed its electromechanical damper on Clearpath's Husky UGV, evaluated it on one such bridge, and was able to generate inertial control forces to counteract the vibrations caused by pedestrian or vehicle activity.[109]

The way this system works is that, once the damper-enabled Husky UGV is positioned on the bridge, it senses the vibrations generated by the bridge due to the corresponding traffic patterns emerging on top of it and develops counterforces that are magnified by the body dynamics, i.e., the rugged profile of the Husky. These counterforces are applied at specific locations on the bridge that are critical to neutralizing structural vibrations. The Husky-enabled mobile damping system autonomously steers itself to appropriate locations on the bridge at regular intervals, based on the emerging traffic footprint, applies these control forces, and stabilizes the bridge.

The choice of the Husky UGV was driven by the university's need for a rugged mobile robot that could carry the required payload, i.e., the electromechanical damper, offer a low profile, and feature lugged tread tires, which were critical for providing the necessary static friction to prevent sliding and enable effective transfer of control forces.

During this project, one of the unique challenges the team faced was localization. The bridge structure appeared essentially as a rectangular tunnel on the ROS map, which made localization extremely challenging. To address this challenge, the team added QR tags along the bridge structure, which enabled the robot to localize itself at regular intervals of time and thus resolved the issue. When tested on the research bridge constructed by the team, this solution led to a 70% reduction in the lateral displacement of the bridge. In effect, the system was able to counteract and reduce bridge vibrations by approximately 70%, making the structure significantly more stable and resilient.

While this research team in Canada was able to test its algorithms to stabilize these mission-critical structures using Clearpath robots, another research team further south wanted to test theirs, this time to enable safe autonomous navigation in pedestrian-heavy outdoor environments. That meant it was time for Clearpath to take its robots to the research testbed at the Massachusetts Institute of Technology (MIT).

MIT Develops Autonomous "Socially Aware" Robot Using Jackal UGV

The research team at MIT was developing an autonomous navigation algorithm to enable safe operation in urban environments that are typically pedestrian-heavy and devoid of lane markers catered toward pedestrians. Their goal was to ensure the robot doesn't collide with pedestrians or vehicles, while still maintaining sufficient speed and battery life to travel from one end of the city to another in a safe and reliable manner.

Although the team initially developed and tested the algorithm on a custom-built, in-house robotic platform, the system proved inadequate for real-world urban testing. It was not robust enough to support high-speed operation in urban environments, respond effectively to emerging traffic patterns, or navigate the city on a single charge. To address this challenge, the team, after evaluating several platforms, settled on the Jackal UGV, the company's low-cost, high-speed, and easily programmable outdoor mobile robot platform.[110]

The robot was outfitted with a Velodyne VLP-16 LiDAR to enable mapping and obstacle detection, an array of webcams for pedestrian detection and classification, Intel RealSense R200 cameras for depth sensing, and ultrasonic sensors mounted to the bumper for close-range obstacle detection. Additionally, given that these robots were ROS-compatible, the team was able to transition to this platform with minimal setup. Not only did the robot do its job, but this research paper was successfully published at the International Conference on Robotics and Automation, and received the Best Student Paper Award at IROS 2017. With this success, the Jackal UGV gained significant visibility in the research community—cementing Clearpath Robotics' reputation as a leading robotics platform provider in academia.

The company had officially etched its name on the NYSE of the academic world, and it was now on the radar of universities across North America. The next stop on this journey was Queen's University, Canada.

Queen's University — Autonomous Mine Mapping Operations

Queen's is the only university in Canada that has a well-equipped explosives test facility and is well known around the world for its research in defense, and especially in the mining vertical. Researchers at the Mining Systems Laboratory (MSL) were building an algorithm to map the internal structure of mines and predict lethal rock falls and rock bursts at these sites, thereby enabling safer mining operations for workers. This process is formally referred to as geotechnical mapping of mines.

At the time, this process was largely manual and conducted by mine workers, who would carry a handheld device, insert it into nearby rock faces, collect data, analyze it, and then repeat the process across different areas of the mine. In addition to being tedious, the work was extremely dangerous, with workers operating under the constant threat of sudden rock bursts that could trigger deadly showers of debris and hazardous gaseous emissions. Given how time-consuming and unsafe the existing process was, the research team sought to automate it using their algorithm.

To test this algorithm, they needed a mobile robotic platform that could carry the required equipment, provide strong battery life to navigate hundreds of miles in these tunnels on a single charge, and enable autonomous operations.

The company's Husky UGV proved vital here, as it not only met these requirements but also enabled the team to deploy their algorithm seamlessly on the platform and test it within mining facilities.[111] Not only did the robot do its job, but the team's algorithm was accepted into the IEEE Transactions on Robotics journal and was widely recognized by geologists around the world as a novel and safer way to conduct these operations.

Building on the success seen by researchers in Canada mapping mines, a corresponding research team at the University of Virginia, already working on a similar effort to map a historic railroad tunnel, decided to adopt the company's cost-effective mobile platform, the Jackal UGV.

University of Virginia — Historical Railroad Tunnel Mapping

Researchers at the University of Virginia were tasked with mapping the Blue Ridge Tunnel, a historic railroad tunnel running through the Blue Ridge Mountains in Central Virginia. Decommissioned in 1944, the tunnel had remained closed to the public for decades due to safety concerns. As part of a county-funded restoration effort to reopen the tunnel as a pedestrian rail trail, project organizers required a detailed assessment of the tunnel's structural condition. For this work, they turned to the research team at the University of Virginia, to explore the possibility of using a robot to carry out this tunnel inspection process.[112]

The primary goal of the research was to generate a high-resolution 3D map of the tunnel and compare it against historical records to understand how the structure had changed over time. At the time, existing inspection methods were largely static and relied on expensive, long-range LiDAR systems deployed at the entrance of the tunnel. While effective at long distances, this static setup in reality produced a lot of anomalies and blind spots on the 3D map, as it could not capture regions occluded by rocks, uneven walls, and other obstructions within the tunnel. Hence, the team needed a rugged mobile robot platform that could navigate this wet and uneven terrain with ease and was ROS compatible.

To address this challenge, the researchers adopted the company's low-cost rugged platform, i.e., the Jackal UGV, and equipped it with two LiDAR sensors to generate horizontal and vertical scans of the tunnel, which they could fuse into a coherent 3D map.[112] Specifically, they used a Velodyne VLP-16 3D LiDAR and a Hokuyo 2D LiDAR. The Hokuyo was mounted vertically to capture the tunnel's cross-section, while the Velodyne was mounted horizontally to generate dense 3D point clouds along the tunnel walls and ceiling. Combined with onboard IMU, encoder data, and a ROS-based SLAM pipeline, the system enabled the team to collect high-resolution spatial data while tele-operating the robot through this one-mile stretch of the tunnel. This data was then processed using ROS to generate a detailed 3D model of the tunnel, which was then shared with the preservation team to kick-start restoration efforts.

The research team confirmed that with the adoption of the Jackal UGV platform, they were able to complete the survey nearly ten times faster, and with greater precision, than traditional inspection methods, thereby providing actionable data to the preservation team to accelerate restoration efforts.

With these deployments addressing the needs of researchers across academia, from stabilizing bridges to automating mining operations and mapping tunnels, the Clearpath team had demonstrated that their robots were capable of carrying a diverse range of sensor payloads and navigating harsh environments. While these robots had proven effective at enabling autonomous mobility solutions and transporting payloads for these researchers, they were not yet capable of physically interacting with their surroundings. They were not equipped with spatial manipulators.

The Next Bottleneck: Spatial Manipulation

This was a major bottleneck for academic researchers at the time, as a vast majority of these research projects had now matured beyond remote sensing and were progressing toward direct physical interaction with their environment. At this stage, many of these teams increasingly needed the ability to extract, transport, and analyze samples from these hazardous environments. Mobility and rugged design alone were not the success criteria; spatial manipulation was.

This surge in demand meant it was time for Clearpath to elevate its offering, i.e., equip its fleet of robots with dexterous and affordable arms. To address this demand in the market, the company inked a partnership with Universal Robots.

Partnership — Robot Arm Integration with Universal Robots (2018)

In 2018, Clearpath Robotics formalized a partnership with Universal Robots to integrate its mobile robot platforms, including Husky and Jackal, with collaborative robot arms developed by the Danish robotics company. This partnership enabled Clearpath to offer academic researchers fully integrated mobile robotic platforms equipped with ultra-dexterous 7-DOF manipulators, specifically the UR3, UR5, and UR10, supporting payloads of 3 kg, 5 kg, and 10 kg, respectively.[113]

The objective was to deliver standardized, pre-integrated, ROS-compatible platforms fully equipped with robot arms to accelerate the adoption of Clearpath's autonomous ground robots across academic and industrial verticals. These newly launched integrated models allowed researchers and early adopters to focus their time and resources on developing applications on top of these platforms rather than being consumed by downstream challenges in the stack, such as autonomous navigation and system integration.

In simple terms, Clearpath's robots were not only capable of traversing hazardous sites and carrying heavy payloads, but could now interact directly with their environments. This integrated model turned out to be a timely boon for the company, as it directly addressed an urgent and unmet need in the market, especially across academia, defense, and agriculture.

The company was now on the radar of researchers at the U.S. Army Research Laboratory and at Clemson University.

Defense & Agriculture Research — U.S. Army & Clemson University (2019)

The U.S. Army is the largest of the six branches of the United States Armed Forces and conducts land-based military operations. Within the Army, the U.S. Army Research Laboratory (ARL) brings together government and private-sector partners to lead defense-focused research programs. In 2019, as a part of one of these programs, researchers at ARL developed a novel autonomous navigation algorithm designed to enable mobile robots to efficiently navigate diverse, unmapped outdoor terrain.[114]

Traditional approaches to terrain mapping were extremely compute-heavy, time-consuming, and inefficient. To address this challenge, and to ensure the robot fully understood the cost of each route it evaluated across a given terrain, including factors such as risk profile and operational efficiency, the team developed a novel machine learning–based approach and started looking for a suitable platform to test this algorithm.

Clearpath's mission-critical outdoor robotic platform, the Husky UGV, proved valuable in this effort, allowing the team to test the algorithm using two of these robots. One robot ran the newly developed terrain classification algorithm, while the second served as a verification platform. By comparing outputs between the two, the team was able to identify errors and refine the model. This setup accelerated the development timeline by eliminating the need to build custom robots and allowing the researchers to focus on the application layer rather than the navigation stack.

Building on this momentum, researchers at Clemson University adopted the Husky UGV to address labor shortages and high equipment costs in the agriculture sector, specifically in cotton farming.[115] Traditional cotton harvesting methods rely on expensive machinery that can cost upwards of $700,000. Most farmers cannot afford these machines and instead opt for manual methods that are both labor-intensive and difficult to scale in a cost-effective manner.

To address this challenge, the researchers developed an algorithm to automate the cotton harvesting process using robotics principles. Given that the task is repetitive and physically demanding, and largely comes down to perception, navigation, routing through the field, and spatial manipulation, namely harvesting the crop, the team set out to build an affordable robotics solution for farmers.

While the team had already developed and tested their algorithm on in-house robotic platforms, deploying it in outdoor, production-grade farms meant the team needed a rugged mobile platform capable of carrying the necessary payload, including harvesting equipment, and operating reliably in wet and uneven terrain. After evaluating platforms available in the market, they chose the Husky UGV, as it directly met these requirements.

Using the Husky UGV as a base platform, the team developed a proof-of-concept mobile harvesting system capable of navigating outdoor terrain through GPS-enabled waypoint navigation, along with a custom suction-based mechanism designed to collect and store cotton from the field. This work resulted in the launch of a low-cost autonomous cotton-harvesting robotic platform for farmers who had previously been unable to scale production using expensive industrial machinery or a large labor workforce.

Across both defense and agriculture applications, Husky UGV served as an extremely rugged and agile platform, one surgically developed to accelerate research projects across industry verticals. This cross-vertical validation implied that these robots had officially matured enough to serve customers looking for outdoor robotic systems across the industry—and that meant it was time for the company to head indoors. It was time to enter the factories and warehouses.

OTTO Motors Expansion — Funding & Industrial Customers (GE, Toyota, Nestlé) (2020)

Having launched its indoor robotic platform under the OTTO Motors division in 2015 to enable autonomous transport of goods in warehouses and factories, the team at Clearpath, over the next five years, inked large-scale deals with enterprise customers including GE, Toyota, Nestlé, and Berry Global. What is important to note here is that by 2020, more than 70% of its deployed robots were operating within the facilities of Fortune 500 companies across the U.S., Canada, Mexico, and a handful of sites in Japan.[116]

While the expansion into Japan was strategic at the time, over the years, after having done a few deployments overseas, the company realized that the farther it moved away from its home base, the more expensive it became to service and maintain its robots for customers. Given this lesson, the team decided, with this round of funding, to focus its expansion efforts in North America.

To support growing demand on the continent and scale its production capacity, OTTO Motors raised a $29 million Series C round that year, led by Kensington Private Equity Fund, with participation from Bank of Montreal Capital Partners, Export Development Canada, and existing investors iNovia Capital and RRE Ventures.[116] This fresh supply of capital was used to scale its footprint, grow its partner network, and accelerate the development of enterprise-grade fleet management software that customers needed to manage these growing fleets of warehouse robots across their facilities.

This was a critical phase for the company. Clearpath, with its fleet of indoor and outdoor robots, was no longer a Waterloo-based provider of robotics platforms for academia and factories. By this point, it had deployed thousands of robots across the continent and across verticals ranging from agriculture to defense, and from academia to factories, railroads, and mines. It had essentially become a core artery responsible for mobilizing resources both in some of the most dangerous sites on earth and within factories and warehouses.

That meant the company had now etched its name across the industry, and with its accelerating adoption footprint, it was now on the acquisition radar of the giants of the manufacturing world—companies providing the core operating system for industrial operators.

Strategic Exit — Acquisition by Rockwell Automation, ~$600M Deal (2023)

On October 2, 2023, Clearpath was officially acquired by Rockwell Automation for approximately 600 million.[98-99] This was a strategic acquisition by Rockwell, a global industrial automation leader providing hardware and software solutions to optimize manufacturing processes for some of the world's top-tier enterprises, including but not limited to Rivian Automotive, Inc., Abbott Laboratories, ExxonMobil, Amgen, Walmart, Amazon, Procter & Gamble, Ford, and others.

These customers were already using Rockwell's control systems in their manufacturing facilities and on their assembly lines to orchestrate workflows across industrial machines and human workers. A good example here is a simple controller that directs a series of machines on the assembly line to weld, bolt, paint, or inspect parts as raw parts move across these stations in the factory. These workflows are orchestrated using controllers provided by Rockwell and this control layer consists of both hardware and software components.

Now, the challenge these customers faced was that the existing solution addressed only one end of their need, orchestrating workflows across static machines deployed on the factory floor. It did not help automate the movement of goods across these facilities, i.e., between worker stations, inventory areas, assembly zones, and other operational areas. Once a part was assembled at one station, it had to be transported to another station, either using labor-intensive methods such as forklifts or static conveyor-based solutions, which were often inefficient.

As a result, the existing solution was missing a critical piece needed to automate the transport of goods within these facilities. The OTTO AMRs provided by Clearpath directly addressed this need by enabling customers to transport goods and parts within their facilities in an autonomous manner using indoor mobile robots.

Beyond autonomous transport, there was another gap that existing solutions did not fully address. These customers were relying on a fragmented array of solutions provided by independent vendors, including Rockwell and others, to orchestrate their manufacturing workflows spanning a wide range of tasks, such as building parts, assembling them, transporting them, and inspecting facilities as a whole. While Rockwell's solutions automated production workflows, and OTTO robots enabled autonomous transport, the inspection layer was still incomplete. These customers still had to rely on independent vendors to automate that layer.

To close this gap, Rockwell repositioned Clearpath's Husky and Jackal robots as mobile inspection systems post-acquisition in 2023. This offering was officially launched as the Husky Observer system. The company, in its publicly released statement, cited that this integrated solution, consisting of Rockwell controllers, OTTO AMRs, and the Husky Observer system, had long been a dream of plant managers across industries ranging from automotive to consumer packaged goods. As a result, operators could now scale production without having to deal with integration bottlenecks and multi-vendor dependencies that had previously constrained these facilities.

Post-acquisition, in 2025, Rockwell launched a large-scale production facility at its headquarters in Milwaukee, Wisconsin, to scale manufacturing of Clearpath robots, particularly the OTTO AMRs, and deliver this integrated solution to its customers across more than 100 countries.[117] With annual revenue for fiscal year 2025 at approximately $8.2 billion, the acquisition was expected to elevate this number by 1% or about $82 million in the near term, a significant contribution for an organization of this scale and caliber.

Summary

For a robotics startup that came to life in a cold Arctic basement in Waterloo in 2009, with the dream of becoming the BlackBerry of the robotics world, this journey was marked by lessons in endurance, focus, and clear execution. Over a period of just 15 years, the company managed to establish its roots across more than 35 countries, spanning academia with universities such as MIT, Princeton, and the University of Coimbra, as well as defense and industrial verticals with Fortune 500 clients including, but not limited to, GE, Toyota, and Nestlé. This trajectory culminated in a strategic and well-positioned acquisition that resulted in a win-win outcome for both the founding team and the enterprise organization.

Given this impact, the journey warrants deeper exploration. These 15 years offer a glimpse into hard-earned lessons and enduring insights that can serve as a guiding light for engineers, researchers, and entrepreneurs — those brave enough to build the next generation of robots on this planet.

Lessons for Founders & Engineers

Lesson 01: No Matter Your Starting Point, You Can Fund Your First Prototype

The most important lesson here is that these founders, fresh out of university, were able to come up with creative ways to fund their first prototype. While raising a seed round at an early stage could be extremely daunting, and often not the best move when you are still validating your thesis, it goes without saying that without capital resources, even this part of the journey becomes extremely challenging. To address this challenge, the founders strategically cut down their expenses by operating between a basement space and the accelerator center, and additionally secured grants from government agencies such as the Ontario Centres of Excellence and the Natural Sciences and Engineering Research Council of Canada. So no matter your context, even before you raise a seed round, there are ways to fund these early prototypes, and government grants are indeed a strategic channel.

Lesson 02: Landing Early Adopters Should Not Be a Challenge for Fresh Grads

The second key lesson here is that landing early adopters should not be a challenge for fresh graduates, and the reason is that since you are fresh out of academia, you deeply understand the problems of your potential customers working in this space, such as researchers and professors. Even if you do not have the full picture yet, you have spent years in this ecosystem and have a strong sense of the challenges and unmet needs within it. This puts you in a position to dive deeper into these areas, study these problems, come up with ways to solve them, and position your solution directly in front of the people who need it most.

The founders of Clearpath took this route and secured early adopters in academia, from MIT to CMU and the University of Virginia to the University of Coimbra. These early adopters proved vital not only to validate the product but also to build strong brand awareness in the market for Clearpath, which led to even more incoming requests, such as one from a professor at the University of Calgary for the M200 maritime robot. Thus, landing early adopters should not be a challenge, especially if you are an aspiring founder coming straight out of academia.

Lesson 03: Brand Awareness & Market Positioning Matters More Than You Think

In the case of Clearpath Robotics, one of the founders received a call on a snowy winter day from a professor at the University of Calgary inquiring about a robotics platform for maritime applications. This inbound request came after the team had launched the M200 maritime robot platform at local trade shows and demonstrated it across a few academic institutions.

The takeaway is clear: as a founder, your approach to positioning your solution and driving awareness in the right communities matters a lot, especially at an early stage. In this case, the founders did an excellent job of positioning their platform for academic research and building brand awareness through local trade shows, a low-cost marketing channel that directly translated into inbound demand.

Lesson 04: You Can Scale Adoption with One Product & Two Variants

In this case, the company made a strategic shift in 2015 to position its robots to support both outdoor and indoor use cases, including warehouse and factory environments. While from the outside these two variants of robots might seem like two totally different platforms, in reality, you are essentially repurposing the same outdoor platform to better function in indoor spaces, which are far more controlled and reliable and offer far fewer variables that can harm your product. This launch of the OTTO Motors division, and the strategy to expand into the indoor vertical with the robot platform, was key for the company to continue scaling its adoption across verticals and generate revenue.

Now that we have distilled the core lessons embedded in Clearpath Robotics' trajectory, from its origins through its evolution and the pivotal decisions the team made along the way, it is time to decode the underlying business model.

In the final section of this chapter, we will break down the business model and illustrate how the product portfolio maps to select primary use cases.

Clearpath Robotics: Product Snapshot I — Academia & Defense

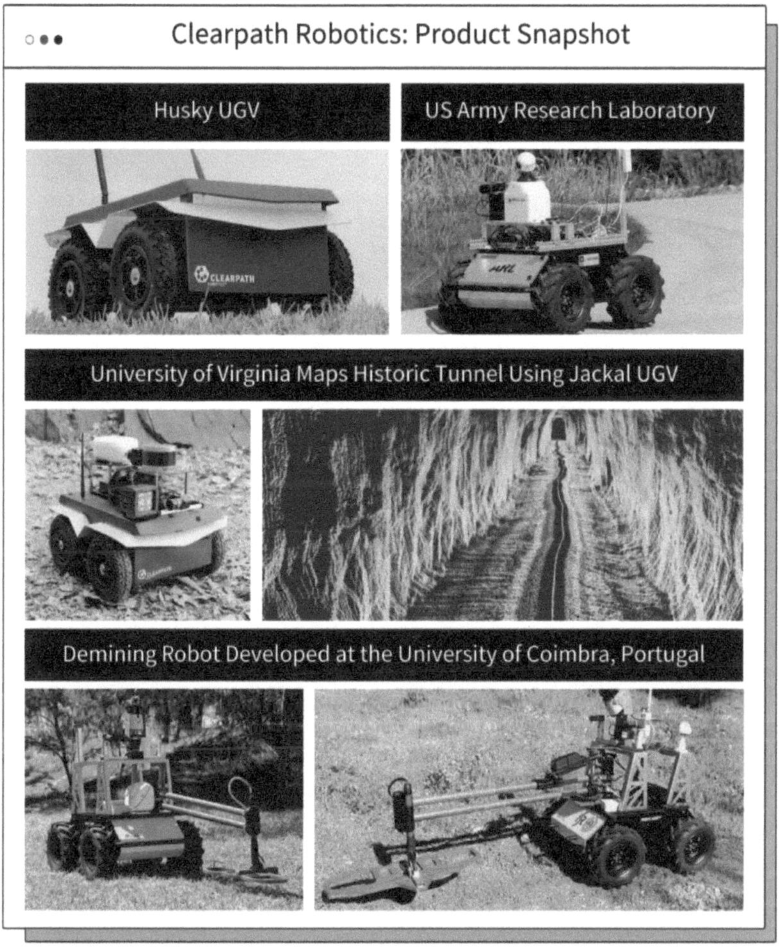

Figure 8.1 — Clearpath Robotics Academia & Defense

Illustration by the author. Robot images from Clearpath Robotics customer stories page.

Clearpath Robotics: Product Snapshot II — Geological Mapping & Cotton Farming

Figure 8.2 — Clearpath Robotics Geological Mapping & Cotton Farming

Illustration by the author. Robot images from Clearpath customer stories page.

Clearpath Robotics' Business Model — Part 1

Figure 8.3 — Clearpath Robotics' Business Model: Part 1

Illustration by the author.

Clearpath Robotics' Business Model — Part 2

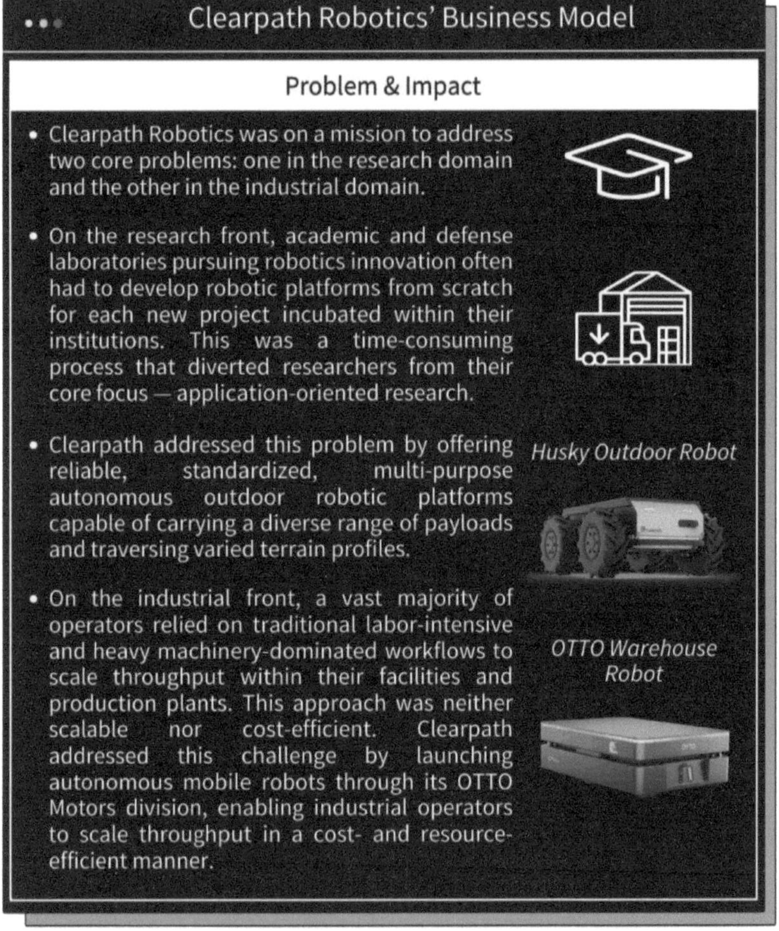

Figure 8.4 — Clearpath Robotics' Business Model: Part 2

Illustration by the author.

Clearpath Robotics' Business Model — Part 3

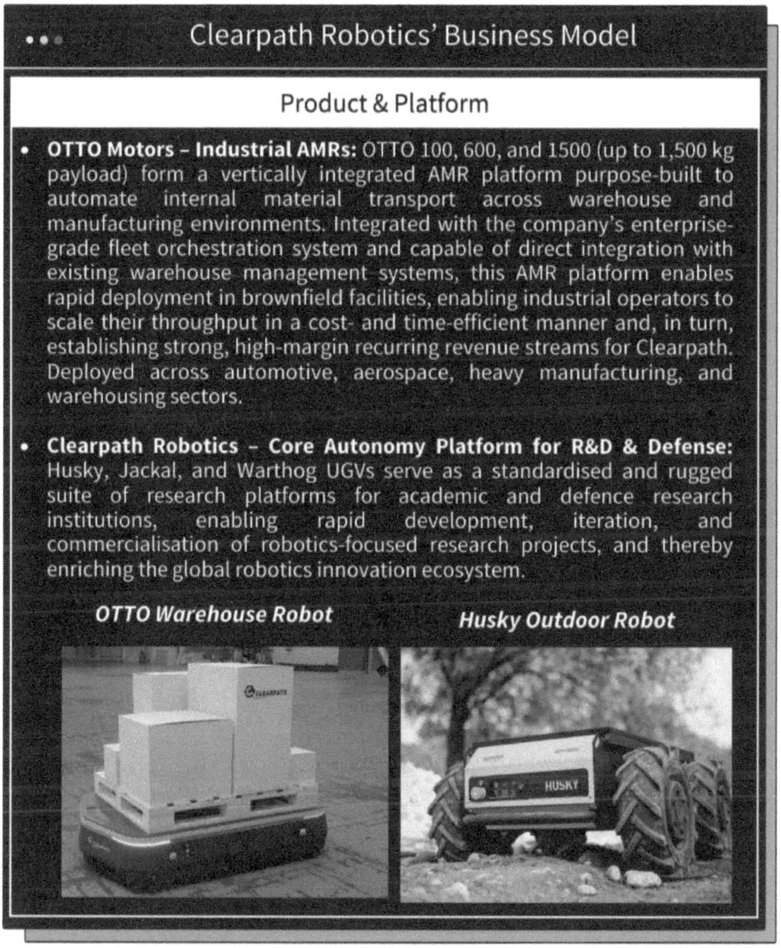

Figure 8.5 — Clearpath Robotics' Business Model: Part 3

Illustration by the author.

Clearpath Robotics' Business Model — Part 4

Clearpath Robotics' Business Model

Customer Segments

- **Primary Verticals:** Automotive OEMs, industrial manufacturing, aerospace, logistics and warehousing, academic research institutions, and defense contractors.
- **Customer Profile:** Leading industrial enterprises and research institutions seeking mission-critical autonomous mobile material transport systems for both indoor and outdoor applications.

Notable Customers & Deployments

- **Global Deployments & Scale:** 75% of robots deployed operate in the facilities of academic institutions and Fortune 500 companies across the U.S., Canada, Mexico, and Japan.
- **Defense Research - U.S. Army Research Laboratory:** Adopted the Husky outdoor autonomous robotic platform to evaluate terrain classification algorithms designed to support the Army in exploring and navigating hazardous, unmapped terrain with the highest levels of safety and reliability.
- **Agricultural Research - Clemson University:** Adopted the Husky UGV to develop a low-cost cotton-harvesting mobile platform designed to help farmers minimize their reliance on both labor-heavy farming practices and capital-intensive agricultural machinery, which can cost upwards of $700,000 and remains beyond the financial reach of many family-owned farms.

Figure 8.6 — Clearpath Robotics' Business Model: Part 4

Illustration by the author.

Clearpath Robotics' Business Model — Part 5

Clearpath Robotics' Business Model

Notable Customers & Deployments

- **Railroad Tunnel Mapping - University of Virginia:** Researchers adopted the Jackal UGV to map a historic railroad tunnel running through the Blue Ridge Mountains in central Virginia. The robotic platform enabled the team to generate a coherent and detailed 3D map of the tunnel, minimising the need to send research personnel inside the tunnel, significantly enhancing the safety profile of the operation and providing the research team with detailed insights into sections requiring immediate structural attention, thereby supporting informed and efficient restoration efforts.
- **OTTO Motors - Industrial AMR Deployments:** From 2015 to 2020, the team secured large-scale deployments with enterprise customers including GE, Toyota, Nestlé, and Berry Global, deploying thousands of autonomous mobile warehouse robots across their facilities and helping these Fortune 500 customers scale throughput in a cost-effective manner.

UNIVERSITY of VIRGINIA TOYOTA Nestlé

Go-To-Market(GTM)

- **Pre-Acquisition Model:** Direct enterprise sales through OTTO Motors, complemented by channel and systems integrator partnerships, with a strong commercial focus on North American automotive and manufacturing sectors.
- **Post-Acquisition (Rockwell Integration):** Following the 2023 acquisition, Clearpath's product portfolio was integrated into Rockwell Automation's global sales and distribution network, enabling scaled deployment across a robust array of Fortune 500 industrial clients.

Figure 8.7 — Clearpath Robotics' Business Model: Part 5

Illustration by the author.

Clearpath Robotics' Business Model — Part 6

Clearpath Robotics' Business Model

Revenue Model

- **Hardware Sales (Research):** Clearpath's Husky and Jackal unmanned ground vehicle (UGV) platforms sold under a CapEx model to academic, defence, and research institutions.
- **Hardware Sales (Industrial):** Autonomous mobile robot (AMR) unit sales, including OTTO 100, 600, and 1500 models, deployed across manufacturing and warehouse environments.
- **Software Revenue:** OTTO Fleet Manager — enterprise-grade fleet orchestration and control platform enabling traffic management, workflow optimisation, analytics, and MES/WMS/ERP integration.
- **Services & Support:** Training, technical support, and integration services.

Average Pricing & Economic Value Proposition

- **OTTO 1500 AMR:** Positioned at approximately 20% of the equivalent operating cost of traditional forklift-based workflows. Company-referenced calculations indicate that a single OTTO unit operating continuously (24 × 7 × 52) costs approximately $40,000–$50,000 per vehicle per year, compared to $200,000–$280,000 annually for a human-operated forklift, including driver labor, equipment lease, and operating expenses.
- **OTTO 100 AMR:** Average unit pricing is typically cited in the $15,000–$25,000 range per vehicle and represents approximately 10% of the labour costs of human drivers when compared on a 1:1 basis.
- **ROI Cycle:** The typical payback period hovers around 12–24 months per customer in large-scale industrial deployments and is governed by fleet size, utilisation rates, and workflow complexity.

Figure 8.8 — Clearpath Robotics' Business Model: Part 6

Illustration by the author.

Clearpath Robotics' Business Model — Part 7

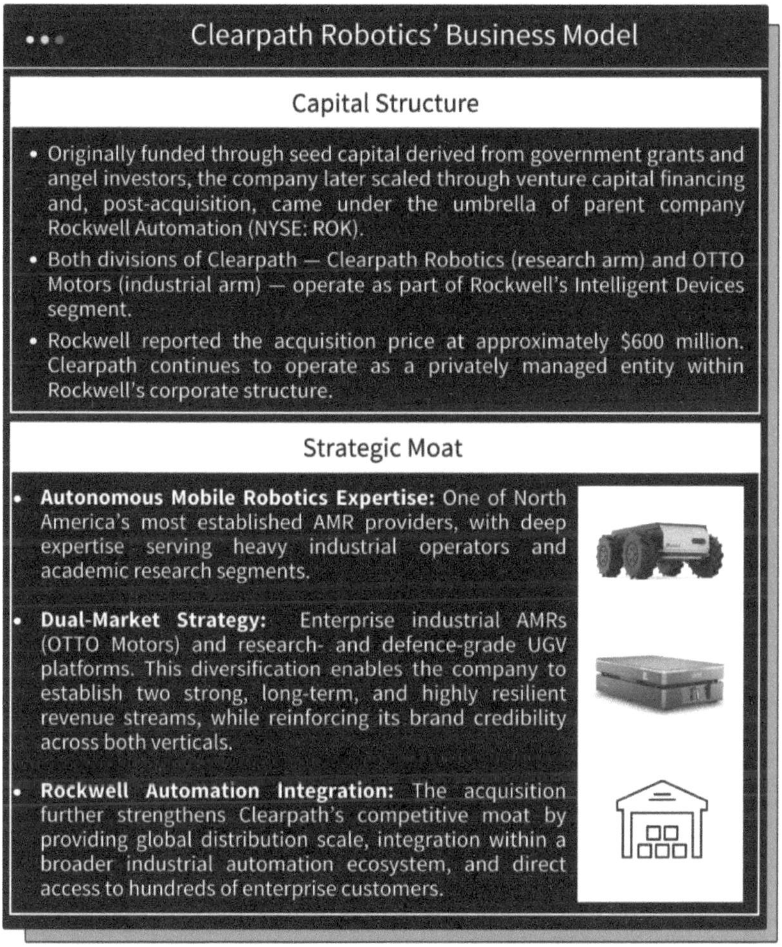

Figure 8.9 — Clearpath Robotics' Business Model: Part 7

Illustration by the author.

Chapter 09: AutoStore — Norwegian Roots to Whole Foods Scale

While most robotics companies are born in the tech arteries of the Silicon Valleys of the world—San Francisco, Boston, Berlin, London, Tokyo, Mumbai, and Shenzhen—there is one that emerged from the shadows of the Northern Lights and the womb of mighty fjords: Norway.

To the western edge of Norway lies a remote village called Nedre Vats that sits on the edge of a 700-meter-deep fjord. This is the home of the Hattelands, village merchants since the 19th century and builders of an old warehouse located nearby. It was here, forty years ago, that Jakob Hatteland, the scion of the family, started a television repair business, which later evolved into a company selling electrical components. Soon, this family-owned business became the largest supplier of electronic parts in Scandinavia and, by the mid-1990s, had grown too big for its warehouse.[6]

Warehouse space thus became both its major constraint and a key opportunity for innovation, forcing the company to fit far more electronic components into a confined footprint while remaining profitable. At the time, an employee of Hatteland's, Ingvar Hognaland, who was deeply familiar with this challenge, decided to address this constraint for the family head-on. He exclaimed, "What's the biggest thing in the warehouse? Air—there's too much air."

Ingvar wanted to replace all that air in the warehouse, along with aisles, with mobile robots that could both stack items and transport them at the same time. In simple terms, he didn't like the idea of a warehouse serving as a static storage facility. Instead, he envisioned it as a system designed from the ground up using mobile robots that could act as both storage and transport units.

It is this idea and their partnership that led to the launch of a private robotics company called AutoStore in 1996 here in Norway. The goal was to eliminate aisles and store inventory in an array of cubical robots that could serve as both mobile storage and transport units.

Over the next twenty years, the company grew from this small village in Norway to a global robotics powerhouse. It went on to serve customers across multiple verticals, including Medline, H Mart, Best Buy, and the FBI in the United States; Albert Heijn in the Netherlands; DHL and Siemens in Germany; Panasonic in Japan; and SoftMoc in Canada.[119] The company also enabled vertical farming for Whole Foods and ultimately partnered with SoftBank, leading to an IPO widely regarded as one of the largest in Norwegian history, valued at roughly $12 billion.[6]

From a remote village in Norway to automating warehouse and logistics operations for customers in 35+ countries around the world with its global fleet of 20,000+ robots, this company has successfully etched its soul onto the robotic canvas of the global warehousing and supply chain industry.[6] Given its unconventional roots, remarkable trajectory, and, most importantly, its global footprint, this company deserves deeper examination.

To better understand the story of AutoStore, we must travel back in time all the way to the year 2005.

Building AutoStore's Robotic System (1996–2005)

After its official inception in 1996, the team building AutoStore invested the next 10 years iterating on its modular robotics concept, refining both hardware and software to build a highly-resilient production-grade robotics system. That raises an important question: how did they design this system and how does it work?

The team designed this robot-led warehouse automation system around a single core concept: maximizing space utilization in the warehouse while aligning throughput with the operator's ability to process orders at high speed. To do so, the company introduced the concept of towers and bins. Essentially, a tower is a group of vertically stacked bins designed to house inventory or goods stored in the warehouse. Thus, goods across the warehouse are organized into an array of these towers, which are placed adjacent to each other.

On top of this grid of towers is a metallic highway, or network, deployed for robots to access each of these towers and fetch bins. The fleet of robots operating directly on top of this tower grid receives the list of items that need to be shipped out of the warehouse for the day. Then, each of these robots navigates across this tower grid, extracting the corresponding bins and the items from those bins, and transporting them to their destination stations—the workstations where items are extracted and consolidated per customer order.

No matter the scale or size of the warehouse building, these tower grids can be installed in a matter of days at any greenfield site around the world, with minimal need for construction or redesign. Thus, this grid of towers filled with bins, and the fleet of robots operating on top of them almost like a crane, enables warehouse operators to double—and in some cases triple—their throughput. That leads us to the next question: with this prototype system solidified, how did AutoStore land its first customer?

Partnership.

First Production System — Elotec, Norway (1996–2005)

In 2005, the company partnered with its first major adopter, Elotec. Elotec is a Norwegian manufacturer and distributor of security equipment and became the first company in the world to install the AutoStore system.[120] The company supplies safety and security devices including but not limited to fire and burglar alarms, emergency lighting, and surveillance systems to industrial and residential customers across Scandinavia and Germany. Given the mission-critical profile of its product portfolio, the company needed a highly reliable system to automate its warehouse. Shipping the wrong alarm to its customers could mean the difference between life and death.

Given this need, in 2005 the company decided to deploy AutoStore's system at its facility, where it remains in use to this day. The company evaluated the solution and its operational impact around the two-year mark, in 2007, and decided to expand its adoption further. Elotec attributes minimal downtime, high picking accuracy, customer satisfaction, and fast delivery times among the many benefits of installing AutoStore's robot-led warehouse automation system. That leads us to the next question: with a production-grade system in place and its first strategic partner secured, how did AutoStore scale its customer base?

That is where the enterprise model comes in.

Hub-and-Spoke Model for Enterprise Distribution (2005–2016)

Rather than scaling through a traditional direct-sales enterprise model, AutoStore expanded through a hub-and-spoke distribution model. The hub here refers to its distribution partner and integrator, Element Logic, and the spoke refers to the scale it unlocked for AutoStore.[121] Element Logic designed and delivered AutoStore's first production-grade system at Elotec and went on to deploy the platform across industries and geographies, acting as both system integrator and commercialization partner. This model allowed the Norway-based robotics startup to scale globally without building a large in-house sales or deployment team, while leveraging integrators' deep operational knowledge of local markets and warehouse workflows.

By the mid-2010s, this strategy had enabled AutoStore to expand from a single Norwegian warehouse into hundreds of installations across approximately 28 countries spanning five continents.

Enterprise customers during this period included Puma, Siemens, Gucci, Lufthansa, and DHL, signaling early adoption by industrial, logistics, airline, and retail leaders. Given this scale of adoption in European markets, it was time for the company to head west across the Atlantic Ocean. It was time to enter the North American market.

That voyage began with a defining partnership—one that would anchor AutoStore's presence in the United States for the next decade.

Medline Deployment — North America (2013–2025)

Medline is one of the largest providers of medical and surgical products in the world, supplying hospitals, doctors' offices, surgical centers, and long-term care facilities across the United States. By 2013, Medline was experiencing a surge in custom orders, requiring the company to handle hundreds of unique items per order. This demand for customized orders meant that the company's ability to meet next-day service commitments, within an operation that was already tightly run, was at risk of being compromised. To keep pace with this demand and evolving order profile, the team at Medline determined that automation was essential.

Following an exhaustive search, Medline selected AutoStore's robot-enabled, grid-based warehouse automation solution to scale its operations.[122] Swisslog was chosen as the implementation partner and as the provider of the warehouse control software that links AutoStore to Medline's warehouse management system. The partnership between AutoStore, the provider of the leading robotic warehouse management system at the time, and Swisslog, a fast-moving systems integrator, enabled this deployment to proceed at remarkable speed.

Medline became the first company in the United States to adopt the AutoStore robotic order fulfillment system. After implementation, the medical-supply giant realized throughput gains of approximately 200%. That meant the company was now able to fulfill four times more orders each day.

Beyond throughput, Medline cited system agility and reliability as key reasons for choosing AutoStore over competing solutions. Agility refers to the grid-based design, which allows the system to be expanded or contracted based on operational needs, while reliability refers to the extremely low downtime enabled by the robust network of dedicated tracks on which AutoStore's robots operate atop its grid-based warehouse structure.

Encouraged by these results, Medline expanded the footprint of its AutoStore-driven warehouses nationwide in 2025. Today, Medline is AutoStore's largest customer globally, with more than 1,700 robots deployed across its 50+ cosmos-scale distribution centers spanning the United States. Given these large-scale deployments in North America and Europe, the company now had sufficient revenue to raise the product bar. It was time to quadruple the throughput of its system.

Launch of Router — Robot Navigation System (2020)

In September 2020, at ProMat, one of the largest global exhibitions for warehouse automation and supply chain technology, the company unveiled its next-generation robot navigation system, called Router. With the launch of Router, the company was now able to officially quadruple throughput for its warehouse operators while retaining the same geographic footprint when deploying the AutoStore grid-based, robot-enabled warehouse automation system.

This raises an important question: how was the company able to quadruple throughput? To enable this, the company connected the robots' navigation and decision-making systems directly to the order fulfillment system. This meant that navigation decisions made by the robots were optimized in real-time based on incoming orders, cancellations, and other variables governing the flow of goods through the warehouse. In effect, the company married the robot navigation stack with the order management stack of the warehouse, resulting in a fourfold increase in throughput for its customers. With this launch at ProMat in the U.S., the company came on the radar of the FBI for good reason.

Government-Grade Credibility — FBI Records Automation (2020)

By 2020, the FBI faced a mounting operational challenge: managing and retrieving more than 2 billion pages of sensitive paper records that had been dispersed across more than 250 field offices. Traditional records management methods were no longer sufficient to handle the scale, security requirements, and retrieval speed demanded by modern law enforcement operations. The agency needed a solution that could consolidate these archives, ensure strict compliance with federal records standards, and accelerate retrieval workflows.

To address this challenge, the U.S. General Services Administration, in consultation with Swisslog, evaluated multiple automation technologies and selected AutoStore's cube-based storage and retrieval system for the FBI's new Central Records Complex in Winchester, Virginia. Installed in 2020, it became the first robot-enabled records warehouse deployed by a U.S. federal agency. Certified by the National Archives and Records Administration, the 250,000-square-foot facility replaced traditional aisle-based storage with a high-density cube-based design, deploying 140 robots operating on a sixteen-level steel grid to retrieve files from more than 360,000 bins.[123-124]

The impact of this deployment was transformational. Automated retrieval significantly reduced search times for individual records, enabling a team of 500 FBI employees to shift their focus from manual file-sorting processes to higher-value investigative and analytical work. This deployment demonstrated that AutoStore's platform was not only commercially scalable but mission-critical. It had earned the level of trust needed to support core U.S. national security infrastructure.

With successful flagship deployments at Medline and the FBI, it was time for AutoStore to expand its footprint beyond medical logistics and the government vertical. The next frontier was the consumer market, and that meant automating the grocery distribution network.

H Mart Micro-Fulfillment Partnership: U.S. Grocery Market Expansion (2021)

In 2021, the company landed a deal with H Mart, the leading Asian-American specialty grocer, to introduce a fully automated micro-fulfillment center supporting H Mart's online grocery operations in Carlstadt, New Jersey.[125] Powered by AutoStore's grid-enabled, robot-led automation system, the facility automated both in-store pickup and home delivery workflows. As a result, the grocery chain was able to deliver orders at twice the speed and half the cost.

With this expansion in North America, AutoStore was now ready to win the capital markets of Norway—and that meant an IPO was next on its radar.

SoftBank-Led $12 Billion IPO on Norway's Capital Fort (2021)

AutoStore's path to the public markets was shaped decisively by SoftBank's entry in 2021, which established the financial and strategic foundation for its IPO later that year. In April 2021, SoftBank acquired a 40% stake in AutoStore through a $2.8 billion investment, valuing the Norwegian warehouse automation company at approximately $7.7 billion at the time.[126-127] SoftBank described AutoStore as a "foundational technology" for global logistics, positioning the company squarely within its long-term vision for automation-led supply chains.

This investment marked a fundamental inflection point for AutoStore. It was no longer a fast-growing robotics company; it had become the core operational backbone for the global warehouse and logistics industry, setting the stage for its public debut later that year. By October 2021, AutoStore launched its IPO on the Oslo Stock Exchange at a valuation of up to approximately $12 billion—one of the largest listings in Norway's history. With this IPO, the company secured a strong capital runway to expand its product portfolio—and it was time to launch its next core product: the R5.

AutoStore's R5+ Robot Launch (2022)

In January 2022, AutoStore launched the R5+ robot to enable its system to handle larger products, a capability that had been limited in earlier robot designs. The R5+ robots are significantly taller than their predecessors and can operate with larger bins, allowing warehouse operators that handle medium- to large-sized items such as shoe boxes or computers to use the same AutoStore architecture to support a broader range of products.[128]

As a result, customers can now store and retrieve larger items using the standard AutoStore system by replacing existing bins and robots with newer versions. For customers across North America and Europe, the R5+ opened up new use cases, enabling AutoStore systems to support a wider range of products, from apparel and electronics to bulkier retail and grocery items, without changing the underlying infrastructure. This allowed operators to use warehouse space more efficiently and scale operations while maintaining the speed and reliability AutoStore's system is known for.

This launch of the R5+ also meant that one of its early adopters was now ready to expand its footprint.

DHL Germany — 1st Deployment for DHL (2022)

In February 2022, AutoStore was inducted into DHL's European operations for the first time at its facility in Braunschweig, Germany. There, the logistics giant installed a fully automated warehouse system based on AutoStore technology, effectively doubling the speed at which it processed 18 million packages per year. The system stores goods in more than 40,000 containers stacked across 16 levels, with 26 robots navigating atop the grid to retrieve products and deliver them to worker stations for packing and dispatch.[119]

By automating much of the sorting and storage-centric workflows, the warehouse effectively cut the labor workforce needed to support this operation in half while achieving extremely high throughput, i.e., order fulfillment rates, within a relatively small footprint. This deployment cemented AutoStore's partnership with DHL for decades to come and demonstrated that the company's robot-led, grid-enabled warehouse automation was ready for large-scale deployment across the EU.

StrongPoint Expansion — Nordics & Baltics (2022)

In October 2022, AutoStore expanded its footprint in Northern Europe through a strategic deployment with StrongPoint, a leading technology provider for grocery retailers and distribution centres operating across more than 20 countries in the Nordics and Baltics. The partnership introduced AutoStore's first solution capable of handling frozen goods, operating across three temperature zones—ambient, chilled, and frozen—and enabling grocery retailers to automate their full product range within a single system.[129] The AutoStore grid can now operate down to −25°C (−13°F), while allowing both fresh and frozen goods to be picked from chilled zones within the warehouse, maintained at approximately 2°C (35.6°F).

StrongPoint designed and installed the solution for the Norwegian food distributor DLVRY at a facility in Sveio, Norway. Until this point, frozen goods were typically picked and packaged manually by workers. By incorporating support for frozen items into the AutoStore grid, the company officially eliminated one of the biggest barriers to adoption for its system by food logistics giants across Europe.

While the StrongPoint deployment positioned AutoStore's robot-led warehouse automation system for large-scale organisations, many small and mid-sized grocers and businesses still lacked the capital and resources to adopt the company's technology. That bottleneck culminated in the launch of its next product: Pio.

SMB Expansion Strategy: Launch of Pio (2023)

In January 2023, AutoStore expanded its market footprint with the launch of Pio, a simplified and modular version of its automated storage system designed specifically for small and medium-sized businesses. Introduced initially in North America, Pio was developed to make the company's robot-led warehouse automation system accessible to retailers that previously lacked the scale or capital required for full-scale AutoStore deployments.[130]

Pio is offered as a modular, plug-and-play model and integrates easily with standard e-commerce and shipping platforms. Unlike traditional deployments, customers do not purchase the robots or software upfront; instead, AutoStore retains ownership of the system and charges a pay-per-pick fee, significantly lowering the barrier to adoption for small and mid-sized retailers operating across verticals such as apparel, sporting goods, crafts, and cosmetics to automate order fulfillment workflows without requiring additional labour or floor space. With only about 15 per cent of the world's warehouses currently automated, Pio positioned AutoStore to address a far broader segment of the retail market by extending the reach of its automation system beyond large enterprises to independent and mid-sized operators.

Continuing this momentum, DHL, which had started working with AutoStore back in 2022, was now ready to expand its adoption footprint.

DHL Partnership Expansion: Enabling 10 Million Orders per Day (2023)

By this time, DHL had already deployed and tested nine large-scale AutoStore systems across Singapore, Poland, Australia, the U.S., and Germany, effectively adopting a fleet of 1,000 robots to boost operational efficiency across its warehouse network. Beyond throughput gains, this system offered DHL a modular and scalable approach to scale its operations. This meant the company could now expand and contract the capacity of each of its warehouses to accommodate peak demand cycles without raising their physical footprint. Given this agility and scale, DHL decided to extend its adoption with four additional large-scale systems in Nov 2023.[131]

With this expansion, AutoStore was effectively helping DHL manage approximately 1.2 million bins filled with goods across its global fulfillment centers spanning Europe and North America. That's roughly 10 million items processed per day by AutoStore's robot-led automation system. This expansion, driven by DHL, meant the country was ready to adopt this technology even further, now to ship its famous Gouda cheese and groceries.

Albert Heijn Expansion — Netherlands, Belgium & Germany (2023)

Albert Heijn, the largest supermarket chain in the Netherlands, with more than 1,000 stores across the country, Belgium, and Germany, decided to adopt AutoStore's technology. This raises a critical question: what was the exact problem this supermarket giant was trying to address? It turns out that, at the time, online grocery orders received by the company were escalating at a rapid pace. This meant that its ability to fulfil those orders across its warehouse network was becoming increasingly constrained.

To address this challenge, the supermarket company adopted a technology-first approach by developing a new warehouse in Barendrecht, partnering with AutoStore to procure its robot-led automation system, and with Swisslog to integrate that system into this new facility. Under this model, Swisslog led the implementation, while AutoStore supplied the robots and software.

In the new facility, everyday products such as jars of peanut butter and tubes of toothpaste are now transported through an automated grid system, with robots travelling across the top of stacked bins to retrieve and consolidate items for dispatch. Powered by over 300 AutoStore robots, the system processes about 45,000 orders per week, significantly accelerating order fulfillment timelines and enabling Albert Heijn to strengthen its e-commerce operations.[132]

At this time, the company had several large-scale contracts in place, and that meant its success now hinged on its ability to deliver on these contracts. Its next major bottleneck was no longer generating demand for its product; it was addressing that demand, which meant manufacturing these robots at scale.

Until then, the company operated a single manufacturing facility in Poland, responsible for fulfilling orders for its customers worldwide. This facility soon became a critical bottleneck, as high labour costs and operational expenses in the region limited its ability to scale effectively.

So how did the company address this challenge? It decided to head east, this time to the land of Pad Thai, to establish its manufacturing hub.

Thailand Factory — Production at Scale (2024)

Why Thailand?

To better answer that question, it helps to understand the context here. This was June 2024, and by this time, the company had implemented its system across more than 54 countries, with over 17,000 robots operating in the U.S. alone for brands such as Puma, Gucci, and Best Buy. With a surge in demand for its robots from North America, and only one core manufacturing facility in place in Poland, the company was hard-pressed to scale its manufacturing operations.

Heading west was not an option, since rising labour costs meant thin margins and potential loss of revenue. To address this problem head-on, the company had to head east. Workforce availability, proximity to major harbours and airports, favourable labour costs, and government incentives for automation companies made Thailand an ideal location for its second robot factory, built to support AutoStore's North American operations.[133-134]

Now, with the new manufacturing facility in place and growing demand from existing customers to support a wide range of products, the company soon encountered its next major bottleneck: its existing robots were only capable of handling dry goods. They were not engineered to operate in frozen warehouse aisles, and this became a significant adoption barrier for its customer base. That raises a critical question: how did the company address this challenge? It redesigned its robots to survive Arctic conditions — quite literally.

Sub-Zero Robots for Grocery Fulfillment (2024)

In September of that year, the company introduced a fully reengineered version of its robots, now capable of surviving the sub-zero temperatures inherent in warehouses primarily handling frozen goods and vegetables. This solution allows warehouse operators to use the same AutoStore Grid across their entire warehouse, even when they include multiple temperature zones.

It is designed to ensure the robots can operate equally effectively in both frozen and chilled zones inherent in these warehouses, with a temperature range from -13°F in the frozen sections to +42.8°F in the cooled areas. This new system was built to help retailers manage the growing demand for online grocery orders by automating the fulfillment workflows associated with fresh, chilled, and frozen goods in one integrated solution, without requiring warehouse workers to manually retrieve items from these harsh environments.[135]

With support for both dry and frozen goods, large-scale adoption across Europe, and the launch of its new manufacturing facility in Thailand, it was time for the company to head west and address the surging demand from its customers in North America.

North America Expansion — Medline, Kenco 3PL & SoftMoc (2024–25)

Medline was the first company in the United States to adopt AutoStore's robot-led order fulfillment solution and remains its largest customer globally. The company operates more than 1,700 robots, the first of which was deployed in 2013 at one of its warehouses. Over the following decade, Medline experienced a sustained fourfold increase in warehouse throughput. Given this performance — and more specifically the agility it offered that allowed the medical supply giant to support orders of all sizes and profiles — the company decided to deploy the system at its newly launched large-scale order fulfilment facility in the U.S.[136]

This scale of adoption by a leading medical equipment supplier brought the Norwegian robotics startup onto the radar of U.S. third-party logistics giant Kenco.

That same year, Kenco, a leading third-party logistics (3PL) provider, adopted AutoStore's automated storage and retrieval system (AS/RS) at its distribution centre in Jeffersonville, Indiana. This became the supply chain giant's largest AutoStore installation to date, featuring 49,000 bins powered by 130 order fulfilment robots. Designed to serve Kenco's CPG customers, the system quadrupled storage capacity at the facility, enabling 24/7 operations. It now allows the company to fulfil approximately 15 million orders per year at four times the speed and with one-quarter of the real estate compared to Kenco's other facilities, delivering F1-grade service levels year-round to its customers.[137]

With these large-scale deployments enabled by Medline and Kenco in the U.S., it was time for the Nordic-born startup to head north.

SoftMoc, a Canadian footwear retailer, selected AutoStore to design and scale its order fulfilment system at its newly constructed facility in Ontario in response to growing demand driven by its e-commerce operations. The new AutoStore system, implemented by Element Logic, was designed to enable the company to fulfil customer orders nationwide with high speed and operational efficiency, reinforcing SoftMoc's commitment to customer service.

The new fulfilment centre in Ontario features nearly 100,000 bins and 42 robots, enabling high-speed, high-accuracy order processing workflows. Integrated with a 3,225-foot conveyor network supporting picking, packing, and shipping, the system delivers an average bin retrieval time of just 27 seconds and industry-leading throughput. Together, these capabilities position SoftMoc to deliver shoes to its customers at unprecedented speed.[138]

Now, as the Nordic robotics company became capable of handling shoes, groceries, frozen goods, and more, it had until this point managed these items primarily using bins and robots. From the robot's perspective, this meant simply extracting and transporting bins across the warehouse, from the inventory area to worker stations. At these stations, workers manually packaged items from the bins and transferred them onto a network of conveyor belts, from where they were routed to outbound trucks.

The key challenge was that the company had automated only half of the warehouse workflow by transferring goods in bins from inventory to consolidation stations. The remaining half, consolidation and packaging, was still manual and quickly became its next major bottleneck. To automate this workflow, mobile robots alone would not suffice. The company had to develop a robot arm, similar to a human hand, capable of identifying, picking, and sorting individual items per customer order.

So, in April 2025, it launched its next product, an AI-powered robotic arm called CarouselAI.

Next Frontier — AI-Powered Robot Arm Launch (2025)

With the launch of CarouselAI, the company's first AI-powered, robot-enabled piece-picking solution, developed in partnership with Berkshire Grey, a leader in robotics, AutoStore automated one of the most labour-intensive stages of the order fulfilment workflow, enabling 24/7 operations for retailers, doubling warehouse throughput, and unlocking a faster path to return on investment.

The company also ensured that existing customers could adopt this robotic arm with minimal disruption, enabling a seamless transition. Now, having developed mobile robots capable of transporting bins filled with groceries and a robotic arm capable of identifying, picking, and sorting individual grocery items, it was time to build a farm — quite literally. It was time to build a vertical farm.

Vertical Farming — Opollo Farm & Whole Foods (2025)

In 2025, the company launched Opollo Farm, a fully automated vertical farming system powered by AutoStore's robotics system. Located in Arizona, the system is already supplying Whole Foods stores in the Phoenix area with leafy greens grown locally in this farm. That raises a critical question: how does this robot-led vertical farm work?

At the heart of Opollo Farm is AutoStore's robotic grid, now adapted to transport living plants instead of inventory bins. Crops are seeded and transferred into this grid for germination, and then repositioned frequently to ensure they receive optimal light, water, and nutrients. This is a closed-loop system and uses up to 95% less water than conventional farming methods and produces harvest-ready greens in as little as 15 days—roughly half the time of conventional methods.[139] Unlike traditional farms, the fully enclosed, climate-controlled structure eliminates the need for pesticides and enables sustained year-round growth, while its compact footprint allows farms to be placed near urban centers or co-located with distribution hubs.

Strategically, Opollo Farm addresses several structural challenges facing modern agriculture: labor shortages, water scarcity, supply chain bottlenecks, and the demand for fresh and locally grown food. In effect, this deployment ensured that crops can now be grown directly inside major grocery outlets, enabling a new model for urban food infrastructure. While still in its pilot phase, this deployment with Opollo Farm signals a strategic shift in AutoStore's trajectory—from automating warehouses for enterprise customers to scaling food production for consumers around the world.

Where do you go next after scaling food production? You deliver it to consumers at lightning speed. After scaling the production process, it was time for AutoStore to ensure these freshly grown vegetables reached consumers on time.

That meant automating the order fulfilment and delivery process for these groceries.

Vienna Veloq — AI-Driven Grocery Fulfilment Centre (2025)

In October 2025, the company partnered with Veloq, which offers an AI-powered grocery fulfilment platform, to enable grocers across Europe to automate their order fulfilment and delivery workflows. In simple terms, it provides the software needed to automate this process end to end, from production and consolidation at distribution centres to fulfilment and delivery at retail grocery stores. This partnership culminated in the launch of what is described as the world's most advanced AI-driven grocery fulfilment centre in Vienna.

At this centre, Veloq's AI-powered order retrieval and delivery system is tightly integrated with AutoStore's robot-driven order fulfilment system, enabling the facility to support up to 20,000 SKUs, or unique items, spanning fresh produce, bakery, pharmacy, ambient, and frozen goods across seven temperature zones.

Furthermore, the system can route orders for dispatch in under 30 minutes once they enter the system and fulfil these orders with a three-hour click-to-delivery promise.[140-141]

By eliminating manual sorting workflows and strategically integrating the order retrieval, fulfilment, and delivery workflows into a single system, this facility has cut labour requirements in half and doubled throughput without sacrificing quality. In effect, it has solved one of the hardest problems in online grocery: scaling speed without compromising accuracy.

This combination represents a major breakthrough for grocery retailers and distributors around the world. Veloq's platform, with its 90 NPS and 94% order satisfaction rate reflecting orders delivered in full, on time, and error-free, combined with AutoStore's robotics solution for automated picking, labour management, and routing, marks the arrival of the next generation of high-speed and high-precision grocery fulfilment systems across Europe. This new model is now positioned for international rollout, including expansion into North America.

Summary

What started as a small warehouse automation company founded by the Hatteland family in the remote village of Nedre Vats, Norway, has evolved into a powerful operating system for modern commerce. Today, the AutoStore system sits at the heart of fulfillment networks for organisations ranging from Medline and the FBI to DHL and Whole Foods, managing everything from medical supplies and fresh produce to documents and home goods. In doing so, the company has become the engineering backbone of modern order fulfilment systems, delivering twice the throughput in half the footprint and setting the performance standard for the supply chain and distribution arteries of the world.

So, what are the core lessons next-generation robotics founders and engineers can learn from this journey?

Lessons for Founders & Engineers

Lesson 01: Solve One A+ Problem That Truly Matters

It may sound repetitive, but the importance of this first lesson cannot be overstated: focus on solving one core problem — and solve it exceptionally well.

The right problem is one you deeply understand and one that is so painful for your customer that, if solved with the correct approach, it should at least double throughput and cut operating costs in half. That is exactly what the Hatteland family did. They encountered this problem firsthand while trying to scale their own warehouse to support a growing electronics business. They wanted to do so without expanding its physical footprint. Their solution maximised aerial and floor-space utilisation rates while enabling high-speed inventory handling and order fulfilment workflows.

In effect, what started as an experiment to devise a solution for their own warehouse became a system powerful enough to serve top-tier customers worldwide, from Medline and the FBI to Whole Foods and H Mart. The takeaway: find one problem that truly matters, and build a world-class solution around it.

Lesson 02: Secure One Long-Term Early Adopter

Once you have the right problem and the right product, your highest priority is to secure a long-term early adopter. The goal is to partner with a customer who allows you to deploy your system in a live production environment and operate it over time, so that the results it generates become a strong and industry-validated proof point for future customers. In the case of AutoStore, the Hatteland Group secured Elotec and Medline as their early adopters and long-term partners, and this bond is as strong today as it was when they started.

It is important to choose this early adopter with a long-term horizon in mind. This ensures you can maintain a standardised, scalable workflow for bringing top-of-funnel prospects to this facility, demonstrating in real time how your robot-led solution solves their problems, and building trust. In effect, this customer benefits from the throughput your system adds to their company and in return serves as your official long-term champion for your future customers. It's a win-win.

Lesson 03: Surgically Evaluate and Address Your A+ Bottlenecks

As you build a robotics company, at each stage your A+ bottleneck will evolve. At the beginning, it will be landing your early adopter; then it will be tapping into the centre of the hub, i.e., an integrator or a partner that helps build trust with customers at scale; and then it will be ensuring you continuously build and improve your product to better address the needs of those customers.

While the A+ bottleneck will shift over time and vary by context, it is critical to identify your primary constraint each quarter and eliminate it decisively.

In the case of AutoStore, the company addressed its first A+ bottleneck by landing early adopter Elotec. The next bottleneck was landing large-scale customers, so they tapped into one long-term customer, Medline. Then it was supporting products beyond medical supplies, which meant grocery, and hence came H Mart, Albert Heijn, and more. Then it was ensuring their system could support both dry and wet goods to better serve grocery distribution centres. That meant launching robots that could survive these harsh conditions. While this list will continue to evolve as you grow your company, it is always important to keep an eye on your A+ bottleneck at each stage and address it surgically.

Lesson 04: Keep Your Burn Extremely Low

Another major challenge in robotics is ensuring you keep your burn rate low at all stages. This is extremely hard to accomplish, since there are so many variables that can raise your burn rate overnight without you even realising it. From manufacturing to supply chain, deployment models to product portfolio, there is more than we can list here. So how do you keep your burn rate low at all stages?

Each quarter, you want to take a look at your expense sheet as a company and identify the top three line items that are elevating your burn exponentially. Once you find them, you want to address at least one of those three each quarter and bring it down. If we look at the case of AutoStore, the company was able to keep its burn rate low in the first phase by tapping into an integrator-led model. This meant it provided the robots, while deployments were scaled using third-party integrators. This worked well in phase one.

Then, as the company expanded its footprint across Europe and North America, manufacturing quickly became the number one line item on its quarterly expense sheet. To address this, the company incubated a new manufacturing facility in Thailand to lower production costs and gain access to affordable parts from Asia without sacrificing quality. In addition, it is important to note that the company maintained a highly disciplined approach to its product catalogue. It remained focused on a single core product for a significant period before introducing its second product. Its first product, the mobile robot-led warehouse automation system, served from 2005 all the way to 2025, when its second product, the robot arm was introduced.

In short, retain a small number of high-quality products and ensure the protocols that govern your manufacturing, engineering, and service delivery workflows are highly capital- and resource-efficient.

Now that we have distilled the core lessons embedded in AutoStore's trajectory, from its origins through its evolution and the pivotal decisions the team made along the way, it is time to decode the underlying business model. In the final section of this chapter, we will break down the business model and illustrate how the product portfolio maps to select primary use cases.

AutoStore: Product Snapshot I — System Overview

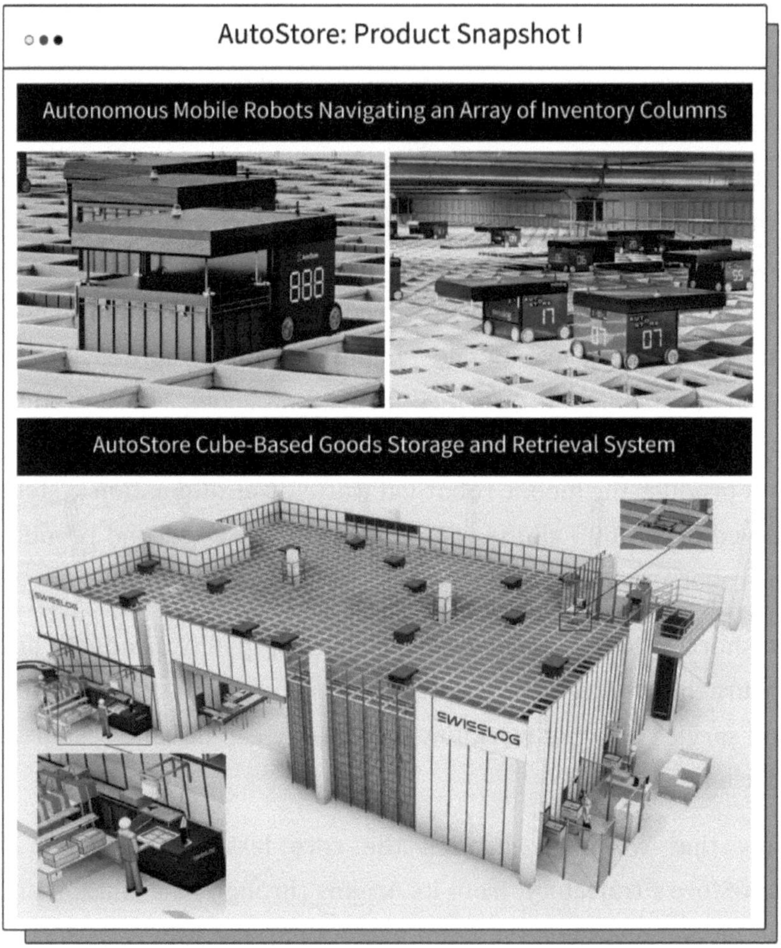

Figure 9.1 — AutoStore: System Overview

Illustration by the author. Robot images from the AutoStore system overview page.

AutoStore: Product Snapshot II — DHL & IKEA Deployment

Figure 9.2 — AutoStore: DHL & IKEA Deployment

Illustration by the author. Robot images from the AutoStore customer cases page.

AutoStore's Business Model — Part 1

Figure 9.3 — AutoStore's Business Model: Part 1

Illustration by the author.

AutoStore's Business Model — Part 2

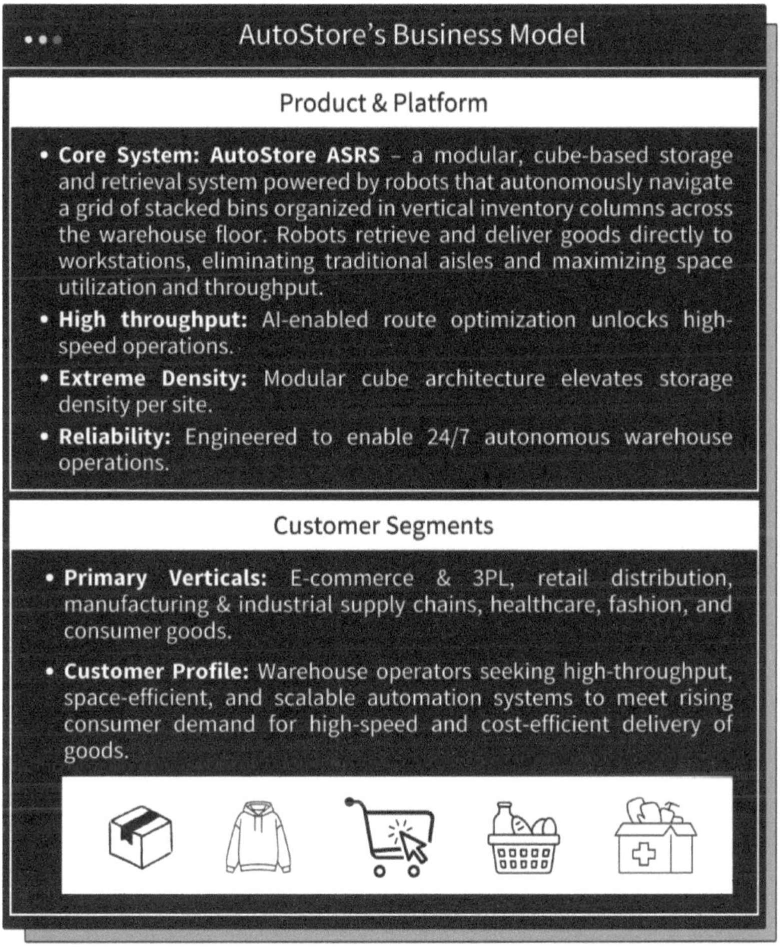

Figure 9.4 — AutoStore's Business Model: Part 2

Illustration by the author.

AutoStore's Business Model — Part 3

Figure 9.5 — AutoStore's Business Model: Part 3

Illustration by the author.

AutoStore's Business Model — Part 4

Figure 9.6 — AutoStore's Business Model: Part 4

Illustration by the author.

AutoStore's Business Model — Part 5

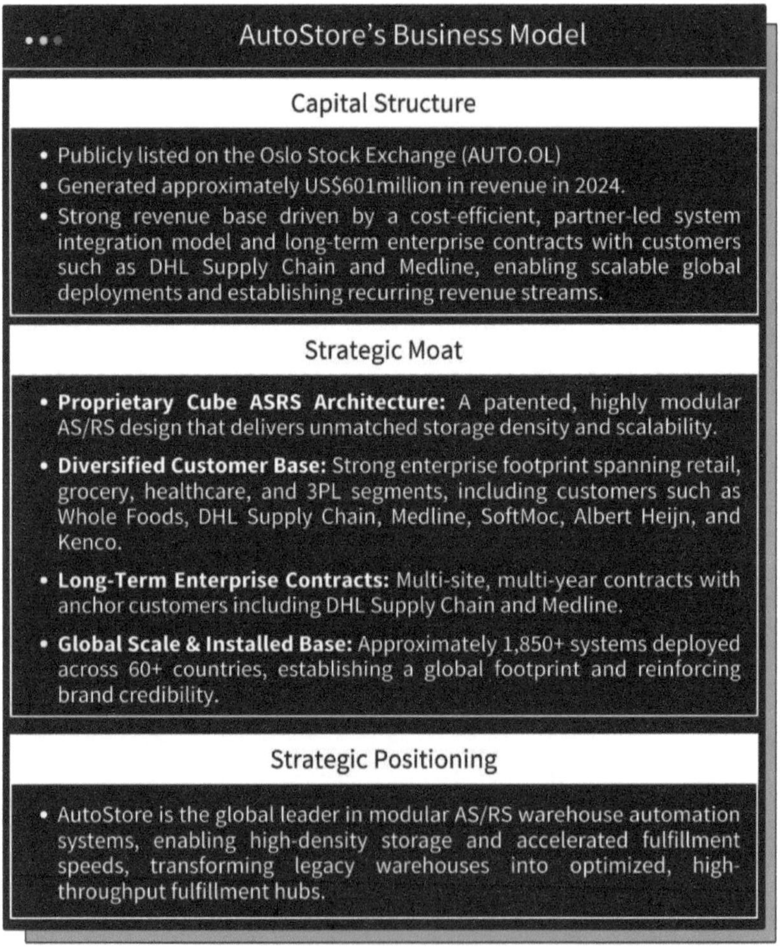

Figure 9.7 — AutoStore's Business Model: Part 5

Illustration by the author.

Chapter 10: Universal Robots — Danish Cobot Pioneer to Teradyne Exit

Odense, a city located on the island of Funen (Fyn), is one of the most historic and third-largest cities in Denmark, the happiest country in the world. The city was an important site in the Nordic Iron and Viking Age and is named after Odin — the chief god in Nordic mythology. Beyond its Viking roots, the city became the birthplace of Hans Christian Andersen, a legendary Danish author, famous for his literary fairy tales, which have travelled the world since the 1800s and have been translated into more than 125 languages to date. Continuing this legacy, in the 1990s the city became known for trade and craftsmanship, giving birth to some of the best shipbuilding craftsmen in the world. While the city boasts a rich shipbuilding heritage, this craft came under serious threat in the 1980s, given the surge of low-cost and highly automated shipbuilding companies that had started emerging in Asia.

Given this fierce competition, at the time, the Lindø shipyard, a local shipbuilding company owned by the Mærsk Group, approached the nearby University of Southern Denmark, then known as the Odense University, for help developing welding robots to make the shipbuilding process more cost-efficient. That collaboration between the local shipbuilding business and the university, seeded in the 1980s, has now turned the city into a robotics fortress.[142-143]

Building on its strong academic roots in robotics, Odense now serves as fertile ground for the next generation of entrepreneurs, researchers, and more than 150 robotics companies that sprouted out of the local university. In 2003, this same university became the nexus for three robotics researchers who are now known to the world as the founders of Universal Robots—a company that pioneered cobots, small programmable and cost-efficient robot arms developed exclusively for small and mid-sized businesses.

From 2003 to over the next two decades, the company managed to secure top-tier customers across the world, including but not limited to Nissan, Mercedes, and Dell. Today, the company commands approximately 40% of the global collaborative robot market, offering a wide range of low-cost robot arms with payload capacities ranging from 3 kg to 20 kg, deployed across more than 50 countries and supporting mission-critical workflows across manufacturing, logistics, medical, automotive, and food production industries.

Within just two decades since its inception, this Danish robotics company managed to establish a strong footprint around the world and was acquired by Teradyne, a leading industrial robotics company in the U.S., in 2015.[144] Teradyne is now on the path to automating the U.S. automotive and manufacturing sectors, kickstarting its efforts in the heart of the country's industrial corridor—Michigan.

Born in the once-remote town of Odense, the Danish robotics company has travelled across the ocean and has found its place in the heart of America's industrial arteries. Given its journey and its significance for both the Danish and the U.S. economies, it warrants deeper exploration.

To better understand its trajectory and the lessons it now has to offer to us, we will need to travel back in time, all the way to the year 2003.

Founder DNA — University of Southern Denmark (2003–2005)

In 2003, three young researchers with a deep passion for robotics met at the University of Southern Denmark, at its Odense campus. The university was formally established in 1998 but emerged from its predecessor, i.e., Odense University, which traces its roots back to 1966, and is now known for its research, specifically in the fields of sports science, biotechnology, and robotics.

At the time, all three were working on a research project and were trying to address one of the biggest limitations inherent in the adoption of traditional industrial robot arms: the high acquisition cost and the specialised programming skills needed to train and operate them. While these robots had until then served companies building automobiles and ships, these two adoption barriers meant they were not accessible to small and mid-sized businesses. This posed a major challenge for local Danish businesses seeking to lower manufacturing costs for their goods while remaining competitive against emerging Asian industry players.

These researchers sought to address this problem and, as part of their research, developed collaborative robots (cobots), or robot arms, that are lightweight, low-cost, and easy to programme, offering small and medium-sized enterprises an automation solution comparable in performance to traditional industrial robots but at significantly lower cost. This research process culminated in the launch of Universal Robots (UR), and those three researchers were Esben Østergaard, Kasper Støy, and Kristian Kassow.[145] This naturally raises the next question: how did these entrepreneurs fund their startup?

They approached Denmark's equivalent of Y Combinator: Syddansk Innovation.

Universal Robots Established — Odense, Denmark (2005)

Funded by Syddansk Innovation, one of the largest innovation incubators in Denmark, and by the Danish State Investment Fund, the company was officially incubated in 2005 in Odense, Denmark, where it retains its R&D headquarters to this day. From its launch in 2005 to 2008, the trio focused on developing their first prototype for this model of collaborative robot, or cobot. This research finally culminated in 2008 with the launch of their first production-grade prototype, called UR5, or Universal Robot 05.

The UR5 cobot was among the world's first commercially sold collaborative robots, designed to operate safely alongside humans without the need for metal safety fences. It was specifically developed for small and medium-sized manufacturers that either could not afford traditional industrial robots or lacked the operational expertise required to adopt them. The launch of this robot arm caught the attention of local Danish enterprises that were in desperate need of automating their manufacturing facilities to compete head-on with merchants from Asia—one of which was Linatex.

First Production System — Linatex, Denmark (2008)

In December 2008, Linatex, a Danish manufacturer of specialised plastic and rubber components used across industries ranging from aerospace to healthcare, became the very first commercial customer of Universal Robots. Linatex needed an automation solution that could be easily reprogrammed on the fly to support the manufacturing workflows governing the production of diverse parts and the corresponding orders it received from customers across multiple industry verticals.

Specifically, the company sought to automate the CNC machine-tending process, which had become a highly tedious task within its workflow and accounted for significant labour costs. To better understand this challenge, it helps to visualise this workflow: CNC machine tending involves workers manually feeding plastic parts into a CNC machine and removing finished parts once processing is complete. Traditional industrial robots were expensive—often costing tens of thousands of dollars—and required highly skilled programmers, creating a barrier that many mid-sized enterprises could not overcome.

Linatex addressed this challenge by purchasing a UR5 collaborative robot and automating the workflow, eliminating one of the most repetitive and labour-intensive steps in its production cycle. Rather than fencing the robot behind metal safety cages—

the industrial norm at the time—the company deployed the UR5 directly on the shop floor alongside human workers, marking one of the world's first commercially viable collaborative robot deployments. Within its first few years of operation, the low-cost robotic arm produced roughly 10,000 parts—an output that would previously have required hundreds of workers operating continuously for months at this production facility.[146]

The company was equally impressed by the return on investment, reporting that it recouped the full cost of deploying the robot within just three months of installation. Ease of programming, rapid payback time, and a low total cost of ownership were cited as key drivers for choosing Universal Robots over alternative solutions. This early adoption by Linatex served as a strong proof point for the company's expansion across Europe. The cobots were now on the radar of Denmark's neighbors — particularly Norway and Germany.

First Adoption Wave — Manufacturing Across Norway, Germany, and France (2010)

By 2010, manufacturers in Norway and Germany, two of Europe's most industrialised economies, were increasingly becoming aware of the potential of UR cobots to automate their production facilities and significantly reduce their operating budgets. Given the value these robots offered, UR was approached by one such manufacturer in Norway, Engmark Meteor AS.

Engmark Meteor, Norway (2010)

Engmark Meteor AS is a small local manufacturer specialising in the production of large griddles (traditional cooking pans) and other heavy metal products. The manufacturing process for these pans is extremely tedious and requires workers to handle raw metal components over extended periods, making it extremely labour-intensive and difficult to automate.

With a workforce of just 5–25 employees, these manual, labour-intensive workflows constrained Engmark's production system, creating operational bottlenecks and limiting throughput. As a result, the company struggled to remain competitive. High production costs within Europe resulted in thin profit margins, making it challenging for the company to compete with manufacturers that had outsourced production to countries offering lower-wage labour. Engmark Meteor therefore needed a way to scale production while retaining its manufacturing facility in Norway, without sacrificing net margins or placing additional strain on its workforce.

In 2010, the company approached UR and adopted cobots to automate core production tasks. Specifically, the company purchased four UR5s (robot arms capable of handling lighter payloads) and two UR10s (robot arms capable of handling heavier payloads), transitioning a range of labour-intensive workflows, from machine tending to packaging, onto a flexible robotic platform. These robots could be programmed by in-house staff and redeployed across manufacturing workflows, providing a flexible and affordable automation layer tailored to the company's evolving needs.

The impact was immediate: productivity doubled, working conditions became safer, and the company was able to retain all production locally while remaining globally competitive. When asked explicitly about the return on investment, owner Håvid Engmark noted that, at a 24/7 utilisation rate, the cobots could pay for themselves within just a few months—an especially compelling return on investment for a small manufacturer.[147]

Following the successful adoption of UR cobots by a Norwegian manufacturer of heavy metal products, automotive parts suppliers in Germany who faced similar production challenges began exploring comparable automation solutions. One such company was Böco Böddecker.

Böco Böddecker, Germany & Boog Machining, France (2010–11)

Böco Böddecker is a local manufacturer in Germany that supplies parts to the automotive industry. These parts must be clearly marked and labelled so they can be traced throughout the supply chain. This marking process must be extremely precise and consistent, as even minor errors can cause serious challenges for car manufacturers.

The challenge Böco Böddecker faced was that its existing approach was labour-intensive and manual. Workers were manually annotating each part, which meant the error rate was high. A part annotated with the wrong number has serious implications for the manufacturer. As the automotive industry maintains strict discipline around how parts are annotated, even a single incorrectly marked part can lead to serious consequences for automakers and their customers. Hence, automotive standards require every part to be marked with the same level of precision at all times. As production volumes surged, it became extremely difficult for human workers to consistently maintain the required speed, accuracy, and reliability needed to meet demand.

To address this challenge and meet customer requirements, Böco Böddecker introduced a UR5 collaborative robot in 2010 to automate this process. The UR5 performed these tasks with significantly higher accuracy rates and consistency than human operators, reducing errors and elevating throughput. Thanks to its collaborative design, the robot could be integrated directly into existing production lines without safety cages or major layout changes, allowing the company to improve productivity without disrupting factory operations.

By automating this process, Böco Böddecker strengthened both productivity and quality assurance, while ensuring defective parts were surgically identified and eliminated from customer-bound shipments.[148] This is especially critical in the automotive industry, where a single faulty component can have serious safety consequences.

Given this adoption in Norway and Germany, metal parts manufacturers in France left no stone unturned to adopt this solution. As a result, Boog Machining, a local metal parts manufacturer, became one of the country's first adopters of UR cobots, automating its production line in a manner similar to Engmark Meteor using lightweight UR5 robots.

Together, these early deployments across Norway, France, and Germany demonstrated that UR cobots were industry-agnostic and capable of supporting a wide range of applications, from metal cookware to large-scale automotive components. This early adoption validated the technology and positioned the company to expand beyond its initial verticals into industries that are operationally demanding and harbor workflows that are significantly more complex to automate.

One such sector was food processing.

Second Adoption Wave — Food Processing & Consumer Goods Packaging Across Germany, Iceland, and Greece (2011–2013)

Minitüb (Germany) – UR5 Automates Veterinary Packaging Operations

Minitüb is a leading supplier of systems that enable assisted reproduction for farm animals. The company produces and packages specialised products used in veterinary applications, operating in a highly regulated and quality-sensitive environment. In 2011, as demand for its products surged, the company started evaluating solutions to automate its packaging workflows, which at the time were labour-intensive and difficult to scale.

To automate these workflows in a capital-efficient manner and reduce strain on its employees, the company decided to test UR5 collaborative robots in the first quarter of 2011. Integrated directly into its packaging line, the robot was positioned at the end of the conveyor belt, where it picked up freshly bagged products and placed them into cardboard boxes. A custom gripping tool equipped with suction cups enabled the robot to safely handle these sealed product bags. The UR5 packaged an average of 40 bags per minute, delivering a stable and consistent throughput while significantly reducing labour requirements for this workflow and elevating operational efficiency for the company.[149]

The successful deployment of UR5 robots at Minitüb demonstrated that the technology had been validated in highly regulated and quality-sensitive veterinary environments. This validation positioned the company to scale its footprint beyond manufacturing, drawing interest from other producers with similar operational requirements, including large-scale dairy processors in Iceland.

Mjólkursamsalan Akureyri (Iceland) — Dairy Production Automated

Mjólkursamsalan Akureyri is Iceland's largest dairy producer, processing and packaging a wide range of dairy products for domestic distribution. The company is known for its cream cheese products in the region. The production process for cream cheese that the company had originally implemented was labour-intensive and required workers to manually pick 1,000s of packages of cream cheese from the conveyor belt as they exited the sealing machine and then place them into large trays. These trays were used to transport the sealed cheese packages to trucks for shipment. This entire workflow was manual and operationally inefficient, limiting the company's ability to scale production.

To address these challenges, the company integrated two UR5 collaborative robots directly into its cream cheese production line.

The first robot was deployed to lift multiple 250-gram packs of cream cheese from a conveyor belt and transfer them into large trays. The second UR5 stacked the filled trays onto pallets for shipment. The robots were integrated directly into the production line and operated safely alongside employees, eliminating the need for safety fencing or major layout changes. Within the first year of deployment, the robots saved the equivalent of three full-time employee years of opex-intensive and physically straining work, enabling employees at Mjólkursamsalan Akureyri to focus on other workflows and significantly elevating operational efficiency for the company.[150]

Now, having successfully automated packaging workflows in highly regulated veterinary facilities in Germany and high-volume dairy production facilities in Iceland, Universal Robots demonstrated that its collaborative robots could operate reliably in highly regulated and hygiene-sensitive environments. This validation meant the company's robots were now ready for adoption by large-scale consumer goods manufacturers, including Johnson & Johnson in Greece.

Johnson & Johnson (Greece) — Consumer Goods Packaging Operations Automated with Cobots

At its facility in Athens, Greece, Johnson & Johnson manufactures consumer healthcare products, including shampoos and skincare items. The packaging process for these products includes two major workflows: one that outputs finished products onto a conveyor belt, and a second that takes these products and packages them. The second workflow related to packaging was manual and labour-intensive. This meant the company was only able to package about 45 bottles per minute at each of these packaging stations. The company wanted to double this throughput and was actively looking for a solution.

Johnson & Johnson, having heard about UR5 deployments in Germany, Iceland, and Norway, decided to test this robot at its plant in Athens. Once deployed, the robot picked up multiple bottles simultaneously from the production line, oriented them, and placed them into the packing machine. Bottles come in different shapes and sizes, which meant the robot had to handle these variations while ensuring no damage was done to these fragile products.

Not only was the UR5 able to handle these variations and package the bottles, but it was also able to do so at nearly twice the speed of the manual workflows previously in place. The robot processed approximately 70 bottles per minute, compared to the 45 bottles handled manually. Furthermore, workers were trained to reprogram the robot as needed, allowing it to be redeployed across different areas of the plant to automate additional workflows.[151]

This adoption by a Fortune 100 client demonstrated that the cobots had effectively passed the enterprise test and that the company was now ready to expand its footprint beyond Europe. It was time to enter Asia.

Asia Expansion — China Market Entry (2011)

Around 2011, the company launched its first office in China with the goal of expanding its production over time and growing its customer base in the region. This raises an important question: why China?

China is the manufacturing hub of the world, which means it has a pressing need to automate the millions of manufacturing plants located across the country that produce goods for Fortune 100 companies around the globe, including, but not limited to, Apple, Google, Microsoft, and others. The country is also home to the largest installed base of industrial robots in the world, with a staggering 250,000 units already deployed in these manufacturing plants, growing at an exponential pace each year. This was one of the primary reasons the company decided to expand into China. Now, having established its new office there, it was time to finalise the launch of its next product — a high-payload robot arm, the UR10.

UR10 Launch & U.S. Market Entry (2012)

In 2012, having secured its first office and a future manufacturing hub in China, the company geared up to launch its next product: a high-capacity robotic arm, the UR10, designed to carry payloads of up to 10 kg. The UR10 was targeted at manufacturers looking to automate workflows involving larger items, such as laptops, electronics, appliances, and other bulky consumer goods, particularly in the European and North American markets. Following its launch, the company secured contracts to expand its adoption footprint at Mjólkursamsalan Akureyri's dairy production facility in Iceland and Engmark Meteor's manufacturing facility in Norway, deploying both UR5 and UR10 robots across their sites.

In addition to expanding its footprint and product line, the company established its American subsidiary in the same year, based in New York, marking its official entry into the U.S. market. With a newly established office in China and a seed stage office in the USA, the company was now ready to sell UR5 and UR10 robots across the Asia-Pacific and North American markets.

It is important to note here that, by this time, the company had deployed around 1,600 robots across Europe, and to serve these next two geographies it had to scale its production engine. That meant it was time to launch its first large-scale factory.

Scaling Manufacturing — New Factory in Odense, Denmark (2014)

In 2014, the company launched a 12,000 sqm production facility in Odense, Denmark, to facilitate the production of more than 4,500 robots on a monthly basis.[152] The facility was established to meet rising customer demand and to clear an order backlog of nearly 2,000 units for the year. With production scaled in Odense and a local office already established in China, the company was now ready to officially expand its customer base. That meant it was time to deliver the first set of orders received from customers in the Asia-Pacific region.

Having incubated its base in Asia in 2011, the team operating in China secured contracts with a local manufacturing company, PLC Industries, over the three-year period from 2011 to 2014. It was now time to deliver on those commitments, execute these contracts, and build a strong fortress in Asia.

Southeast Asia Expansion — PLC Industries, Singapore (2014)

PLC Industrial is a precision engineering manufacturing company headquartered in Singapore, producing components used in photonic devices, biomedical and imaging systems, and healthcare equipment. This meant the company operated dozens of CNC machines to turn raw materials into finished products before shipping them to customers.

In 2014, a shortage of skilled labour, coupled with rising material and real estate costs, had become a major challenge for manufacturers like PLC Industries in Singapore. To address this challenge, the company decided to test UR10 high-capacity robot arms using a phased approach. First, the company conducted a safety assessment to ensure workers were comfortable both handling and operating alongside the robots. It also ensured that these robots did not damage any of the existing equipment or the components being manufactured.

Once the product passed the safety assessment test, the company ran a pilot program to objectively evaluate the value these robots offered. In simple words, it was now time to evaluate the throughput gains.

Did the UR10 pass this test? It did.

By deploying just two UR10 high-capacity robots, the company was able to scale throughput at two of its CNC stations by 40%. If that was not enough, it also greatly improved labour efficiency. Each worker was now handling four CNC machines, whereas previously multiple workers were required per CNC machine. In addition to this, the company was able to recoup the product and installation cost in just 12 months — an impressive payback period for a domestic mid-sized manufacturer adopting robots in the industry.[153]

With this deployment in Asia, Universal Robots had installed about 2,500 robots globally, operating across industries including food, manufacturing, dairy, and consumer goods. From Johnson & Johnson in Greece to businesses across Iceland, Norway, Germany, Singapore, and France, the company demonstrated that its products were validated and adopted at scale across multiple geographies.

Given this scale of adoption and its ambition to expand its footprint in North America, the company was now positioned squarely on the radar of industrial robotics giant Teradyne.

Teradyne Acquires Universal Robots (2015)

In 2015, Teradyne acquired Universal Robots for approximately USD 285 million, with a mission to re-energize the soul of manufacturing and the automotive industry in North America. The acquisition also led to the establishment of Universal Robots' first major development center in Michigan, the heart of America's automotive industry, as well as the inception of a strong partnership with Ford Motor Company, the local automotive giant.[144]

From its modest beginnings in the town of Odense, Denmark, the company evolved to serve dairy production facilities in Iceland, automotive parts suppliers in Germany, and Fortune 100 customers in Greece, while establishing offices in China and deploying more than 2,500 robots worldwide. What started as a small European robotics venture matured into a truly global enterprise, ultimately securing a long-term strategic partner across the Atlantic.

Today, the company produces robotic arms supporting payloads from 3 kg to 20 kg and serves customers in more than 50 countries, retaining a strong foothold across Europe, North America, Japan, and China. From manufacturing and logistics to electronics, medical, automotive, and food, its cobots are automating production plants across industries.

While building a successful robotics company is an extremely challenging endeavour in itself, scaling it globally, progressing, ensuring a strategic and successful exit, and ultimately becoming the very engine that now powers one of the world's largest economies is even more challenging. The lessons from this journey warrant deeper examination.

Lessons for Founders and Engineers

Lesson 01: A Founder's Ability to Identify Opportunities Is Often Underrated

To begin with, one of the most important lessons in this journey lies at the very root of this startup. If you notice, the founders entered an industry that was heavily dominated by leading robot-arm players such as Kuka, ABB, Fanuc, Yaskawa Motoman, Omron, Kawasaki, and others. Even with this lineup of A-grade industrial players manufacturing robot arms, the company identified an opportunity in the market — a need for small and mid-sized manufacturers to automate their production with affordable robotic arms. A need that was not addressed at all. It identified, validated, and addressed this unmet need for businesses around the world by building affordable and agile industrial robot arms.

The lesson here is to remember that even though a market appears saturated with solutions similar to those that you are planning to build, in reality customers operate at different scales. A solution that works for enterprise customers may not work for small and mid-sized businesses, and it's this crevice of unmet need that creates opportunity for startup founders. Do not be afraid to enter markets that appear saturated; they are often hidden gems filled with opportunities.

Lesson 02: The Pie Is Much Bigger Than You Think

Most founders in the robotics arena tend to chase the same or a similar set of customers during the initial stages, and this often creates a heavy layer of resistance to product adoption. In many cases, startups immediately target enterprise customers — hospitals, mall owners, oil refineries, or large-scale production facilities. While these customers do bring scale and are important to secure your long-term runway, in the early stages you need to optimize for speed of adoption, because you are operating in a constrained-resource mode.

You need to build credibility with limited resources and demonstrate that your product works and generates value. To do this, you must optimize for organic speed of adoption, and one of the simplest ways is to go where no one else is going. Go outside the city you are based in. Head towards the outskirts. This is often where the opportunities lie. The pie — the total addressable market — is far bigger than most founders realise, yet many do not take the time to meet and understand customers who fall outside the traditional startup lens.

Think small and mid-sized farms, family-owned metal and recycling businesses, the leather industry, or even the fishing industry. There are countless sub-verticals that remain largely un-roboticised. Just like the founders of Universal Robots landed opportunities with dairy farmers in Iceland, metal product manufacturers in Denmark, and small-scale automotive parts suppliers in Germany, you can do the same and secure your early adopters with ease. By establishing early adopters in these arenas, you naturally reduce direct competition and maximise focus on product improvement. At the same time, you are able to land deals faster with customers who genuinely need and value your solution.

Lesson 03: Build One Core Product, Then Scale It Across Verticals

One of the important lessons from the journey of cobots is that the founders focused on building one strong core product at the beginning — the UR5. Only after several years did they launch the UR10 to address customers who needed higher payload capacity, followed later by the UR16e and UR20 to support payloads ranging from 5 kg to 20 kg. The product curve here remained lean and focused.

Instead of diversifying too early or adding multiple unrelated products to the portfolio, the company stayed committed to one core product and continuously improved it to address evolving customer needs. The core problem never changed — customers needed a robotic arm that could act as an additional hand to automate operations. What changed was the scale of automation customers wanted as they began to see value, which naturally created demand for higher payload capacity.

This was organic product growth. In reality, the company was solving one core problem with one foundational product, yet over time it created a portfolio that appeared diverse but was fundamentally unified. Build one strong product for your customers, not ten disconnected variants.

Lesson 04: Local Expansion First, Global Expansion Next

One of the most important lessons from this journey is the founders' ability to stay committed to Europe and grow their customer base in their local market for several years before even considering expansion into Asia or North America. From 2005 to 2011, during the first six years, the founders remained focused on the regional market in Europe and, within that, concentrated on the periphery of the market, namely small and mid-sized businesses with unmet needs.

Only after the product was well adopted across industries, from dairy and automotive to metal manufacturing and consumer goods, did they launch their first office in Asia. The upside of this approach is that you can continue to operate lean, maintain a lower burn rate, and sustain a higher focus rate when you prioritise local operations for an extended period of time.

It also means you require a relatively smaller workforce, since you are operating in the same time zone as your customers, and only minimal support is needed outside local working hours. Additionally, being just a few hours away from most customers allows you to deliver high-quality service at speed, which naturally boosts customer adoption and retention rates. The key takeaway is simple: focus on regional markets before making a move into international ones. International expansion may bring scale, but it often comes with significantly higher burn rates and operational bottlenecks that can erode your runway.

Now that we have distilled the core lessons embedded in Universal Robots' trajectory, from its origins through its evolution and the pivotal decisions the team made along the way, it is time to decode the underlying business model.

In the final section of this chapter, we will break down the business model and illustrate how the product portfolio maps to select primary use cases.

Universal Robots: Product Snapshot

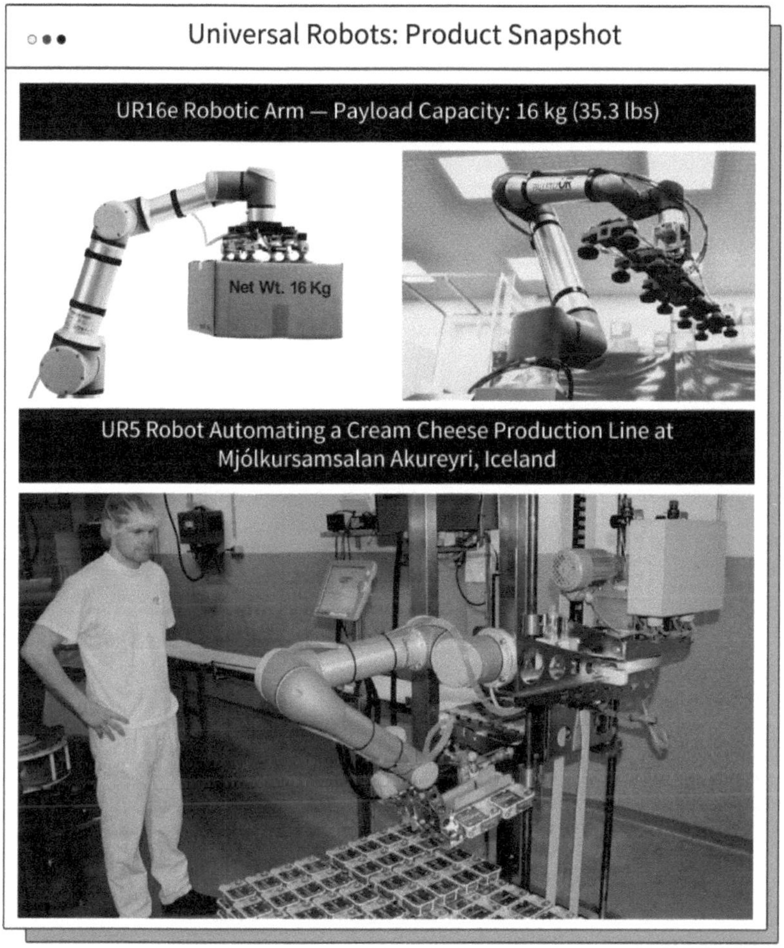

Figure 10.1 — Universal Robots (U.R.): Cream Cheese Production at Mjólkursamsalan, Akureyri, Iceland

Illustration by the author. Robot images from the U.R. case study pages

Universal Robots’ Business Model — Part 1

Figure 10.2 — Universal Robots’ Business Model: Part 1

Illustration by the author.

Universal Robots' Business Model — Part 2

Universal Robots' Business Model

Product & Platform

- **Core System:** Collaborative robotic arms (UR Series), designed for safe human-robot interaction, ease of integration, and rapid deployment across diverse industrial environments.
- **Product Portfolio:** UR3e–UR20 | **Payload Range:** 3–20 kg
- **Agility:** Lightweight and modular arms that can be easily reprogrammed for multiple use cases, such as pick-and-place, welding, inspection, and packaging.
- **Ease of Operations:** Intuitive graphical programming interface that minimizes the need for specialized robotics expertise.
- **Ecosystem (UR+ Platform):** A certified partner ecosystem offering end-effectors, software, vision systems, and accessories that integrate seamlessly with UR cobots.
- **Safety & Compliance:** Integrated force sensing and advanced safety architecture enable safe human-robot collaboration, eliminating the need for traditional cages in most industrial environments.

Customer Segments

- **Primary Verticals:** Manufacturing (SMEs and large enterprises), Electronics & Semiconductors, Automotive, Metal Fabrication, Packaging, Food & Beverage, Healthcare & Research.
- **Customer Profile:** Small and mid-sized manufacturers pursuing agile, reconfigurable, and cost-efficient automation solutions to mitigate labor constraints, elevate throughput, and scale operations with minimal investment.

Figure 10.3 — Universal Robots' Business Model: Part 2

Illustration by the author.

Universal Robots' Business Model — Part 3

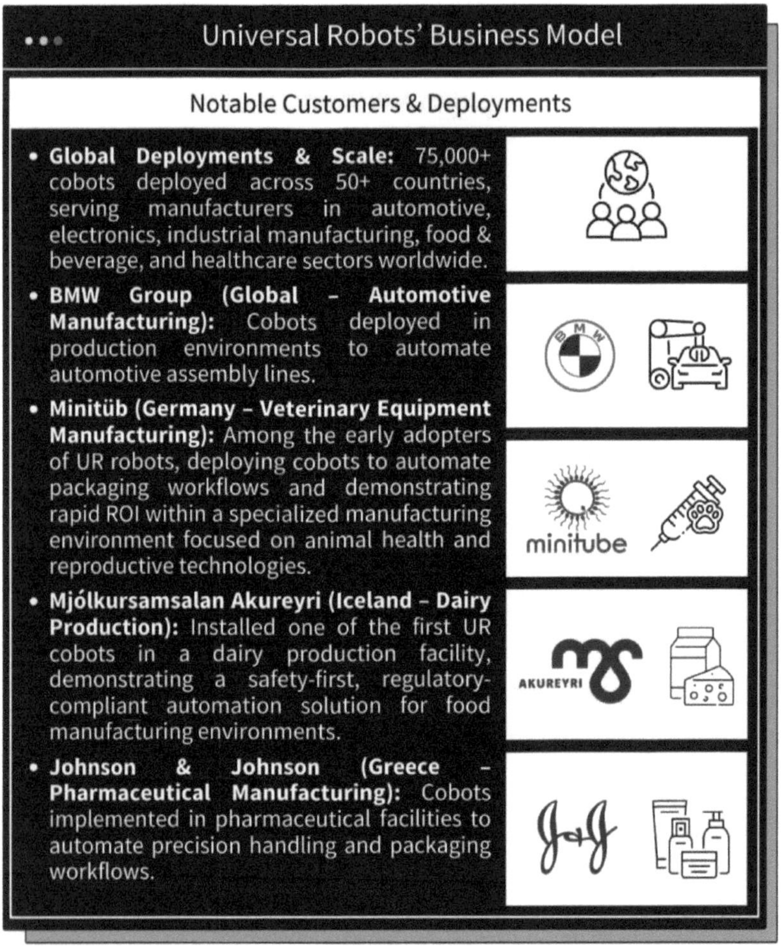

Figure 10.4 — Universal Robots' Business Model: Part 3

Illustration by the author.

Universal Robots' Business Model — Part 4

Figure 10.5 — Universal Robots' Business Model: Part 4

Illustration by the author.

Universal Robots' Business Model — Part 5

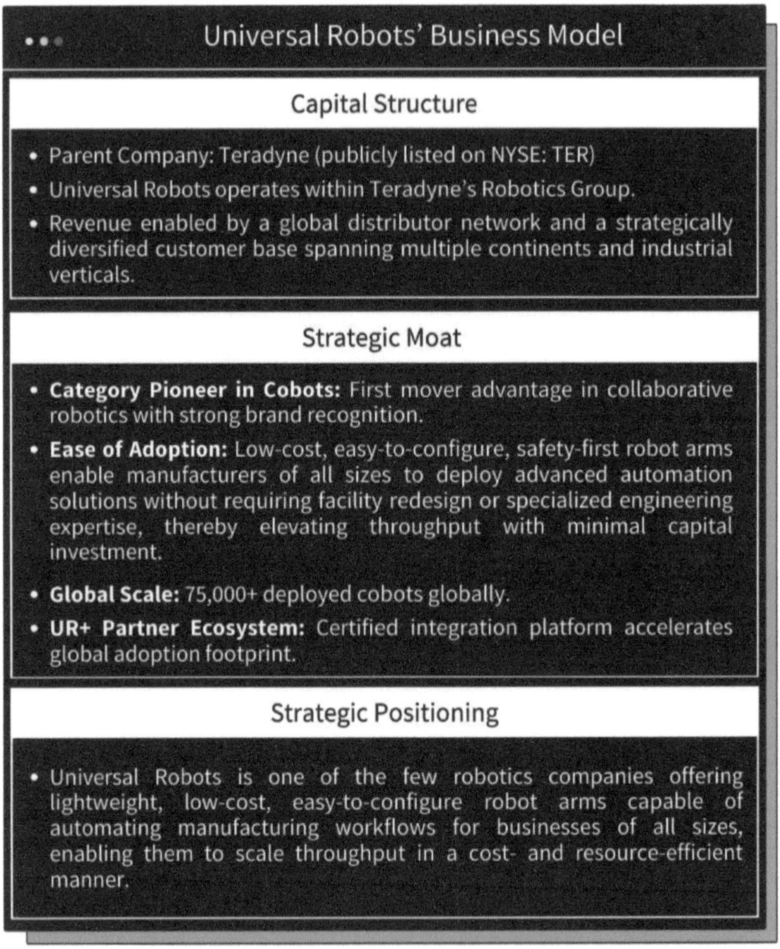

Figure 10.6 — Universal Robots' Business Model: Part 5

Illustration by the author.

Chapter 11: ANYbotics — ETH Zurich Labs to Shell and Vale

Around 5,000 years ago, nomadic settlers arrived on the shores of one of Europe's largest natural lakes. Over the centuries, this hamlet blossomed into a vibrant modern city — one known for sweeping views of the Alps, delicious chocolates, and rich fondue made from creamy Alpine milk sourced from some of the happiest cows in the world, grazing peacefully in the lush green pastures on its outskirts. This city is Zurich, in Switzerland — a country of 9 million people known for its monastic devotion to nature and fresh, organic food.

What do you build when you get to study in college with such stunning views, savor fresh mountain-rich food, and keep a pulse on time with some of the most exquisite watches ever made on our planet? Well, it turns out that if you are one of its nine million citizens who have access to this rich, mind-nourishing ecosystem, you start exploring ways to make the city and the country more productive. You solve the problems that matter most to the citizens of your country.

What kind of problems?

While the country may have access to mountains, cows, cheese, and chocolate in abundance, labor has been its biggest bottleneck.

You need a skilled workforce to turn that creamy milk into fine cheese and delicious chocolates. That means you need a skilled workforce to scale the country's net economic output — its GDP. Rising wages, persistent labor shortages, and a shrinking workforce meant the citizens had to address this problem to ensure the country could sustain its trajectory. In response, a group of citizens-turned-engineers chose to confront this challenge head-on — a quest that ultimately led to the birth of what many now refer to as the "Boston Dynamics of Europe": ANYbotics.

A company with deep expertise in legged robotics, its robots are now being adopted by industrial giants around the world, from Shell and BASF to Petronas and Siemens. From its first product launch in 2016 to commercial deployments spanning multiple industries in less than a decade, this robotics company has established a global footprint with an exceptionally lean operating model: a team of around 200 passionate robotics professionals, $22M in Series A funding followed by $60M in Series B, and a top-tier customer base — a trajectory that certainly warrants deeper examination. To better understand the route this startup took and the lessons it offers, we must travel back in time to the year 2015.

Founder DNA — ETH Zurich Robotics Systems Lab (2015)

Back in the mid-2010s, the founders of ANYbotics, many of whom were working on a research project at the ETH Zurich Robotics Systems Lab, were trying to address a core challenge in robotics. The challenge was to develop perception and motion-planning algorithms for legged robots to ensure they could traverse harsh and challenging environments — hills, mountains, tunnels, mines, oil rigs, and chemical factories. This, in fact, was the core research thesis of one of its principal researchers and now co-founder and CEO, Péter Fankhauser.[154]

The goal was to build a robot that could perceive and navigate these environments and offload tasks from workers conducting inspection, extraction, and repair work at these sites. While the mission was clear in terms of what the robot should do, the team had to decide what form it should take. Should it be a humanoid, a mobile robot with four wheels, or a quadruped robot like Boston Dynamics' Spot? When faced with the choice of form factor, the founders drew inspiration from mountain goats. Mountain goats are among the few lean animals that can both navigate and retain balance in extreme alpine terrain, and this quickly became the decisive factor that led the researchers to choose quadrupeds as the form factor for their project — their robot, called ANYmal, in 2015.[155]

ANYmal is a rugged four-legged robot designed to automate routine inspection patrols that are conducted by workers across facilities such as tunnels, chemical plants, mines, oil rigs, and manufacturing plants to maximize uptime, improve safety, and reduce operational costs. Customers today include Petronas, Shell, Siemens Energy, BASF, and Outokumpu, one of Europe's largest producers of stainless steel.[156]

That raises an important question: how did the company land its first customer? Well, if your startup has its roots in the academic soil, chances are you might find your early adopters in the same soil. This was true for ANYbotics.

First Commercial Delivery (2017)

In 2017, the company delivered its first production-grade robot, ANYmal, to the University of Edinburgh. This robot was designed to ensure mission-critical reliability: it could survive rugged and hazardous environments, was waterproof, and was equipped with autonomous navigation capabilities. Researchers at the Edinburgh Centre for Robotics used ANYmal to continue their work on legged locomotion, perception, and motion planning systems. In effect, the robot became a research platform for the university and its students, enabling them to test their theses and algorithms in some of the most challenging environments imaginable.[157]

Needless to say, the robot is capable of navigating stairs, doors, and slopes while conducting autonomous inspection patrols. It can read sensor gauges located on factory floors or in chemical plants and transmit the data back to operators, all while navigating around obstacles in its path, including humans and equipment.

Having successfully landed its first customer in the academic cosmos, the company started absorbing structured feedback from these researchers, i.e., its early adopters. That meant it was time for product iteration. From 2017 to 2019, the team rigorously refined the design, hardware, and software stack of the robot to address gaps identified during these early-stage deployments.

In addition to these researchers in academia, around September 2018 the company also secured a pilot project with another early adopter, the Dutch-German transmission system operator TenneT. TenneT connects large-scale offshore wind farms to the onshore grid through high-voltage DC connections and, in effect, enables reliable and low-cost energy transmission and distribution for energy operators. The company partnered with ANYbotics to enable robotic inspection on one of its offshore platforms, which also happened to be the world's largest offshore converter platform in the North Sea — making ANYmal the world's first autonomous offshore inspection robot. This deployment allowed TenneT's onshore teams to remotely inspect offshore energy infrastructure, significantly minimizing the need to send workers into the middle of the North Sea to conduct inspections, thereby ensuring safer and more reliable year-round operations.[158]

This phase was critical for ANYbotics, as the company was gearing up to spread its roots beyond academia and the North Sea. It was now ready to demonstrate its product on an international stage, and that moment arrived in 2019.

First Major Product Launch — ANYmal C, World Robot Conference Beijing (2019)

In 2019, the company launched the next version of its robot, ANYmal C, at the World Robot Conference in Beijing, China. This model was the culmination of the feedback loops and corresponding efforts the company had undertaken from 2017 to 2019 with its early adopters, such as the University of Edinburgh and TenneT, to refine its product and ensure it was optimized for production-grade environments — specifically for conducting autonomous inspection patrols across chemical plants, oil refineries, mines, and factories, while delivering a mission-critical solution with extremely high availability and a strong safety and reliability profile.[159]

With this launch, customers could now define and execute customized missions for each of these robots to autonomously patrol their sites and collect data that mattered to them based on their use case. A mission, in this context, implied a predefined route the robot would follow while navigating these facilities, along with specific checkpoints at which it would pause and capture gauge readings using its suite of sensors, including but not limited to thermal cameras, gas-detection sensors, and standard AI-enabled cameras, all while identifying and circumventing obstacles using its LiDAR-enabled ROS algorithms. The goal was twofold: to conduct autonomous routine inspections and to identify hotspots that could potentially lead to disasters. Hotspots in this context referred to gas leaks, chemical leakages, burst pipes, and other critical anomalies that typically require human intervention.

To enable this, ANYmal C was equipped with LiDAR and depth cameras that provided 360-degree vision, a tele-operation module that allowed operators to assume manual control when required, and a rugged, water- and dust-resistant IP67-rated body. It offered a payload capacity of approximately 10 kg, although its operational runtime remained limited to roughly two hours on a single charge. So, in short, while the company had made significant progress optimizing autonomous navigation algorithms that enabled remote inspections in these challenging environments, there were still areas that emerged as bottlenecks for its adoption across verticals. This phase could best be described as Phase 2 — the robot was well-suited for supervised industrial adoption, while a truly unsupervised model was yet to come.

In effect, the robot was, in a way, ready to help energy, chemical, and oil plant owners conduct autonomous routine inspections in a safe and reliable manner, but enabling fully unsupervised, 24×7 continuous operation was not yet possible. That brings us to an important lesson on managing product trade-offs, especially at early stages. When you are a startup, you don't always get to perfect every aspect of your product; instead, you optimize the features that matter most to your customers, and there is always a trade-off. In this case, the trade-off was deliberate. Customers could deploy multiple robots at a single site to enable near-continuous 24×7 operations, so battery life was not an immediate bottleneck for the startup. It was not its A-plus bottleneck.

The real A+ bottleneck in this case was production and the scale of adoption. The company had to secure its financial runway while serving its existing customers, which meant it had to manufacture these robots at scale and expand its adoption footprint in parallel. To do so, it needed capital — and that meant it was time for a Series A.

Series A Funding – $22.3M (2020)

In 2020, the company raised $22.3M in a Series A round led by Swisscom Ventures, with participation from co-investors such as Ace & Company and EquityPitcher Ventures, to accelerate the commercialization of its autonomous inspection robots.[160] At the time, the company was extremely lean, maintaining a footprint of about 50 employees who had undergone a rigorous selection process. As I started researching the trajectory of this company, one of the key patterns that became evident was its consistent focus on maintaining a lean profile — a small team of roughly 50 engineers, lean funding rounds, and an intense focus on its mission while keeping an exceptionally low burn rate and high clarity of purpose. That was one of the important lessons here, and as we continue to decode its trajectory, it will surface again and again.

As a robotics founder, or an engineer-to-be founder, it is important to remember this lesson because running a robotics company can at times feel like running multiple companies in parallel, which can lure founders into aggressive hiring and oversized funding rounds. In reality, however, it is eye-opening to see how one can maintain a lean footprint and still build a successful robotics company that creates exponential value for employees, customers, and investors alike.

With the Series A raised, the company now had the funds to scale its production and expand its adoption footprint.

First Industrial Adoption Wave (2020–2022)

Having launched its production-grade model, ANYmal C, at the conference in Beijing in 2019 and secured its financial runway through Series A funding in 2020, the company embarked on a cross-industry voyage to land some of the world's top-tier customers operating in the energy, chemical manufacturing, and heavy metals sectors. The first in line here was Petroliam Nasional Berhad (PETRONAS).

PETRONAS — Autonomous Offshore Inspection, Malaysia

PETRONAS is Malaysia's national oil and gas company with upstream and downstream operations spanning offshore platforms and refineries across multiple geographies. In simple terms, the company operates oil rigs at sea and on land in several countries and extracts crude oil and gas (upstream activities), then refines these into finished fuel and energy products (downstream operations) that are later sold to customers around the world. Routine asset inspections on offshore facilities such as the company's Dulang B platform in the South China Sea traditionally required deploying specialized trained staff members into hazardous, rough-weather-exposed environments via boat or helicopter.

These inspections were capital- and resource-intensive and hence could not be scaled by the company, despite the need to conduct them on a frequent basis. In addition to this, the majority of these sites required explosion-safe (Ex-certified) equipment, which further constrained the company's ability to seek automation solutions and raised health and safety risks for its workers.

To make this process safe, reliable, and scalable, PETRONAS partnered with ANYbotics to deploy ANYmal C for autonomous offshore inspections. Starting with onshore assessments, the company graduated the platform to a multi-week live offshore deployment on the Dulang B offshore platform in the South China Sea, Malaysia. The robot performed visual inspections, digital and analog gauge readouts, thermography, and acoustic analysis of offshore equipment located at the site while leveraging LiDAR-based autonomous navigation and docking capabilities to ensure safe and secure on-site routing.[161]

This deployment led to a significant reduction in the workforce required to conduct offshore inspections, elevated the quality and frequency of these offshore inspections, enabled automated reporting, and helped the company establish a reliable year-round operational safety profile. This collaboration with the nation's largest oil and gas company further culminated in the co-development of an officially Ex-certified ANYmal, purpose-built for commercial deployments in 2022 across industrial inspection environments where stringent safety certifications are mandatory.

Given this successful adoption by one of the largest oil companies in Asia, it was now time for ANYbotics to test its robots in environments as hazardous as offshore oil rigs but on land. The next stop was chemical manufacturing plants in Germany.

BASF — Chemical Plant Inspections, Germany

BASF is one of the world's largest chemical manufacturers, operating massive and highly complex production facilities that must be monitored constantly to ensure safe, reliable, and regulatory-compliant operations. At its Ludwigshafen site in Germany, which is the largest (~10 km^2) integrated chemical complex in the world, the company had to conduct inspections on a regular basis to keep track of all its assets and their health status. This process was largely manual, with some level of automation enabled using static IoT sensors. The problem the company ran into was scale. The existing inspection process was not scalable, since adding more workers to the facility meant doubling the operational expense at each site. In addition to this, each of these sites carries a heterogeneous terrain profile, meaning they are not as flat as a warehouse floor and include stairs, narrow passages, and multi-level equipment, which limited the company's ability to adopt off-the-shelf solutions. To address this challenge, BASF reached out to ANYbotics.[162]

The company worked with ANYbotics to evaluate the autonomous robot ANYmal as a potential scalable mobile inspection solution for its site. The robot conducted inspection patrols, reporting acoustic sounds from pumps and compressors to help remote operators detect anomalies, capturing thermography images to help detect fire and smoke hazards, and transmitting sensor readings from each asset that helped operators ensure all on-site equipment was healthy, while navigating autonomously across these plants. This deployment greatly benefited the company as it automated the entire process of conducting inspections, collecting data, and reporting anomalies in parallel during a single patrol with a single robot, minimizing the need to send teams of workers for hours at each site to conduct these patrols. This meant time and cost savings, and overall throughput nearly tripled for each site. This was the scalable solution the company had been looking for, and given this deployment by BASF in Germany, and earlier by PETRONAS in Malaysia, ANYbotics was now on the radar of oil and chemical manufacturing companies around the world. The next stop was Brazil, in South America.

Vale — Autonomous Inspection in Mining Operations, Brazil

Vale is one of the world's largest mining companies and a leading global producer of iron ore and metals, operating in nearly 30 countries with a workforce of approximately 125,000 employees. Given the scale of its operations, workforce safety and asset reliability are critical business priorities. At its Cauê Mine facility in Minas Gerais, Brazil, the company was required to conduct frequent inspection patrols and that meant deploying teams of workers into narrow, dark, wet, and machinery-dense corners of these facilities, exposing them to heightened health and safety risks. In addition, the process was largely manual and capital-intensive, which limited the company's ability to scale inspection frequency.

To address this challenge, Vale partnered with ANYbotics to evaluate the autonomous robot ANYmal as a long-term mobile inspection platform for this facility. Following an initial commissioning phase that resulted in the development of detailed 3D facility maps and inspection-point databases (patrol routes), the robot was deployed to perform fully autonomous inspection patrols. It conducted full-site scans using onboard high-definition visual cameras, thermal imaging sensors, and acoustic sensors, providing operators with multi-layered insights for all on-site equipment and components, including but not limited to conveyor belts, gauges, electrical panels, lubrication points, drainage systems, and large milling equipment, while navigating stairs, slopes, mud, and wet terrain. Thus, through this deployment, the company was able to fully automate routine inspections in hazardous environments, reduce reliance on manual labor, and scale inspections in a capital- and resource-efficient manner while maintaining a strong safety and compliance profile.[163]

This collaboration, alongside earlier deployments with global energy and chemical industry leaders, further culminated in the co-development of an officially Ex-certified ANYmal, purpose-built for commercial deployments.

ANYmal X Commercial Launch — Certified for Hazardous Environments (2022)

In March 2022, ANYbotics introduced ANYmal X, the world's first Ex-certified legged inspection robot approved under IECEx and ATEX Zone 1 standards. This release marked a decisive engineering inflection point: until then, robots capable of navigating stairs, narrow passages, and multi-level industrial structures were not legally permitted to operate on explosion-prone sites. ANYmal X closed that gap by marrying a quadruped mobility platform with explosion-proof electronics, dual-redundant drivetrains to ensure uninterrupted navigation even in the event of a subsystem failure, and batteries engineered with extremely high temperature thresholds to withstand both extreme heat and cold.

This launch of an explosion-proof certified solution meant the company's robots were now officially ready to conduct autonomous inspections on sites prone to explosions, including oil rigs, chemical plants, and nuclear facilities—sites where even a minor spark could lead to a full-scale explosion and major operational loss. The certification provided formal assurance that the company's robot was not only fully autonomous and capable of conducting remote inspection patrols, but also rigorously validated for mission-critical operations through stringent industrial safety certification processes.

The launch of this certified mobile inspection robot squarely positioned the company on the radar of major energy, oil, and chemical plant operators worldwide.

Second Wave of Adoption (2023–2024)

With the Ex-certified platform validated and early industrial pilots completed, the company officially ignited the second wave of adoption for its robots. From Outokumpu in Finland to Equinor in Norway, and even nuclear facilities in Finland, the company started actively positioning its solution for leading global industrial enterprises. First in line was Outokumpu.

Outokumpu — Heavy Industry & Metals | Finland, Germany, Sweden (2023)

Outokumpu is one of Europe's largest stainless-steel producers, operating high-temperature, chemical-intensive manufacturing plants across several countries. In 2023, the company initiated a multi-site robotic inspection program spanning its plants in Krefeld (Germany), Tornio (Finland), and Sweden to assess the possibility of automating inspection patrols at these live production sites using mobile robots. This decision raises an important question: what challenge was the company trying to address?

The primary challenge was ensuring safe and reliable inspection protocols across its facilities. Workers conducting patrols at these hazardous sites were frequently exposed to hydrofluoric and nitric acids, molten metals, dust, and extreme temperatures. This meant the existing approach adopted by the company to conduct these patrols was neither safe nor scalable. However, discontinuing these patrols was not an option, as it would leave the company without any insight into the health status of its on-site equipment and components. Given the occupational and regulatory risks involved, the company chose to address this challenge head-on and partnered with ANYbotics to deploy ANYmal D under a robotics-as-a-service model.[165]

Under this model, the company adopted a phased approach to test and roll out this robot-led site inspection service. In Phase 1, the robots underwent several weeks of on-site training, during which they developed a deep understanding of the layout of each facility, along with the corresponding patrol routes and inspection checkpoints defined by the company. This was followed by Phase 2, in which the robots started executing fully autonomous inspection patrols. During these patrols, each robot conducted a three-layered inspection assessment that included insights captured through high-definition visual cameras, thermal imaging sensors, and acoustic and frequency analysis of on-site components and equipment such as motors, panels, gauges, and lubrication points. The IP67-rated quadruped navigated stairs, narrow passages, and wet floors with ease.

The impact of this deployment was immediate and tangible: the robots were able to conduct nearly double the number of patrols with roughly half the number of workers required, significantly increasing operational throughput. In effect, these robots not only made this process operationally efficient and safer, but also helped advance Outokumpu's stated goal of zero workplace incidents across its steel manufacturing plants. Given these multinational deployments spanning Finland, Germany, and Sweden, the company was now on the radar of one of the largest oil producers in the world, Equinor.

Equinor — Offshore Energy Operations, Norway (2023)

Equinor is headquartered in Stavanger, and the city is also known as the oil capital of Norway. The company operates in about 30 countries worldwide and is approximately 67% owned by the Norwegian government. This should not come as a surprise, as the company serves as a leading producer and supplier of oil and gas across the European Union, with the sector contributing roughly 22% to Norway's GDP.

In a global context, Norway produces about 2% of the world's oil, with Equinor leading production in the region at around 2.07 million barrels per day. Considering that approximately 81 million barrels of oil are produced globally each day, and that less than 5% of this originates in the European Union, Equinor's contribution as a major European energy company to this sector is particularly significant.

Given the role the company plays in the country's and the region's GDP, along with the shrinking workforce in this sector, the team at Equinor was actively looking to scale production across its oil and gas plants. However, to scale the production, the company needed to scale its ability to inspect these sites and proactively address critical safety anomalies. This proved challenging, as the majority of inspection workflows were manual and heavily reliant on on-site personnel.

As a result, Equinor had to deploy teams of skilled professionals across multiple shifts to maintain 24/7 operations at these sites. The existing model for conducting inspection patrols, addressing critical anomalies, and scaling production was neither operationally sustainable nor safe and scalable. To address these constraints, Equinor decided to partner with ANYbotics to evaluate its solution at one of its production facilities.

In 2023, the two companies initiated a pilot to test the Ex-certified quadruped robot ANYmal X at the Kårstø gas-processing facility on Norway's west coast. This is essentially a full-scale production site containing active explosive-certified (Ex) zones where hydrocarbons are processed. The objective behind this pilot was to scale the company's ability to conduct inspection patrols at these hazardous and explosion-prone sites without disrupting plant operations or altering existing physical infrastructure. To evaluate the ANYmal X robot, the company meticulously developed a full-scale assessment plan spanning over a month.[164]

During this deployment, ANYmal X completed a five-week structured evaluation program that included more than 30 autonomous inspection missions spanning multiple levels of the facility. Using onboard sensors, the robot scanned for critical anomalies while navigating stairs, narrow passages, and multi-level structures independently. It operated reliably in adverse weather conditions—including wind speeds of up to 20 m/s, heavy rain, and temperatures near 0 °C—and continuously transmitted inspection data captured from this site directly to the company's centralized plant monitoring dashboard.

This pilot demonstrated that ANYmal robots can help energy operators automate inspection patrols while minimizing risk for human workers and upholding strict safety standards. In effect, it proved that production at these facilities can be scaled in a cost-efficient manner without compromising operational integrity. The success also strengthened ANYmal's regional reputation and sparked interest from the European government, which was seeking a solution to conduct similar inspections in extremely hazardous environments—nuclear sites in Finland.

European Commission Joint Research Centre — Autonomous Nuclear Facility Inspection, Finland (2024)

Finland is a country many remember for Lapland or the Northern Lights, but few may be aware that it is also a nuclear powerhouse. Nearly 40% of the country's energy comes from five major nuclear reactors located across the country, making Finland one of the larger producers and consumers of nuclear energy today.

Given this reliance on nuclear power, it goes without saying that, like all forms of energy, the full cycle, from production to consumption, generates waste—and in the case of nuclear energy, it is hazardous waste. That means, for a country that experiences Arctic temperatures for year-around, managing this nuclear waste had become a major challenge. After all, no one wants to work in sub-zero temperatures at a normal facility, let alone at a nuclear site to clean nuclear waste. This presented a major challenge for the government and, at the same time, a strategic opportunity for ANYbotics to deploy its quadruped robot to clean this waste.

In collaboration with the local company Posiva Oy, responsible for cleaning nuclear waste at the Olkiluoto Nuclear Power Plant, as well as the European Commission and its Joint Research Centre, ANYbotics deployed its robot at the Onkalo spent nuclear fuel repository, located adjacent to the plant. Onkalo is the world's first long-term disposal facility for spent nuclear fuel, characterized by harsh, wet, and radioactive tunnel networks extending more than 400 meters below ground. Conducting manual inspections and managing nuclear waste in this radiation-active environment deep below the Earth was an extremely hazardous and unsafe process for employees.

To address this challenge, the team deployed a quadruped robot that, in phase one, conducted routine inspections of the facility by navigating safely through the tunnels and transmitting site-critical information—such as radiation levels, equipment health status, and other anomalies—back to operations teams on the surface. This ensured that the company was able to reduce the amount of time workers spent at the site and minimize their exposure to radiation, while still ensuring that nuclear waste generated by the nearby power facility was processed under safe and regulatory-compliant standards.

In addition to this site-critical information, the robot enabled the nuclear site owners to develop a detailed 3D map of the site and its underground infrastructure, which now serves as a safe and tested reference for workers to conduct on-site work while minimizing safety hazards. In effect, this IP67-certified rugged robot provided the company with a scalable and safe solution to ramp up operations, support the ongoing production and consumption of nuclear fuel at the nearby facility that powers the country, and minimize risk to human workers. In Phase 2, the company will evaluate the robot for applications beyond inspection, including site-specific nuclear waste management tasks.[166]

Phase 1 validated the robot in one of the most hazardous and safety-critical industrial environments on the planet. This validation served as a global proof point that the product was mission-critical and ready for deployment worldwide, placing the company directly on the radar of major energy and chemical producers around the world, including Petrobras, Shell, BASF, and driving long-term contracts exceeding 300 units over subsequent years.

Given this demand, the company needed to extend its financial runway to scale its production and operational footprint, and that meant it was time for a Series B.

Series B Funding (2023)

In May 2023, ANYbotics announced a $50M Series B financing round led by Walden Catalyst and NGP Capital, with participation from Bessemer Venture Partners, Aramco Ventures, Swisscom Ventures, and other institutional investors. This round was specifically positioned to accelerate production capacity and global deployments of its autonomous inspection robots across the energy, chemicals, and manufacturing sectors. It is important to note that, at the time of the Series B, the company was still maintaining an extremely lean employee base of approximately 100 employees.[167]

With this lean team and a fresh infusion of capital, the company was able to scale its sales engine and reported several multi-site industrial contracts over the following year. Given this pace of adoption of its robots by customers, it became clear that it was time to extend the Series B.

In December 2024, the company closed a $60M Series B extension, led by Qualcomm Ventures and Supernova Invest, with participation from TDK Ventures and existing investors. This extension brought total capital raised in this round to over $130M, capital the company desperately needed to execute on the multi-site contracts it had secured, each with demand for more than 300 units that were set to be deployed across some of the largest energy production sites around the world.

Given this funding round and the scale of adoption in Europe (EU), the company was now on the radar of energy and chemical producers in the United States (U.S.). And that meant it was time for the ANYbots to cross the Atlantic and build a strong foothold in the U.S. market.

U.S. Expansion — San Francisco Office (2024)

In November 2024, ANYbotics formally announced the incubation of its first office on U.S. soil, in the heart of Silicon Valley, i.e., San Francisco, establishing a permanent commercial hub in the region. This was a strategic move made by the company and was in line with the rising demand from North American oil & gas majors, utilities, and heavy-industry operators seeking to automate on-site work at these sites, given the shrinking workforce in the sector and associated safety hazards. This incubation also meant the company had effectively transitioned to a bi-continental model for its operations, launching sales, engineering, and customer success teams across these two regions, i.e., Europe and the U.S.

Once again, it is important to note here that as of November 2025, a year after this expansion, the company still maintained a lean footprint of about 200 employees spanning these two offices in

San Francisco and Zurich. Given this lean mode of operations and the high pace of adoption by leading energy operators across the two continents, the company's efforts were recognized by Switzerland Global Enterprise, which awarded ANYbotics a 2025 Export Award. This recognition positioned the company as one of Switzerland's leading deep-tech innovators, specifically citing its ability to penetrate into highly regulated and hazardous industrial environments while maintaining a lean employee base and successfully establishing a bi-continental model.

Present State — ANYbotics Today

Today, ANYbotics positions itself as a leading provider of AI-driven robotic inspection solutions for autonomous asset monitoring in hazardous and large-scale industrial environments. Its core product portfolio centers on ANYmal D, an industrial-grade inspection robot, and ANYmal X, an Ex-certified quadruped designed for explosion-prone facilities—both purpose-built to enable mission-critical autonomous inspections.

These robots generate high-fidelity 3D maps of industrial sites, perform thermal and acoustic-based analysis, and produce detailed inspection insights that can be directly integrated using APIs developed by the company, into the asset management systems widely used across the energy sector. This capability stack allows operators to scale their inspection workflows in a safe, compliant, and cost-efficient manner.

Today, primary industry verticals served by the company include Energy, Chemicals, Heavy Industry, Mining, and Utilities—verticals that demand safe, reliable, and predictive inspection protocols under hazardous conditions, enabling high uptime and low operational expenditure for leading operators.

Geographically, the company demonstrates a strong operational footprint across Europe, North America, the Middle East, and Latin America, with customers ranging from GE, Equinor, Outokumpu, European government entities, Vale, BASF, and Petronas.

From its incubation in 2015 to today, the company has, in less than a decade, established itself as one of the leading producers of mission-critical autonomous inspection robots designed exclusively for heavy industry and energy operators worldwide. This trajectory offers several important lessons on strategy, execution, and operational discipline for robotics founders and engineers—lessons that we will now cover in detail.

Lessons for Founders and Engineers

Lesson 01: Talent Density Beats Headcount

One of the most important lessons that surfaced over and over again throughout the timeline of ANYbotics from 2015 to date is that the company managed to secure multiple top-tier customers across Europe and the U.S. with a lean team of around 200 employees. Talent density always beats headcount. In the first phase of its journey, which was largely centered around its growth in Europe, the company operated with a base of roughly 100 employees, each surgically hand-picked by its founders.

This lean employee footprint, which it retained throughout its journey from Europe to the U.S., also implies that the company was highly selective in curating its workforce. The founders and the board likely placed as much attention and focus on hand-picking each employee as they did on selecting customers and investors. This lean model is extremely effective, especially in robotics and deep-tech, given the high burn rates inherent in these sectors. One of the most effective ways to address this challenge is to maintain a lean employee base and a focused product portfolio. ANYbotics' journey serves as a strong proof point of this lesson.

Lesson 02: Only Raise the Capital You Actually Need

In addition to maintaining a lean team, ANYbotics applied the same discipline to its funding strategy. Despite operating in a market aligned with strong global demand for automation, the company raised capital only when it was needed to address clearly defined bottlenecks. It did not try to ride the hype cycle or raise more funds out of fear or worry. Instead, it stayed true to its principle of need. You should only raise the amount of capital that you actually need.

In robotics, more capital does not imply adoption. An inflated war chest has little correlation with success and, worse, can backfire and cause more damage to your startup. You are far better off meticulously calculating the exact amount of capital required to address your bottlenecks over a 3–4 year horizon and raising exactly that amount. Anything beyond that is a waste of time, energy, and capital.

Lesson 03: Own One Most Painful Use Case in the Market

ANYbotics made an intentional decision to position its product around hazardous environments. Across customers ranging from Vale and Equinor to nuclear facilities in Finland and chemical plants operated by BASF, the pattern was consistent: the company focused on environments where the operational risk was highest and the cost of failure was significant.

Although the underlying platform was capable of operating in more general-purpose settings such as warehouses, the company chose not to compete in saturated markets. Instead, it concentrated on a small number of verticals where the need was acute and the value proposition was strongest.

This approach made it easier for the right customers to understand the product, for the team to sell that product, and for the company to stand out in an ocean of robot-led solutions. Strong positioning comes from focus. Products that are designed and marketed around a specific, high-stakes use case tend to see faster adoption than those positioned broadly across many markets.

Lesson 04: One Product Done Well

If you look at the trajectory of ANYbotics, the company remained laser-focused for nearly a decade on one core product: a rugged quadruped designed exclusively for hazardous sites. It didn't diversify into mobile robots or humanoids, even though it could have served many of the same customers with those other forms of robots.

The lesson here is that long-term orientation is in short supply in today's market, yet it is essential to succeeding in capital- and hardware-intensive spaces such as robotics. To succeed, you need to focus on one product and execute it exceptionally well.

This level of surgical focus also gives you the time and resources you will need to work closely with each customer, better understand their needs, and continuously optimize that single core product. Building this one core product is only step one. As we often say here at BoringSage, speaking with your customers, deeply understanding their needs, and iterating on your product with surgical precision is what distinguishes a successful company from the rest.

This approach demands a significant investment of time and the ability to maintain both operational and emotional proximity to your customers. Being in the field and iterating on real-world feedback is what allows products to improve in meaningful ways. Focusing on a single product creates the space to do this well, enabling teams to deeply understand customer needs and systematically incorporate that feedback into the product.

Now that we have distilled the core lessons embedded in ANYbotics' trajectory, from its origins through its evolution and the pivotal decisions the team made along the way, it is time to decode the underlying business model.

In the final section of this chapter, we will break down the business model and illustrate how the product portfolio maps to select primary use cases.

ANYbotics: Product Snapshot I — Onkalo Nuclear Repository & BASF Chemical Plant

Figure 11.1 — ANYbotics: Onkalo Nuclear Repository & BASF Chemical Plant

Illustration by the author. Robot images from the ANYbotics case studies page.

ANYbotics: Product Snapshot II — System Overview

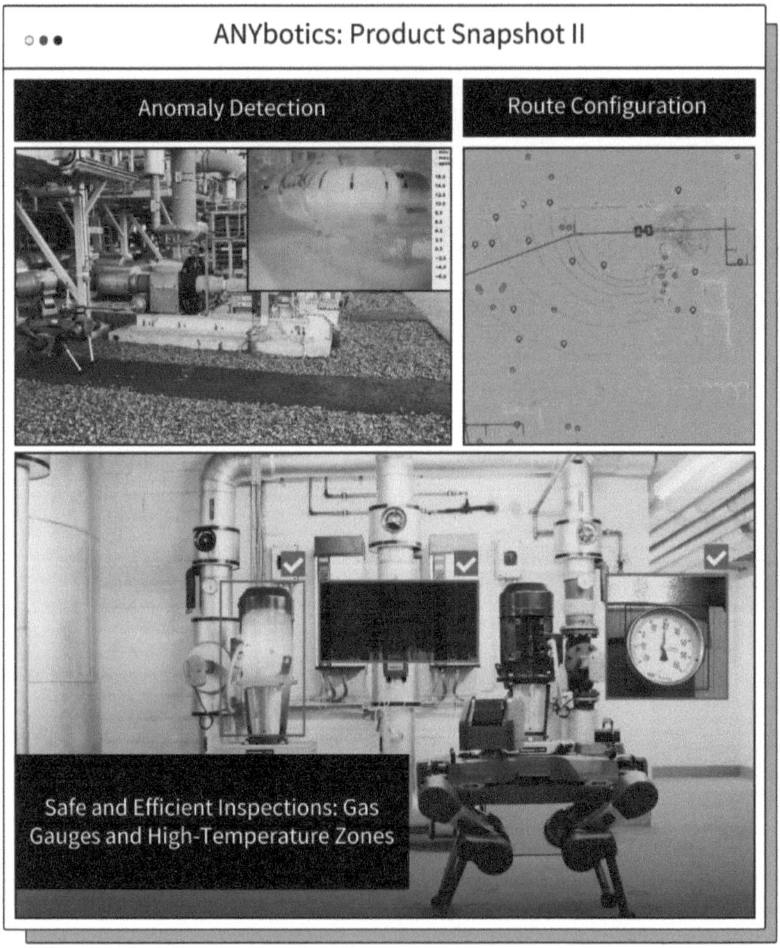

Figure 11.2 — ANYbotics: System Overview

Illustration by the author. Robot images from the ANYbotics capabilities page.

ANYbotics' Business Model — Part 1 of 3

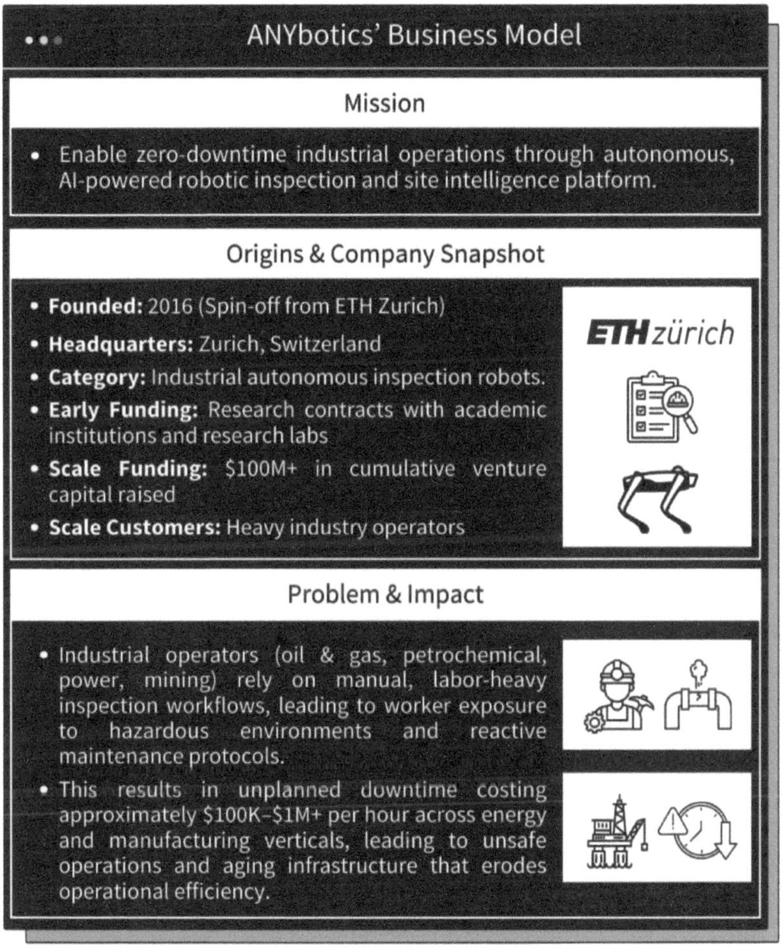

Figure 11.3 — ANYbotics' Business Model: Part 1

Illustration by the author.

ANYbotics' Business Model — Part 2 of 3

Figure 11.4 — ANYbotics' Business Model: Part 2

Illustration by the author.

ANYbotics' Business Model — Part 3 of 3

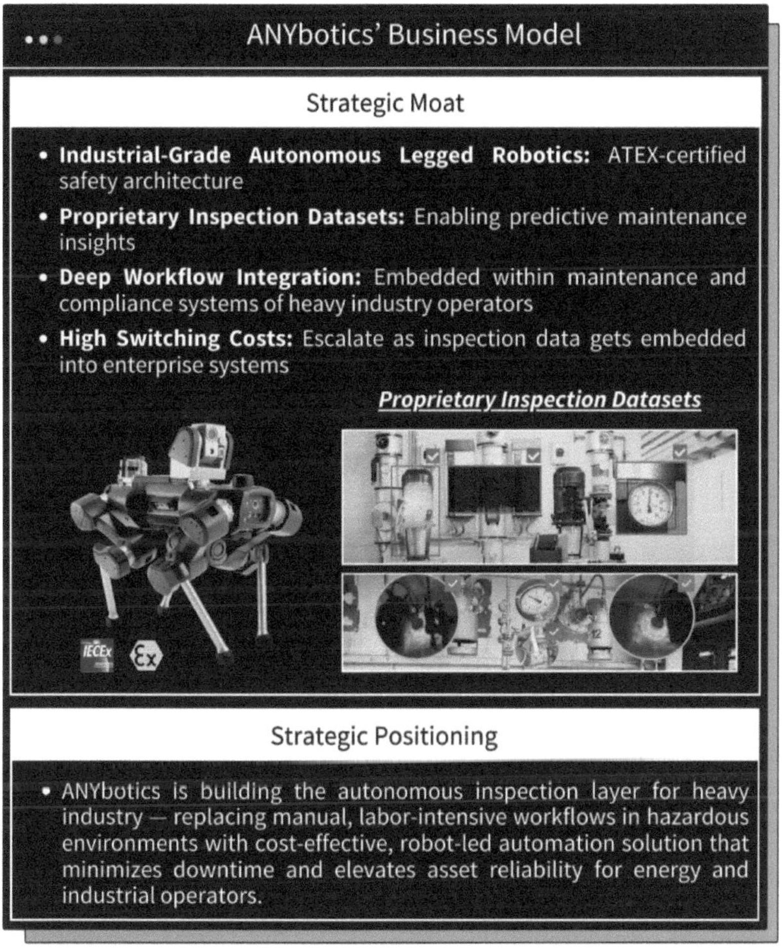

Figure 11.5 — ANYbotics' Business Model: Part 3

Illustration by the author.

Chapter 12: Exotec — French Engineering to UNIQLO Scale

La Ville Rose, i.e., "the Pink City," is the name given to Toulouse, a city located in the southwestern region of France and known for its warm terracotta brick architecture. On a sunny day, the streets here glow with shades of pink and orange, casting a spotlight on the iconic Capitole de Toulouse, the city hall, and offering wanderers a glimpse into the city's rich cuisine and cultural heritage at the bustling Place du Capitole market. Along the nearby Garonne River, scenic riverside promenades only add to Toulouse's enduring charm.

Yet Toulouse is far more than a picturesque historic destination—it is now one of the vital arteries of the French economy, nourishing local talent and fueling startups. The city stands as the country's leading aerospace hub, boasts one of the nation's largest student populations, and is home to one of its most prestigious engineering schools: the Institut supérieur de l'aéronautique et de l'espace.

It was here, at this renowned institution, that an engineer turned roboticist started his journey in 2004. After graduating, he took a job as a sales engineer at a local company, helping drive early programs focused on roboticizing forklifts. Over the years, he became a critical bridge between customers and the engineering team—translating frontline operational needs into product capabilities and, in the process, developing a deep understanding of both the market and the technology. This rare combination of customer insight and technical depth, alongside a clear view of the limitations of existing automation solutions, ultimately led him to serve as the company's CTO.

With this foundation established, he later joined GE Healthcare, where he worked on advanced robotic imaging systems. It was here that he met another engineer equally passionate about robotics. Together, they were responsible for ensuring the mission-critical reliability for these robotic imagers and, in effect, an experience that taught them firsthand what it takes to engineer, commercialize, and deploy hardware products at a global scale.

With this expertise under their belt, their passion for robotics, and a deep desire to solve challenges in the logistics industry rooted in their early work automating forklift systems, they decided to start their own company in 2015. That company was none other than Exotec, now a French industrial unicorn deploying precision-engineered autonomous robots in warehouses around the world. The engineers behind this vision were Exotec's founders, Romain Moulin and Renaud Heitz.[168-171]

Today, the company positions its warehouse robots, known as Skypod, as autonomous mobile robots (AMRs) designed to transport goods within warehouses and elevate operational efficiency. In just ten years since its inception, the company has established a global footprint across 200+ customer sites worldwide, generated $1B+ in revenue, and delivered 10,000+ robots, serving customers such as Carrefour, Gap, UNIQLO, and Decathlon. To date, Exotec estimates it has eliminated 90 million kilometers of walking for warehouse workers, equivalent to roughly 119 trips to the moon and back. Given this scale, impact, and global adoption, the company merits a deeper examination.[172-173]

To understand the founders' trajectory and the lessons behind Exotec's success, we must travel back to the year 2015.

Origins & Seed Stage (2015–2016)

In 2015, after having officially graduated from GE Healthcare, and with a deep passion to address challenges in the logistics industry with robots, the two French engineers, Romain and Renaud, started developing their earliest prototypes, i.e., a small fleet of autonomous mobile robots designed specifically to enable automated transport of goods within the warehouse.

This project had emerged from Romain's earlier work automating forklifts with robots when he worked at a local company called BA Systems, fresh out of university back in 2004. During this time, he understood that while automating forklifts was a good problem to solve in order to raise the speed at which goods were transported within the warehouse, the approach itself was not quite right. In other words, automating a forklift turned out to be a far harder and more complex approach to addressing the underlying problem than developing custom autonomous robots from scratch for these warehouse operators.

Given his engineering background and this clarity of thinking, Romain decided to build a dedicated robot to address this challenge, resulting in the launch of the first Skypod prototypes, autonomous mobile robots purpose-built for warehouse environments. These robots, no doubt, had to be able to navigate seamlessly within existing facilities and integrate with established warehouse management systems (WMS). To enable this, Exotec developed its own software layer, DeepSky, which served as the orchestration backbone connecting Skypod robots with customer WMS platforms. Together, Skypod and DeepSky meant the company now had a full-stack prototype composed of both the hardware and software needed to enable real-world deployments.[168-171]

Their mission was clear: eliminate strenuous work for warehouse employees while elevating throughput for operators by enabling the autonomous transport of goods within these facilities.

With a working prototype in hand and a clear mission, the company now had to secure capital to bring this vision to life. So, in December 2016, Exotec raised $3.5M in seed funding, led by 360 Capital Partners and Breega, refueling its oxygen supplies to power its first commercial deployment, 130 miles northwest in the city famous for its rich French wine, Bordeaux.

First Customer — Cdiscount Bordeaux (2017)

Bordeaux is one of the most incredible cities in France, known for its UNESCO-blessed historical architecture, world-class wine, elegant 18th- to 19th-century mansions, and charming art museums such as the Musée des Beaux-Arts de Bordeaux. Beyond its cultural heritage, the city is also home to the headquarters of

Cdiscount, one of France's largest e-commerce companies, often regarded as the country's answer to Amazon.

Cdiscount sells tens of millions of products each year to its 10 million active customers, ranging from electronics to home goods, personal care, and packaged foods. In addition to its e-commerce marketplace, its business-to-consumer (B2C) arm, the company ships roughly 20 million packages annually to 27 countries through its logistics division, its business-to-business (B2B) arm, C-Logistics, on behalf of its vendors. Given its scale and the exploding order volumes it witnessed in 2017, the company was in dire need to scale its fulfillment engine and deliver goods to its customers at lightning speed with far greater operational efficiency than that provided by traditional conveyor- and forklift-based warehouse operations systems.[174]

To address this challenge, the company decided to evaluate Exotec's Skypod goods-to-person robotics system, autonomous warehouse robots, at one of its facilities in Bordeaux. During the pilot, Exotec's system delivered four times the throughput of the existing setup the French e-commerce giant had in place. In particular, the solution enabled the autonomous pickup and transport of roughly 400 packages per hour, effectively quadrupling order fulfillment velocity overnight.

Building on that success, Cdiscount expanded the system at another large-scale facility, a 60,000 m^2 warehouse located in Réau, France. Here, the company deployed around 43 of these Exotec AMRs, responsible for autonomously ferrying thousands of goods per day between four core zones using about 61,000 bins. This system enabled the autonomous transport of about 1,000 bins per hour, each with the capacity to carry dozens of items, thereby raising storage density at the site by 5× and order fulfillment velocity by 4×.[174]

This rollout demonstrated that Exotec's technology was now capable of not only delivering higher operational throughput for warehouse operators, but also dramatically improving storage density within these facilities, eliminating the need to expand their physical footprint. That meant Exotec was no longer simply a robotics provider for warehouse operators; it had become the very engine they needed to double both capacity and throughput. Given the success of this pilot, the demand for the company's product accelerated rapidly across the region, and that meant it was time for Exotec to scale its execution engine.

It was time for Series B.

Series B: Scale Capital (2018)

In June 2018, Exotec closed a €15 million (~$17.7 million) Series B round led by Iris Capital, with participation from existing backers 360 Capital Partners and Breega. This funding round was explicitly aimed at strengthening its production, sales, and go-to-market engines to meet the accelerating inbound demand it was now experiencing from e-commerce and retail operators across Europe.

With this fresh injection of capital, the company set an ambitious goal: to sell at least 1,000 of these robots over the next 12 months. While the target was etched on paper during the raise, Exotec's team blew past these expectations within just a few quarters. Over the following year, the company secured multi-site contracts across an array of industry verticals, spanning e-commerce, retail, apparel, fashion, grocery, and more.

Customer Base Expansion: Grocery & Global Fashion Logistics (2019)

By 2019, building on its success with Cdiscount, the company had landed multi-site deals with a wide range of customers, starting locally with Carrefour, one of Europe's largest grocery retailers, and extending all the way to Japan with UNIQLO, the country's most admired clothing brand.

Carrefour — Grocery Micro-Fulfillment, Reims & Le Plessis-Pâté

In 2019, Carrefour, one of Europe's largest grocery retailers, deployed Exotec's Skypod goods-to-person robotics system across multiple fulfillment hubs to scale its e-commerce business focused on grocery delivery.

At its Reims micro-fulfillment site, Carrefour implemented a small-scale version of Exotec's AMR system within a physical footprint of just ~619 m^2. Powered by 26 robots, two core stations, and roughly 8,100 bins, the system enabled the automated processing of up to ~630 orders per hour with supervision from just two operators. The deployment significantly reduced order fulfillment times while supporting over 165,000 SKUs i.e., unique items in a space smaller than a school gymnasium. Importantly, this deployment was completed in just six weeks with minimal operational disruption.

Carrefour then scaled the Exotec solution at its larger Le Plessis-Pâté ambient and chilled grocery facility. There, a large-scale system composed of 148 robots, 18 picking stations, and ~30,940 bins delivered a fourfold increase in throughput, enabling the grocery giant to process up to ~4,000 orders per day.[175-176]

By this time, Cdiscount, Exotec's first flagship customer, had developed a strong appetite for these AMRs and fully understood their impact on the company's operational throughput and revenue. Given this impact, the company decided to scale its adoption footprint.

Cdiscount: E-Commerce Warehousing, Réau, Paris

Building on its 2017 pilot, Cdiscount implemented a second large-scale Skypod system in 2019 at its 60,000 m^2 logistics facility in Réau near Paris to handle surging order volumes across a wide range of product categories.

The system included 43 robots, 4 picking stations, and ~61,000 storage bins, organized in high-density racks up to 9 m tall. This Exotec-enabled automation system allowed the French e-commerce giant to achieve a fivefold increase in order fulfillment throughput, enabling just four operators to handle roughly 900 bins per hour across four stations. The company had officially quadrupled its throughput, and given this impact, Exotec was now on the radar of organizations around the world looking for solutions to deliver goods to customers at the fastest speeds possible with minimal resources.[177]

Around the same time, a leading Japanese apparel brand, admired globally for its craftsmanship and its strong brand loyalty, was looking to expand its footprint in international markets and scale its logistics engine with a lean automation solution. Given Exotec's proven ability to deliver step-change throughput gains for European operators, the Japanese giant approached the company later that year, in November, to evaluate its technology.

Fast Retailing (UNIQLO) — Strategic Global Partnership

In November 2019, Fast Retailing Co., Ltd., the Japanese multinational parent company of UNIQLO, announced a strategic global partnership with Exotec to scale its global supply chain and warehouse automation engine in support of UNIQLO's rapidly growing international footprint.[178]

At the time, UNIQLO was expanding aggressively across the world, having established a strong base in Japan while adding new stores across Greater China, South Korea, Southeast Asia, Europe, and North America. With this accelerating store expansion, the company urgently needed to scale its logistics and fulfillment engine while maintaining a lean, capital-efficient operating model. To meet this need, Fast Retailing chose Exotec to transform its warehouse operations globally and elevate throughput across its growing network of store outlets using the company's proven autonomous robotics systems.

Given the pace at which UNIQLO was scaling, Exotec was now entering a phase of explosive growth itself, and sustaining that momentum would require additional capital.

It was time for Series C.

Series C: International Expansion (2020)

In September 2020, Exotec raised a $90M Series C to accelerate international adoption of its Skypod goods-to-person robotics system. The round was led by 83North, with participation from Dell Technologies Capital, and support from existing investors Iris Capital and Breega.

This fresh injection of capital was aimed at fueling growth beyond France, explicitly to establish strong operational bases in Atlanta, North America, and Tokyo, Japan, and to scale its manufacturing engine to support the production of 4,000 robots per year, up from 1,000 in 2019. Having secured this new supply of capital, it was now time to cross the Atlantic Ocean and head west.

First Big Adoption Wave: North America (2021)

In 2021, the company officially established its North America headquarters in Atlanta and began positioning its solution as a system capable of quadrupling throughput for local retail and logistics giants such as Gap, Decathlon, and GEODIS, with the goal of growing its revenue base to €100M that year.

Gap Inc. (United States) — First Major U.S. Retail Deployment

In 2021, Gap Inc. became one of Exotec's first major North American retail customers, selecting the Skypod goods-to-person robotics system to modernize its warehouse network supporting high-volume apparel orders for the company across its brands namely Old Navy, Gap, Banana Republic, Athleta, Intermix, and Janie and Jack. Exotec announced that the deployment would be executed through its U.S. integration partner AHS, positioning Skypod as a scalable, next-generation alternative to traditional conveyor-heavy automation systems commonly adopted in the retail distribution vertical.[179]

For Exotec, this early U.S. enterprise contract marked a critical milestone. It represented one of the company's first large-scale validations on American soil. This adoption by a Fortune 500 giant signaled that these French AMRs had started illuminating the radar of global fashion, apparel, and sporting goods leaders. One of them was Decathlon, the world's largest sporting goods retailer, with 1,800 stores across 79 countries.

Decathlon (France & Canada) — Cross-Border Retail Automation Expansion

During this time, Decathlon, the crown jewel of the sports retail world, with a fortress of 1,800 stores, 101,000 employees, and annual sales of roughly 1.18 billion products, generating a staggering €17.5 billion in revenue, was in a similar position to UNIQLO. With global expansion as its primary goal and a lean operating engine as a core requirement, Exotec's AMRs became a natural choice for the retail giant. The company started with a large-scale deployment at its Montreal distribution center, along with two major distribution hubs in France.[179]

Given this level of adoption by Gap and Decathlon, two major apparel and sporting goods leaders spanning North America and Europe, Exotec was now on the radar of logistics giants in the U.S., including GEODIS.

GEODIS (Nashville, Tennessee) — 3PL Adoption in the U.S. Logistics Core

GEODIS is one of the largest logistics operators in the world, with 1,000+ warehouse sites enabling the transport of goods for more than 91,000 enterprise customers globally. With persistent labor shortages in the industry and growing demand from its enterprise customers for cost-efficient goods transport systems, the company was actively evaluating automation solutions to scale its operations engine. By this point, Exotec's AMRs had already been validated by major retail and e-commerce leaders such as Gap, Cdiscount, and Decathlon, making Skypod a natural fit for a global 3PL looking to modernize its operations engine.

In 2021, GEODIS selected Exotec's Skypod system for deployment at a site in Nashville, Tennessee. The system included 39 AMRs operating across 12,700 square feet of warehouse footprint, transporting approximately 29,500 bins, each filled with dozens of customer orders. The impact of this deployment was immediate: the 3PL giant officially doubled throughput at the site.[180]

This deployment marked a critical turning point for Exotec. Its technology was no longer being adopted solely by retailers; it was now becoming the core operations engine of choice for the world's largest logistics providers. Skypod was increasingly establishing itself as core infrastructure for U.S. and global distribution networks.

Given this deployment with GEODIS, the company experienced a surge in inbound demand for its system from across the world, and once again, it needed to scale its own operations engine to keep pace with this demand. It was time for Series D.

Unicorn Milestone — U.K. & North America Expansion (2022)

In January 2022, Exotec raised a $335M Series D led by Goldman Sachs at a $2B valuation, officially becoming France's first industrial unicorn in its category.[181] This fresh supply of capital ensured the company was now well positioned to accelerate large-scale deployments across North America, Europe, and Asia. Specifically, the company laid out quantified goals for the following year: plans to hire 500 new R&D engineers by 2025, and to pursue commercial expansion into the U.K. and Asian markets in 2022. It also confirmed that it had achieved its previous set of goals, including quadrupling its manufacturing capacity to 4,000 robots per year and tripling its revenue since 2020.

With these milestones, and a strong footprint established in Europe and North America, it was now time for the company to head East to Asia.

Commercial Milestone & Asia Expansion: Korea, Seoul (2023–24)

By October 2023, the company had already established its footprint across 100+ customer sites globally and was on a mission to enter Asia. To bring this mission to life, the company adopted a partnership model by establishing strategic relationships with local engineering and IT services providers, such as POSCO DX in South Korea, to introduce Exotec solutions in-market.[182]

This partnership proved vital, as in the coming year, by 2024, Exotec announced it had sold $1B+ in systems globally across all regions, including Asia, Europe, and North America, while providing 99% uptime for its systems—ensuring mission-critical reliability across the globe. By 2025, Exotec had fully transitioned from a fast-growing European robotics startup into a global automation leader. Following its rapid international expansion across continents, the company reached a major inflection point with the commercial launch of the Next Generation Skypod System in 2025.

Next Generation Skypod Commercial Launch (2025–26)

In February 2025, Exotec launched the Next Generation of the Skypod System, positioning it as an all-in-one warehouse automation platform capable of delivering 50% higher throughput and up to 30% greater storage density compared to its previous generation of AMRs, enabled by a more compact robot design. This momentum continued with the launch of Exostudio, its first permanent demo site for the next-gen platform in North America, signaling the commercial rollout of the system across the region.[183]

As of 2026, Exotec confirmed its systems were operating at full scale across 200+ sites globally, with more than 10,000 robots manufactured and deployed worldwide. In February 2026, Exotec launched the "Imaginarium," a new 25,000 m^2 global headquarters in Northern France intended to support the next leg of its journey through a training-first commercial adoption model. The company plans to bring customers here not only to demonstrate its technology, but to offer prospects a glimpse into the operational value the Skypod system has created for leading warehouse operators across the globe over the years in a visually immersive manner, unique to the company's philosophy that is forged on the roots of elegance and impact.

Given this trajectory and the remarkable impact this French robotics startup has made across continents in only ten years, Exotec's journey offers powerful lessons that we will now unpack one by one.

Lessons for Founders and Engineers

Lesson 01: Solve the Right Problem with the Right Approach

If you note carefully, one of the founders actually started their career in robotics, automating forklifts for a local company in France. Turns out, this was a much harder problem to solve than building an autonomous robot from scratch to transport goods within a warehouse. The lesson to remember here is that you need to be extremely clear about the problem you are trying to solve.

The real problem was not automating the forklift, but enabling autonomous transport of goods within the warehouse. The approach adopted by the local company where the founder worked in the early stages of their career was not only inefficient, but also extremely complex and conceptually muddled.

You've got to work on the right problem and adopt the right approach—one that helps you solve it in a time- and resource-efficient manner. Approach matters as much as the solution you will build, because the former will have far greater long-term consequences for your startup and its trajectory than the latter.

Lesson 02: Fully Integrated Solution from Day 01

Another important lesson here is that the founders not only developed robots for warehouse operators, i.e., AMRs, along with a software stack to manage the workflow for these robots, but more importantly ensured from day one that this software could be seamlessly integrated into the operators' existing infrastructure, i.e., warehouse management systems.

Skypod AMRs and the DeepSky software system serve as a strong reminder of this lesson. That meant this robotic system was not only production-ready, but also optimized for adoption from Day 01. The founders anticipated the bottlenecks they would encounter when it came to deployment and adoption, and addressed them decisively from the outset. A fully integrated solution from Day 01 was their mantra, and they remained deeply committed to it.

Lesson 03: Early Adopters Are Everywhere—If You Know Where to Look

One of the first major deals that this French robotics company landed was with CDiscount, i.e., the Amazon of France. This was not long ago, in 2017. The fact that an e-commerce giant was still operating with conventional conveyor- and forklift-based solutions to transport goods within warehouses, in the era of robotics, should signal that many enterprise organizations continue to rely on tedious and highly inefficient workflows. These are the very workflows that inflate their OPEX and deflate their revenue.

Hence, you, as an emerging and high-velocity startup in the robotics space, should identify these gaps in the market and address them using your robotics-enabled solutions. These gaps exist across verticals all around you.

Early adopters are everywhere. You've got to take the time to find these prospects, speak with them, and understand their challenges. Listen 99% of the time, speak only 1%, and deeply identify and analyze these gaps. That's one of the key rules you've got to remember to land early adopters. If Exotec was able to identify and secure deals with early adopters in a market saturated with robotics vendors, then your challenge is no different — and it can be solved.

Lesson 04: Quantified Milestones Is the Language of Investors

One of the most important lessons we can learn from this team is that each investment round they raised was driven by quantified goals that the team promised and delivered to its customers and investors.

The seed fund of $3.5M in 2015 was structured around the objective of landing one large-scale deal in the industry, and that goal was achieved in 2017 with the CDiscount deployment at the Bordeaux, France site. Similarly, Series B in 2018 was raised for a total of $17.7M, with the goal of selling at least 1,000 robots in the coming year. That goal was achieved, as in 2019 the company secured large-scale contracts with Carrefour, CDiscount Expansion, and UNIQLO.

Likewise, Series C of $90M was raised in 2020 to establish a strong footprint in North America, and by 2021 the company had already secured contracts with Gap, Decathlon, and Geodis. Thus, it is essential to raise capital with quantified milestones, remain transparent about these goals with investors, and deliver ahead of schedule to build enduring trust, and respect in your brand from Day 01. Trust signals your ability to deliver on commitments, while respect comes from doing so in a manner aligned with your values.

Now that we have distilled the core lessons embedded in Exotec's trajectory, from its origins through its evolution and the pivotal decisions the team made along the way, it is time to decode the underlying business model.

In the final section of this chapter, we will break down the business model and illustrate how the product portfolio maps to select primary use cases.

Exotec: Product Snapshot I — System Overview

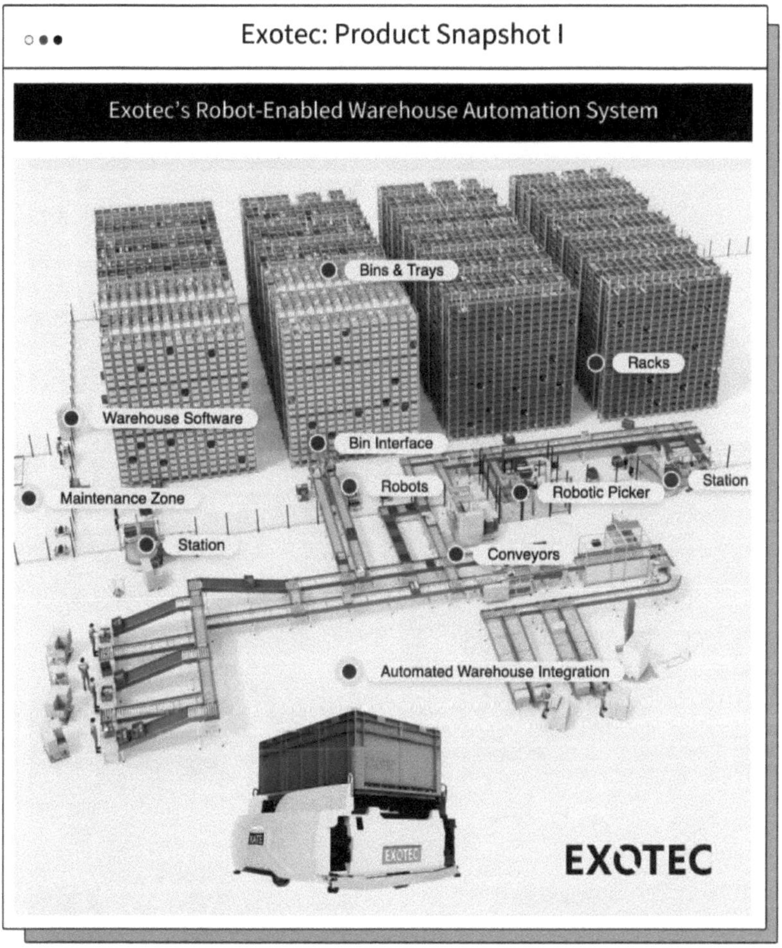

Figure 12.1 — Exotec: System Overview

Illustration by the author. Robot images from the Exotec Skypod automated storage and retrieval system page.

Exotec: Product Snapshot II — Customer Case Studies

Figure 12.2 — Exotec: Customer Case Studies

Illustration by the author. Robot images from the Exotec customer case studies page.

Exotec's Business Model — Part 1 of 6

Figure 12.3 — Exotec's Business Model: Part 1

Illustration by the author.

Exotec's Business Model — Part 2 of 6

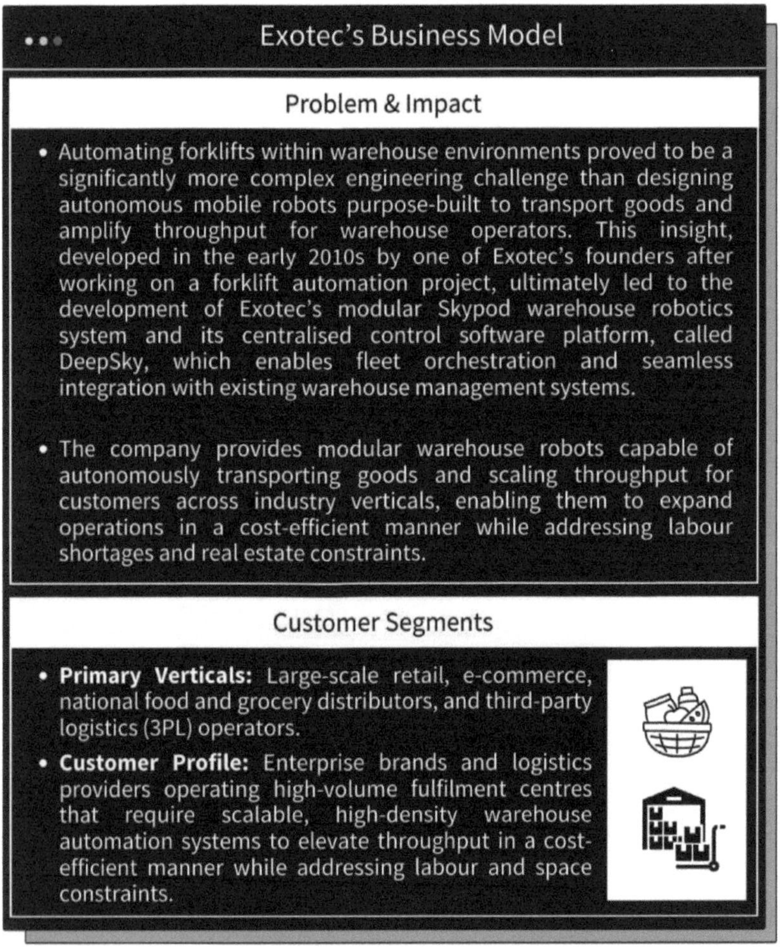

Figure 12.4 — Exotec's Business Model: Part 2

Illustration by the author.

Exotec's Business Model — Part 3 of 6

Figure 12.5 — Exotec's Business Model: Part 3

Illustration by the author.

Exotec's Business Model — Part 4 of 6

Figure 12.6 — Exotec's Business Model: Part 4

Illustration by the author.

Exotec's Business Model — Part 5 of 6

Figure 12.7 — Exotec's Business Model: Part 5

Illustration by the author.

Exotec's Business Model — Part 6 of 6

Figure 12.8 — Exotec's Business Model: Part 6

Illustration by the author.

Chapter 13: iRobot — MIT Lab to Roomba Scale

iRobot was founded in 1990 by MIT roboticists Colin Angle, Helen Greiner, and Rodney Brooks, with a mission to make practical robots a reality. In simple words, they wanted to translate the research work conducted in the robotics space at the time into real-world products. With this mission in mind, from 1990 to 2001 the company kickstarted its efforts and started developing robots for defense and space exploration applications.

These early-stage efforts resulted in the launch of a robot named Genghis, developed exclusively for space missions in 1991, and later PackBot, a tactical mobile robot developed for defense applications with funding from DARPA in 1998. Continuing this momentum, these first 10 years of research work ultimately resulted in the development and launch of its first consumer product, the Roomba vacuum-cleaning robot, in 2002.[184]

Launch of Roomba and the IPO

Roomba, the autonomous floor cleaning robot, quickly became iRobot's flagship consumer product. This launch gave the company much-needed credibility in the consumer markets. Over the period of the next three years, the company experienced an exponential surge in sales of Roomba.

This raised an obvious question: how did iRobot fund this demand? It did so through a public offering—an IPO. In 2005, the company was listed on the NASDAQ stock exchange and thus raised the capital it needed to fuel its trajectory.[184] That leads to another question: what did the company do with this influx of capital?

Expansion.

Product Portfolio Expansion

From 2005 to 2012, during this seven-year period, iRobot launched a series of robots namely Verro, Looj, and Dirt Dog.[184] Each of these robots was surgically designed to automate specific tasks such as pool cleaning, gutter cleaning, shop sweeping, and more. Despite these launches, the majority of the company's revenue during this time continued to come from Roomba, its autonomous floor-cleaning vacuum robot.

While iRobot's revenue chart reflected growth, it also revealed emerging challenges, most notably competition. Low-cost competitors from Asia started flooding the market, significantly impacting the company's sales trajectory. Because iRobot was now a public company, the ripple effects of this competitive pressure started to surface in quarterly briefings.

So how did iRobot respond to these emerging challenges?

Acquisition.

Acquisition of Evolution Robotics

To counter the ripple effects that emerging competitors were having on its sales trajectory, the team decided to acquire

Evolution Robotics in 2012, bringing the Mint floor-cleaning robot into its portfolio.[185] The goal was to maintain leadership in the floor-cleaning robotics segment by acquiring companies with a strong market adoption footprint. While this approach appeared to be a sound strategic move on paper, in practice it failed to address the underlying problem. Why was that the case?

Retaining dominance in the autonomous vacuum-cleaning robot space alone was no longer enough to justify the value of the brand. From 2012 through 2016, the autonomous robot vacuum segment became fully saturated with low-cost alternatives from across the globe. As a result, the robot that had long served as the company's core differentiator had lost its edge. The market had matured faster than the sales trajectory of iRobot.

This effect was again clearly visible in the company's sales chart in 2016. Being a public company, the company had to make decisive moves to counter this competitive pressure and regain a strong hold on its sales trajectory. If it did not do this in time, it would quickly deplete its funding pipeline. So, what did the company decide next?

Global Expansion to Drive Sales

Between 2016 and 2021, the company pursued an aggressive expansion strategy, scaling its product portfolio while extending its global footprint. It launched its first international office in Shanghai, China, and later acquired Robopolis, its largest European distributor, extending its presence across both China and the EU.

This momentum continued with the acquisition of Sales on Demand Corporation (SODC), which enabled the launch of its new offices in Japan. During this period, the company also introduced two new products: the Braava jet mopping robot, exclusively designed for kitchens, bathrooms, and other small areas in the house, and the Wi-Fi-enabled Roomba 690 and 890.[184]

By this stage, in 2021, iRobot had sold roughly 40 million robots worldwide. Yet even with this scale of adoption and its growing global footprint, the company struggled to secure a sustainable financial runway. So what did it do next?

iRobot decided to acquire yet another company in this space—this time in the air-purification segment, called Aeris Cleantec. The goal was to expand its portfolio beyond floor care. In simple terms, the underlying principle was diversification. The thinking was that diversifying products while remaining deeply embedded in the floor-care space, might help the company reignite the flame of growth.

Did this strategy work?

Unfortunately, it did not.

The acquisition failed to reflect meaningful growth in the company's sales chart, and public markets responded accordingly. Even with these continued expansions and acquisitions, the company was unable to secure a sustainable financial runway.

So what did the company do next?

Amazon's Interest and Regulatory Constraints

Yet another acquisition followed—but this time in reverse.

In August 2022, Amazon announced plans to acquire iRobot. The rationale appeared straightforward. Amazon had spent years building a smart-home ecosystem through Alexa, Echo, Ring, and related services. A mobile robot operating inside the home fit naturally within that vision. For iRobot, the deal offered scale, pricing power, and a strong operational base it could no longer build on its own.

The transaction, however, quickly attracted regulatory scrutiny, particularly in Europe. Concerns were raised about Amazon's potential dominance as a core technology provider for consumer devices in the home, as well as the broader implications for surveillance.[186-189] This scrutiny slowed the acquisition process and effectively froze the funding runway iRobot had been relying on.

So what did iRobot do next?

Debt Financing from the Carlyle Group

Given the uncertainty around funding, iRobot secured a $200 million senior loan from Carlyle in 2023, providing the company with a much-needed financial runway. The goal was to ensure this runway remained intact until the acquisition with Amazon was successfully closed. However, amid rising concerns over competition from Amazon and the market power it could wield following the acquisition, the deal was formally abandoned in 2024. Amazon paid a $94 million termination fee, but iRobot was left alone—yet one more time.

While the $200 million funding from Carlyle provided iRobot with a much-needed supply of fresh oxygen, the collapse of the $1.7

billion deal with Amazon made it clear that this supply would not last long. The consequences were immediate. The company announced significant layoffs and warned publicly that its ability to survive was under threat. That threat proved real in the last quarter of 2025.[186-189]

Chapter 11 Bankruptcy

By late 2025, iRobot filed for Chapter 11 bankruptcy protection. Control of the company subsequently passed to its largest creditor, Picea Robotics (Shenzhen Picea Technology Co., Ltd.), a Chinese manufacturing partner.[186-189] This outcome followed a pattern some of you might be familiar with the hardware space: once suppliers become creditors, ownership is usually next.

For a company of iRobot's caliber, it is worth remembering that robotics as a field is still in its early stages. While iRobot's journey ultimately ended in bankruptcy, the company's impact should not be measured solely by that outcome. Both the path it took and the lessons it offers remain deeply valuable for the next generation of robotics founders and engineers. We will now unpack these lessons one by one.

Lessons for Founders and Engineers

iRobot's story is filled with extremely valuable lessons for the next generation of engineers and founders.

Lesson 1: Focus Relentlessly on Your A+ Problem—but Don't Stop There

Every company has an A+ problem—the one problem that matters most to its survival and growth. For iRobot, that problem was Roomba, and more specifically, sustained sales of Roomba. While acquisitions and expansion strategies can help expand a company's footprint, nothing delivers higher returns than deeply understanding the needs of your customer and continuously improving your core product to address those needs with near-surgical precision. In iRobot's case, Roomba was the engine, and everything else was secondary. Consumers cared about the problem Roomba solved, and they were willing to pay for a solution.

The lesson here is clear: start with a real customer problem—but never stop there. Competitors often enter the market and can dilute your offering overnight. This naturally leads to the next lesson.

Lesson 2: Sustained Innovation Is Your Only Defense Against Competition

Product success is never permanent. Whether you are in hardware or software businesses, if you fail to continuously innovate your product, in simple words, sustain the principles that make your product valuable and novel to users—it will not last long in the market. Competitors will knock you out of the market way before you even realize. If you stop pushing your product forward, the market will push you out.

Now the challenge is this: to enable sustained innovation in your company, you need sustained research, and that requires a sustained source of capital. This naturally leads to the next lesson.

Lesson 3: Consumer Robotics Is Deeply Capital-Intensive Business

Consumer robotics is an extremely capital-intensive business. From Amazon's Astro to Kuri, Vector, and Cozmo, the pattern is consistent: building, launching, and scaling a consumer robotics business is an exceptionally hard endeavor. Why is that the case?

The reason is simple. It takes a significant amount of capital to both build a consumer robot and build trust in your brand. In the absence of trust, consumers will not buy the product. This makes both the product development phase and the market adoption phase, that is, the trust-building phase, highly capital-intensive.

As a result, founders must plan for capital needs from day one. Capital structure and fundraising strategy cannot be afterthoughts. You should surgically define them on day one. How do you do that? That leads us to the next lesson.

Lesson 4: Capital Structure Determines Survival

Capital structure often shapes outcomes, especially in the hardware-centric business segments, like robotics. Being capital-intensive, if you do not define your funding strategy with clarity from the very beginning, you will struggle to secure a sustainable financial runway.

That is why it is almost always important to define a hybrid capital structure from day one. This means not relying on a single class of investors to fund the company. This is a mistake many founders make, often turning to high-interest debt-based channels at the last minute. Debt buys you time but also adds unreasonable deadlines that are hard to meet, and rightfully so. For this reason, founders should think through and define a hybrid capital structure for their startup from the outset. How do you do that?

You need to adopt a hybrid strategy. Put simply, if you want to build a financially resilient robotics company, you should start with accelerators to establish early-stage credibility in the market. Next, expand into retail through crowdfunding channels to build a community of loyal users who genuinely love and value your product. Take feedback from this community seriously and use it to refine your product. Then engage institutional customers to scale the business. Many startups skip the retail hop, but doing so often negatively impacts your runway in the long run.

Even with these two ingredients in place—sustained innovation and a strong capital base—success is not guaranteed. Amazon has a strong brand presence and an equally strong capital base, yet customers failed to adopt Astro, its consumer security robot. That means, your focus on your product matters a lot , and that leads to the next lesson.

Lesson 5: Product Focus Is a Double-Edged Sword

Product focus is a double-edged sword. Focusing on a single product—or expanding into too many—are both double-edged swords. In the case of iRobot, if you look at its history over nearly four decades, the team ended up building a series of robots. While this may seem like the right approach in the early stages, when a company is experimenting and searching for a product–market fit, the challenge is that this process is capital-intensive. If not managed carefully, it can erode the very financial runway it is meant to secure.

Ideally, companies should build a small set of highly curated products that can scale and serve clearly defined ideal customer profiles within the same domain. Here, it is often more important to know what *not* to build—and why—than what to build. Every product you choose not to build buys time and resources that can be redirected toward what truly matters for the company. Once you lock down this set of products, your next job is to continuously innovate and build a strong and organic user base for this product. How do you do that? That leads us to the next lesson.

Lesson 6: Acquisitions Rent Trust, Innovation Builds Trust

There is no magical formula for organically building a user base. At the end of the day, you are building trust at scale in your product, and that means your product has to demonstrate its value. It has to solve the core problems of your customers. You must continuously innovate your product and expand its ability to address these core needs with almost surgical precision.

A big mistake many companies make here is acquisition. Acquisitions may look attractive on paper, but in most cases, they are a dangerous shortcut to building a user base. An acquired user base is a rented user base, and it will not last. Trust can never be rented; it has to be earned. Trust is earned through product value and customer experience. It cannot be rented.

To summarize, build your user base organically by clearly solving your customers' problems and being transparent about it from day one. As you solve these problems, you will inevitably need to source components, whether hardware or software, from suppliers—and that leads us to the next lesson.

Lesson 7: Suppliers Hold Power—Choose Them Carefully

Finally, in hardware, suppliers are strategic actors. If you depend on them for survival, they will eventually decide the outcome. This pattern is not unique to iRobot—it is a recurring pattern in hardware companies. Choose your suppliers wisely, and diversify wherever possible to minimize reliance on a single supplier—or, in this case, a supplier that has effectively become a creditor.

That summarizes the lessons we can learn from the journey of iRobot.

Now that we have distilled the core lessons embedded in iRobot's trajectory, from its origins through its evolution and the pivotal decisions the team made along the way, it is time to decode the underlying business model.

In the final section of this chapter, we will break down the business model and illustrate how the product portfolio maps to select primary use cases.

iRobot: Product Snapshot — System Overview & Evolution

Figure 13.1 — iRobot: System Overview & Evolution

Illustration by the author. Robot images from the iRobot product page.

iRobot's Business Model — Part 1 of 2

iRobot's Business Model

- **Mission:** To empower people to do more by building thoughtful, practical robots that automate everyday chores—freeing consumers to focus on what matters most and ultimately improving their quality of life.
- **Target Customer Base:** Mass-market households globally, with strong penetration among dual-income families, pet owners, and tech-forward consumers seeking time-efficient home automation solutions.
- **Primary Use Case:** Autonomous floor care solutions (robot vacuums and mops)
- **Products:** Roomba robotic vacuum platforms and Braava robotic mopping systems.
- **Model:** Consumer robotics company built around the philosophy of a "home cleaning automation" platform, anchored on a hardware-led adoption model and reinforced by proprietary, software-enabled advanced mapping, navigation, and smart home integration services.
- **Funding:** Founded in 1990; publicly listed on NASDAQ under ticker IRBT in 2005.
- **GTM:** Omnichannel consumer distribution strategy spanning direct-to-consumer e-commerce, major retail chains, and global online marketplaces.
- **Growth lever:** Category leadership in robotic floor care driven by a tiered good-better-best portfolio strategy, strong brand equity, ecosystem lock-in via app integration, and recurring revenue streams rooted in consumables (bags, filters, brushes, batteries).

Figure 13.2 — iRobot's Business Model: Part 1

Illustration by the author.

iRobot's Business Model — Part 2 of 2

Figure 13.3 — iRobot's Business Model: Part 2

Illustration by the author.

Chapter 14: Social Robots — Why They're So Hard

From Japan to South Korea, and from Greece to Italy, birth rates are declining at a rapid pace. A recent study confirmed that more than half of the countries in the world now have birth rates significantly below the required level of 2.1 children per woman, which is needed to maintain current population levels and sustain economies around the world.[190] Simply put, a declining population implies a shrinking workforce and, ultimately, lower GDP.

Beyond GDP, these labor shortages also have a direct impact on healthcare systems. If we do not have enough elderly-care workers, the cost of healthcare for the elderly rises, which in turn negatively affects the quality of life for this generation. In fact, it is estimated that the United States alone will face a shortage of about 139,000 physicians by 2033.[191]

In addition to declining population levels, another major contributor that is significantly impacting productivity levels in society today is the internet. It is estimated that the average person spends about 6 hours and 38 minutes each day online, and that number is essentially a core vector for rising unproductivity levels in society.[192] With the continued proliferation of modern chat-based AI models, this vector will continue to surge in the coming years.

Given the declining population, our growing reliance on the internet, the rising need for elder care, and worsening labor shortages, researchers working on this challenge laid out two core options for society to address the problem. One option is for us, as humans, to take a step back, reflect on how we spend each day, and evaluate how we can free up time to better support our loved ones and elders in society. The second is to augment the existing workforce with robots designed to assist with this process. This underlying challenge, and the urgency around it, has led to the launch of a series of social robots since the early 2000s.

Some of you may ask: what about chat apps and voice assistants? While voice assistants and chat apps have often been positioned as solutions to this societal challenge, humans are inherently social creatures. Our ability to connect with one another goes beyond the words we type on a screen or the audio waves produced by our glottis and modulated by our tongue. Our mode of communication with each other extends far beyond these audio waves and words. We enrich it with our emotions, our gestures, and our facial expressions. Chat apps and voice-based systems do not fully meet these needs. Hence, researchers around the world have pursued social robots as a rational medium to address this challenge and help us reinvigorate these bonds in our society. This rationale has led to the launch of a series of social robots such as Anki, Kuri, Jibo, Cozmo, and Vector over the last 25+ years.

If the thesis indeed holds strong rationale, then one might ask: why did these robots fail to gain traction in society? What aspects of these robots were well received, and what weren't? In simple words, where did they succeed in addressing market needs, and where did they fall short?

In this chapter, we will answer these questions and explore the lessons we can learn from the journey of each of these social robots, and from the startups that brought them to life. To begin with, we must first understand why social robots are among the hardest products in the industry to build, finance, and attain mainstream adoption.

Why They're So Hard — 3 Core Reasons

To begin with, it helps to understand that there are three core bottlenecks that make it extremely difficult for startups in this arena to both build social robots and sustainably integrate them into society.

Unforgiving Consumer Hardware Economics

The first core bottleneck that is hard to address is the underlying economics that drive the consumer hardware startups, such as those building social robots. To build one, you need a wide range of components. A single high-quality LIDAR sensor can cost upwards of $500. Then there is the battery, the onboard computers, the GPUs, the underlying software stack, the data that must be stored in the cloud, and the security chip required to encrypt this data in order to retain compliance with SOC and GDPR standards. The list of hardware and software components adds up in no time.

Then there are operating expenses: the talent density required to build and continuously refine these robots, combined with manufacturing and shipping costs. These all add up fairly quickly—more than most realize. We don't have to look far to do a reality check on pricing. Social robots such as Amazon Astro launched at a retail price of around $1,499, Jibo at $899, and Kuri, by Mayfield Robotics, at $700.[193-195] At those prices, it's hard to argue that consumers are not better off paying an hourly wage worker to offer direct assistance than purchasing a robot. That's the first bottleneck, but not the only one.

Homes Are the Hardest Test Case

The second core bottleneck is ensuring that these robots truly deliver on their promise—that is, that they can both navigate and function reliably in our homes, which are likely among the most unstructured environments on Earth, with no two quite the same. This means the conditions under which these robots must provide consistent, high-quality service are almost infinite.

How do you even optimize robots to navigate reliably in such environments, where, beyond the physical footprint and layout, there are Wi-Fi dead zones, variable ambient lighting conditions, narrow hallways that can create localization bottlenecks, and countless other factors that can ultimately strain the navigation and perception stack? In simple terms, how do we ensure that these robots can see, locate themselves, and navigate these spaces in a safe and reliable manner while delivering consistent service to customers? That becomes another major challenge.

Our Expectations for Robots Are Too High

The third core bottleneck is the expectations of us humans. Naturally, when you position a robot as a social robot to a human, we unconsciously set the bar so high in the minds of consumers that there is no doubt most companies will find it extremely hard to meet that bar. From response time to quality of reaction, intensity of emotion, and anonymization of data captured using microphones and cameras located onboard, it becomes extremely difficult to meet expectations across the verticals for these social robots.

Why is that the case? Because consumers will no doubt want a quality of interaction that matches that of a human assistant working with them. At the current stage, it is extremely hard to meet that threshold. It is not impossible, but it would certainly require far more research and work than has been conducted so far.

These three bottlenecks make it extremely hard for consumer robotics startups to succeed in this world. That leads to another question: if these three vectors are well known, and the underlying labor constraints required to enable elder care services at scale are so clear, why did Kuri, Anki, Jibo, and many others make this bet?

The Core Bet Everyone Made

The underlying thesis here was that, given the only two solutions we humans had in place, i.e., to evaluate ways to free up the time of billions of young people around the world to enable better elder care services in society, or two, to add social robots, the former requires significant change in our habits across the planet. The latter is a problem that boils down to changing one habit among elders, i.e., getting people used to using robots.

Unfortunately, whether we like it or not, the former, even though it is a rational and obvious choice, i.e., to free up our time to take better care of elders, in reality, given the way our society has evolved and our responsibilities, is a harder problem to solve at scale than it might seem on paper. There are multiple habit and ownership vectors involved here, from a person's ability to financially support their family, to their health, and more. Unless everyone takes full ownership for all verticals of their lives, the first solution is almost impossible to enable at scale.

Recognizing this challenge, these robotics startups decided to pursue the path of social robots, and hence the bet was made on Kuri, Anki, Jibo, and more.

Jibo, for example, was launched as the world's first family robot, designed to interact with family members, capture their pictures, and share them among them. It was intended to serve as a relationship catalyst within households. The emphasis here was on relationship-driven interaction vectors. Similarly, Kuri was deliberately designed as a friendly home companion, a robot "with personality," by Mayfield Robotics. Anki, with Cozmo and Vector, and Amazon's Astro followed a similar market positioning pattern.

Even with this positioning, some of these startups secured reasonable early adoption, while most still failed to gain traction in the market. For example, while Jibo did manage to launch a successful crowdfunding campaign and raised upwards of $2 million, the company unfortunately went bankrupt and had to shut down all its operations in 2019.[196] Similarly, Kuri, priced at $699, failed to gain traction in the market and had to shut down its operations in July 2018. Anki, with Cozmo and Vector, and Amazon's Astro, even with its strong brand, were victims of this same response from the market.

The challenge here was that while the underlying thesis was rational enough, people were only willing to buy robots if they solved a real problem that mattered to them. While these robots did try to solve part of that problem, they were not a substitute for a human assistant. That meant they didn't fully address the real needs of consumers. There was a large delta between the perceived utility value and what was actually offered in reality. That delta was the cause of failure here. To better understand this delta and the corresponding failure vectors, we will now examine each of these startups and their trajectories one by one in the following chapters.

Consumer Robotics Startups Business Model — Shared Core Parameters

The Business Model of Consumer Robotics Startups

- **Mission:** To elevate quality of life for consumers worldwide by delivering warm, small-form mobile robots that autonomously navigate indoor environments and serve as trusted companions, providing assistive services such as reminders, telepresence, home monitoring, and wellness support out of the box, thus enabling safer homes, greater peace of mind, and deeply unified families.

- **Target Customer Base:** Across these companies, the primary customer base consisted of tech-savvy early adopters willing to embrace a more intuitive and engaging interface than a standard smartphone app, one designed to serve as a true companion for their loved ones.

- **Primary Use Cases:** Core use cases clustered around three segments:
  a. Companionship and wellness services for elderly users, including medication reminders and guided exercise prompts
  b. Educational guidance for children
  c. Home and family assistance features designed to elevate household security posture. Collectively, the platform aimed to deliver peace of mind and practical utility for family members across age groups within a single, integrated solution.

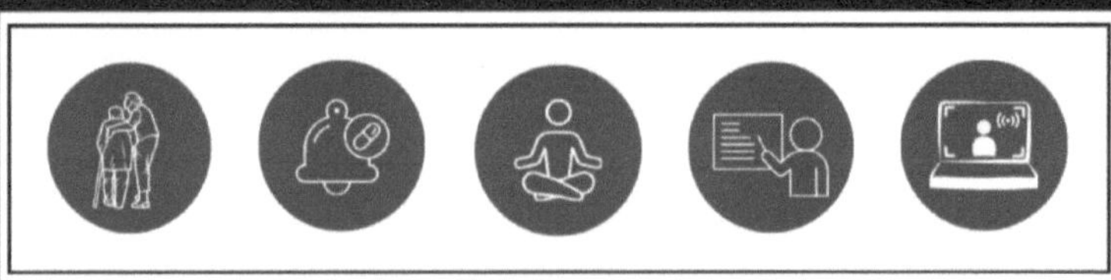

Figure 14.1 — Consumer Robotics Startups Business Model: Shared Core Parameters

Illustration by the author.

Chapter 15: Astro, Anki, Jibo, & Mayfield — Vision to Market Reality

The 2010s marked the first great wave of consumer robotics startups. From Anki to Jibo, from Kuri to Astro, this generation of social robots offers enduring lessons across the full spectrum of the startup ecosystem. From fundraising and product design to economics, market timing, and human behavior, their trajectories offer a blueprint, both cautionary and catalytic, for the next generation of founders, researchers, and engineers building in the field. We start with Anki and examine each of these startups one by one.

Anki — Demand Was Real, the Economics Weren't (2010)

Anki was one of the first robotics startups to pursue social robots with a dual mission: to democratize robotics education while cultivating enduring trust between humans and robots in society.

Incubated in 2010, the company drew its roots from the fertile research grounds of Carnegie Mellon University (CMU). The founders, Boris Sofman, Mark Palatucci, and Hanns Tappeiner, were all students in CMU's robotics program and shared a deep passion for making robots personal, warm, friendly, and most importantly, organically interactive. Over the next several years, they raised a total of $182.5M from top-tier Silicon Valley investors, including Andreessen Horowitz, Index Ventures, and Qualcomm Ventures, making Anki one of the most well-funded consumer robotics startups of its era.[197-198]

To bring their mission to life, the team launched two social robots: Cozmo and Vector, both inspired by the character WALL-E from Pixar's iconic 2008 animation film. Cozmo, released in 2016, was positioned as an educational robot designed to encourage social interaction amongst students by offering an interactive platform to develop programming skills. Vector, launched in 2018, was the company's attempt to evolve that vision into a more sophisticated AI-enabled companion robot that could extend these interactions beyond academia and into our homes. Equipped with an interactive screen, voice assistant, and semi-autonomous navigation capabilities, Vector was designed to enrich these interactions and build a strong sense of trust between humans and robots. Together, with Cozmo and Vector, Anki offered consumers some of the most ultra-compact social robots of our time, paired with a pricing model that strongly resonated with consumers.

Cozmo, launched at a price point of around $179, came equipped with a wide range of programming tools, ranging from MIT's Scratch Blocks, a simple drag-and-drop interface designed for school children, to a full-scale software development kit designed exclusively to help college students learn robotics using Python-based programming models.[199] This functionality ensured it gained adoption even in the robotics classes conducted at Carnegie Mellon University. Cozmo also proved to be a commercial success, becoming the best-selling consumer educational robot by revenue on Amazon in the U.S., U.K., and France in 2017. That traction ultimately paved the way for Vector.

Vector was designed to extend these social interactions beyond academia and into everyday life, which meant the company had to evolve both its design and capabilities. To do so, Anki made strategic efforts to hire designers from Pixar and DreamWorks Animation Studios to make these robots as humane as possible, while maintaining an ultra-compact footprint to minimize adoption friction. Given the design, the compact form factor, and positioning around education and social interaction, the robots managed to win the hearts of consumers and were sold out many times, especially during the Christmas seasons of 2016 and 2018. From 2010 to 2018, Anki managed to sell about 1.5 million robots in total.[200-201]

However, the economics didn't quite work out.

The Limits of Anki's Business Model

Manufacturing these robots and securing components at a price that would make the venture sustainable was probably the hardest problem for the founders to solve. In addition to that, robots like Cozmo and Vector were not fully consumer-ready and required ongoing software support and continued access to the company's cloud-based servers, which inflated operating expenses. Given that the company adopted a hardware-first sales model, a one-time purchase without any recurring revenue layer, these robots could not generate enough lifetime value to cover the capital expenditures and operating costs required to support them at scale, even in the face of strong consumer demand. Thus, while these robots won the hearts of consumers, the business model failed to win the confidence of investors and suppliers.

That ultimately led to the collapse of Anki in 2019, as the company failed to extend its financial runway and align its business model. Post-shutdown, Anki's assets, including Cozmo and Vector, were acquired by a private company called Digital Dream Labs in late 2019. The company attempted to revive these robots but ultimately had to offset the costs of maintaining cloud servers by migrating the project to an open-source, community-based model now managed by global fans of Vector and Cozmo.

While Anki proved that the demand for consumer robots is indeed real, the economics in this arena are indeed unforgiving. Even when you get the economics right, getting them right in time is another challenge in itself, and Jibo's journey is evidence of that principle.

Jibo — A Vision That Never Became a Need (2012)

Jibo was one of the earliest social robots designed with an exclusively humane form factor and with the sole mission of offering a rich, organic human-robot interaction experience to consumers. The robot was the result of years of research led by renowned MIT professor Cynthia Breazeal, who founded and heads the MIT Media Lab's Personal Robots Group. Jibo was conceived to extend the research conducted in her labs in social robotics into a consumer product capable of fostering natural interaction between humans and robots in our society.

Although Jibo as a company, was established in 2012, it was not until 2014 that it formally raised capital through a record-breaking crowdfunding campaign launched on Indiegogo. Through this campaign, the company raised about $3M from 6,000+ supporters, far exceeding its initial $100,000 goal and making it one of the most successful crowdfunding robot campaigns ever. This scale of trust in a consumer robot emerging from the fertile labs of MIT proved that there was strong demand for a thoughtfully-designed, family-oriented social robot. Jibo could be trained to recognize family members, understand their language and accent, deliver personalized information to them, and enable a more engaging experience than the smart devices of the time.[202-203]

Building on this momentum, and faced with the challenge of manufacturing and delivering thousands of units to these 6,000+ early adopters, the company raised approximately $70 million over multiple funding rounds from a series of investors including RRE Ventures, Samsung Ventures, CRV, Flybridge Capital, and Two Sigma. With capital and early adopters secured, the company started turning the prototypes used for the crowdfunding campaign into a commercial product. That meant sourcing components, negotiating with a series of global vendors, finalizing designs, and conducting a series of tests.

The process took nearly three years. By the time Jibo was ready to be shipped to its backers in 2017, it carried an $899 price tag. And that was the problem.

The Limits of Jibo's Business Model

During those years, big players such as Amazon, Google, and others had started proliferating this space with voice-assisted smart speakers that were introduced at about one-tenth of the price of Jibo and provided reasonably similar functionality, albeit without the same embodied form factor and social experience. Turns out consumers didn't care much about the experience, but prioritized affordability and convenience. Amazon alone ended up selling about 71 million smart speakers during those years, while Google sold an additional 52 million.[204]

This shift in the market posed an existential challenge for Jibo. The market was now saturated with these smart-speaker-driven home assistants, and the resulting scale of adoption plunged the demand for premium social robots like Jibo. In retrospect, the three-year gap between the launch of the successful crowdfunding campaign and commercial shipment proved long enough for industry giants to dominate the space. By mid-2018, the company had burned through the majority of its funding and decided to shut down its operations. In June of that year, the company sold its assets and intellectual property to a New York–based investment management firm, i.e., SQN Venture Partners. The robot's servers and cloud infrastructure, which supported much of its social and interactive functionality, were subsequently shut down; in early 2019, existing Jibo units displayed farewell messages as services were discontinued.

While Jibo may not have become a commercial success, it still mattered. It was one of the first serious attempts to bring a thoughtfully designed social robot into the home, and the response to its crowdfunding campaign reflected organic consumer interest in the company's mission. This team had a strong vision that resonated with consumers, but the market matured faster than the product could keep pace, and mounting competition ultimately made the business unsustainable.

So far, we have learned that Anki confirmed economics matter deeply in consumer robotics, and Jibo demonstrated that getting those economics right in time is an even harder challenge. But even when a company manages to get those two right, there are still risk vectors that may not seem obvious at first—yet, when they surface, they can deliver a devastating blow to a startup. The story of Kuri, one of the most elegantly designed robots in the history of personal robotics, is a powerful example of this principle.

Kuri (Mayfield Robotics): Loved by Users, Limited by Scale (2015)

Kuri, a social robot launched in 2015 by Mayfield Robotics, is probably one of the most elegantly designed compact social robots ever made. The company was incubated in Redwood City in Silicon Valley by Mike Beebe (CEO), Kaijen Hsiao (CTO), and Sarah Osentoski (COO), with seed funding from its parent organisation, the German electronics giant Bosch, as part of Bosch's Startup Platform. The two co-founders, Kaijen and Sarah, previously worked at Bosch's research centre in Palo Alto and are among the most well-known robotics professionals in the field. Kaijen, prior to Mayfield, was a robotics research scientist at Willow Garage and is widely recognised for her contributions to ROS-era robotics.[205]

With a strong background in robotics research, the team set out with a mission to build social robots that were explicitly intelligent, joyful, inspiring, and, most importantly, useful for those alone at home—especially the elderly. With this mission, and an elegant design matching the calibre of Japanese craftsmanship, the company debuted Kuri at CES 2016. Later that year, this adorable robot was celebrated at TED, SXSW, The Economist Innovation Summit, and many other high-profile events.[206] This scale of attention was not accidental: the underlying engineering that brought the product to life was equally remarkable.

Equipped with a 1080p camera enabling a telepresence feature, microphones and dual speakers for interaction, and sensors for site mapping and object detection, Kuri was as elegant on the engineering front as it was on the design front. In simple words, it was a product that married Japanese design philosophy with German engineering. Launched at a price point of $699, with a $100 preorder deposit, the robot managed to win the hearts of many around the world through a design that did justice to its positioning: a friendly home companion.

Speaking of design, what made Kuri feel warm and approachable was its form factor, its expressive eyes—perhaps the hardest element to get right—its facial expressions, and the quality of its speech. The company left no stone unturned in making it as humane as possible. It would hardly be an exaggeration to say that Kuri was one of the most humane social robots ever created. So why did it fail to gain traction in the market? More importantly, what led to its demise?

To answer that question, it helps to zoom out and look at the timeline. The launch itself was done right. It garnered significant attention, and the team delivered on its promise. The challenge came the following year, in 2017. Bosch, the parent company, decided to cut the financial lifeline, as it could not figure out how to incorporate Kuri into its broader business units.[207-208] While the reasons behind this decision may be outside the scope of this analysis, what is clear is that this was not a failure of the robotics team. It was, fundamentally, a failure that boiled down to funding and investor understanding.

The Limits of Mayfield's Business Model

The team had secured support from its parent company, but if that parent decides to cut the lifeline at the last minute, there are not many options left for a capital-intensive consumer robotics startup to survive, let alone thrive. The second lesson here is that consumer robotics, unlike software startups, operates on exponentially longer adoption curves. You cannot expect a consumer robot to gain widespread adoption within one year. Why is that so?

Because consumer robotics is a habit-change problem, and habit change takes time. First, companies must build trust with consumers, which Mayfield achieved through design. Second, they need time for consumers to adopt the robot and incorporate it into daily life. The need was clear, and the trust was established—but the runway was not long enough. To buy that time, consumer robotics startups need sustained funding, and once that lifeline was cut, it became nearly impossible for a startup with such calibre and potential to succeed.

Reflecting on this journey, it goes without saying that when Bosch withdrew support in mid-2018, Mayfield soon had to shut down operations later that year. And that meant we lost Kuri, at least for now. While it might appear that Kuri has achieved a state of nirvana, in reality, the lessons that this robot and its creators offer us remain deeply valuable. Hopefully, someone reading this will reinvigorate this mission and bring her back to life. For Kuri was not just a product—it was one of the most humane social robots ever designed, marked by unparalleled elegance and fine engineering.

To summarise, the journey of Kuri reminds us of the importance of retaining a hybrid capital stack and eliminating single points of failure when it comes to funding. The story of Anki reminds us of the unforgiving hardware economics inherent in this field, while Jibo reminds us of timing and the competitive landscape. These are the risk vectors that all consumer robotics startups should keep in mind.

Yet even when these lessons are accounted for, and even if a company is fortunate enough to have a strong brand presence from the outset, ultimately, if the market does not see a strong utility value in your product, then that becomes your major obstacle.

The trajectory of Astro, the home robot launched by Amazon, serves as a strong proof point here.

Amazon Astro: Trust Was the Product Constraint (2021)

Astro was Amazon's first major consumer home robot, announced on September 28, 2021, as an attempt to enable safety and security services for homes. The company knitted the autonomous navigation stack, built with ROS, Alexa voice assistance, and the home monitoring layer into a single mobile device, i.e., Astro. In simple words, the goal was to help family members keep an eye on their home while they were away from it.

Astro was launched through an invite-only "Day 1 Editions" program with an introductory price of $999.99, with an eventual retail price of $1,449.99.[193] But Astro never became a mass-market product. Consumers did not trust it as much as they did Echo and Ring. That raises the question: why was this the case? After all, it was the same company, with a similar use case, but a mobile form factor equipped with a camera. That exactly was the problem here. The problem was not the technology. It was trust.

The Limits of Astro's Business Model

A smart speaker sits in one place. A security camera is fixed to a wall. Astro was different: it was a mobile camera roaming through private living spaces. To enable it to work in your home, you had to give it full rights to map your home and scan it 24/7. That resulted in immediate privacy-led friction, which throttled mainstream adoption for this robot.[209]

Traction appeared minimal. Bloomberg noted that six months after launch, "hardly anyone was talking about Astro online," and only "a few hundred units, at most," seemed to have shipped.[201]

Given this low traction on the consumer end, Amazon repositioned it as a business product and launched Astro for Business in November 2023, priced at $2,349.99. Given similar challenges, in less than a year, the company had to discontinue this product and shut down the project.

Astro's trajectory reinforces a recurring lesson in consumer robotics: even with strong engineering and one of the most trusted consumer brands in the world, home robots do not scale without a must-have use case that outweighs privacy and trust concerns. In the home, trust is often the limiting variable.

Now that we have covered the trajectories of four consumer robots and the teams behind them, it helps to reflect on the lessons we can derive from this study—lessons that shed light on some of the most important risk vectors founders must keep in mind in this arena.

Founder Takeaways

Lesson 01: Hardware Economics Are Unforgiving (Anki)

Anki proved that even when consumers love the product, like the social robot the company designed for its customers, the business can still collapse if the unit economics do not work. Robotics is a capital-intensive domain: sensors, motors, batteries, manufacturing, returns, and support costs all compound quickly. If you do not factor these expenses into your pricing model, you will erode your own runway in no time.

To prevent this from happening, you need to retain sustainable margins in your product, and that means designing your pricing model accordingly. There are at least two pricing models you can consider here, and you should evaluate which one aligns best with your startup, based on your context and the maturity of your brand and product. The first is a hybrid model, which covers static revenue from the first sale and establishes a dynamic revenue pipeline through recurring software or services that you can offer to your customers. The second is essentially a premium pricing model, where you price your product high enough to accommodate both capital expenditures and operating expenses from day one.

Either way, your pricing model must account for these expenses, as hardware costs often eat up a significant portion of the runway for most robotics startups, especially during the early stages. Now, even if you get the hardware costs right, you also have to get them right in time. Timing matters in robotics, and Lesson 02 is a proof point of that principle.

Lesson 02: Timing Matters as Much as Vision and Economics (Jibo)

Jibo had a strong mission and organic demand validated at scale, with 6,000+ backers contributing a total of $3 million in the crowdfunding campaign. In this case, the vision, product, and economics were aligned, but the company failed to get the product out to its users in time. The multi-year gap between the announcement of its first prototype in 2014 and commercial shipment in 2017 gave Big Tech time to flood the market with cheaper substitutes. Thus, timing matters as much as vision and economics in robotics. In simple words, a great product launched too late in the market is often indistinguishable from a bad one and often gets caught in Big Tech's nets.

Now, even when you get the hardware economics and the timing right, if you are a startup relying solely on a single investor or a handful of investors, you are creating a single point of failure. Lesson 03 illustrates this principle well.

Lesson 03: Capital Structure Can Make or Break a Robotics Startup (Kuri)

A single funding lifeline is a fatal vulnerability, especially if you are a founder running a robotics startup. Kuri, designed by Mayfield Robotics, was an elegant, beloved, and well-engineered social robot, but it had only one financial lifeline, namely Bosch. Once the parent company pulled the plug, the startup had no runway left.

This is especially painful because Kuri was a product of extraordinary calibre, with design and engineering that rivalled the craftsmanship of the finest Japanese masters and the precision of German technologists. And yet, none of that mattered once the funding lifeline disappeared overnight. Thus, consumer robotics startups need resilient capital structures that strategically marry at least four diverse sources in harmony: accelerators, retail investors, VCs, and strategic partners. Fragility here is not an option.

Lesson 04: Trust Is a Product Constraint, Not a PR Derivative (Astro)

In the case of consumer robotics, you are putting a robot in people's homes and in their lives. Hence, trust is the biggest barrier to adoption, especially if this robot is equipped with a device capable of recording every moment inside the home, and rightfully so. Astro demonstrated that even Amazon, with its massive distribution scale and brand power, could not easily convince consumers to accept a mobile camera roaming inside their homes.

In the home, privacy and trust are the most important risk vectors for adoption. Privacy is your ability to convey to your customer how and where the data captured by your robot is stored and secured. Trust is the outcome of this process. If you cannot relay this information with clarity to consumers, then trust becomes your major bottleneck for adoption.

In retrospect, this problem is not unique to Astro. There are cameras designed to monitor pets and homes, and consumers have adopted them. The challenge here was how these risk variables, namely privacy and trust, were addressed and conveyed to consumers. It is important to offer consumers insight into where the data is being stored, who has access to it, and, more importantly, their right to erase it at any moment, in return for a service that keeps the home and family members safe and secure. This is easier said than done, and that brings us to the last lesson.

Lesson 05: Adoption in Consumer Robotics Boils Down to Habit Change

Consumer robotics is fundamentally a habit-change problem, and habit change requires time. It takes time for us humans to develop trust in a robot to execute tasks that were previously done by a human assistant. More importantly, it takes time for consumers to incorporate a robot into their daily routines.

This is the shared and most critical risk vector each of these robotics companies encountered: behavioural adoption. The time required for us humans to change our habits in order to adopt these robots far exceeds the patience threshold of a significant number of investors in the arena. Put simply, it takes time for people to adopt robots—often far more time than it took them to adopt software apps, phones, or laptops. This adoption curve has direct implications for the funding model that founders should design for their startup. From the get-go, founders should be transparent about adoption timelines for their product, their implications for ROI, and accordingly align expectations with investors to ensure sustained financial runway.

In summary, Anki teaches product economics; Jibo teaches market timing; Kuri teaches capital resilience; Astro teaches trust—and together, they reveal one core lesson: consumer robotics is a long game.

While these consumer robotics startups encountered bottlenecks that weren't surmountable and failed to thrive, that doesn't imply that the next generation of consumer robotics startups won't make it through. In fact, given the current shifts in demographics, labor patterns, and the technological breakthroughs unfolding around the world, the conditions have rarely been more favorable to reinvigorate the field. The stars may be better aligned now than ever for the second wave of consumer robots.

Why the Second Wave May Be Different — Demographics, Labor Gaps, and AI Breakthroughs

The first wave of social robots, which started in the early 2010s and ended with Astro in 2021, was probably a bit ahead of both the market and the technology. While social robots were designed to address a clear use case—serving as companions and assistants for those who are alone at home (especially elderly people)—beyond a certain point, that use case was still more of a desire than a true need. At the time, human assistants were available at a price point far lower than these robots, and with a quality of service significantly higher than what robotics could deliver.

That landscape is undergoing a shift right now with the tectonic shifts we are witnessing across economies—from Asia to Europe, America to Africa—on both the technological front and the demographic front. Several forces are converging that may finally drive meaningful adoption of social robots in the coming years. We will cover the most important ones here, one by one.

The first and most fundamental driver is demographic. The rate at which populations are aging across major economies is accelerating each year. Take the U.S., for example: residents aged 65 and older reached 59.2 million in 2023, a 9.4% increase since 2020. This demographic is expected to grow to 82 million by 2050, representing roughly 23% of the total population.[211] This shift has direct economic consequences.

An aging population constrains the available workforce in the market, and therefore intensifies labor shortages. This is an especially acute challenge for healthcare systems. If there are not enough people to support the elderly, the problem only compounds over time for each economy and its GDP.

At the same time, the technology landscape has improved dramatically. AI has finally caught up. Today's LLM and VLM models make it far more feasible to deliver natural speech, contextual interaction, and human-robot experiences than what was possible in the era of Jibo and Kuri—often at a fraction of the cost. Along with lower software costs, compute power continues to get cheaper. The cost of sensors and batteries is going down, and efficiency is going up. That means there is more compute power per millimeter that we can fit into these robots than ever before.

Beyond demographics and technology, the utility—i.e., the use cases for these robots—is becoming clearer than ever. Social robots are no longer positioned purely as companions, but increasingly as value-driven home assistants with specific roles. Humanoids like Neo, developed by the company 1X, are an early example of this shift. Equipped with LLM-enabled speech and VLM-enabled vision, these systems are now able to offer a quality of service that is approaching that of a human assistant.

In effect, these forces suggest that while the first wave of consumer robotics startups faced immense challenges, the conditions today are far better aligned to drive adoption of home assistants in the truest sense. Market pull is stronger, the labour gaps are real, and the technological foundations are finally converging to make home robots a practical reality in society.

Now that we have distilled the core lessons embedded in the trajectories of these social robotics startups, from their origins through their evolution and the pivotal decisions their teams made along the way, it is time to decode the underlying business model of each.

In the final section of this chapter, we will break down their business models and illustrate how their product portfolios map to select primary use cases.

Anki Product Snapshot — Vector & Cozmo

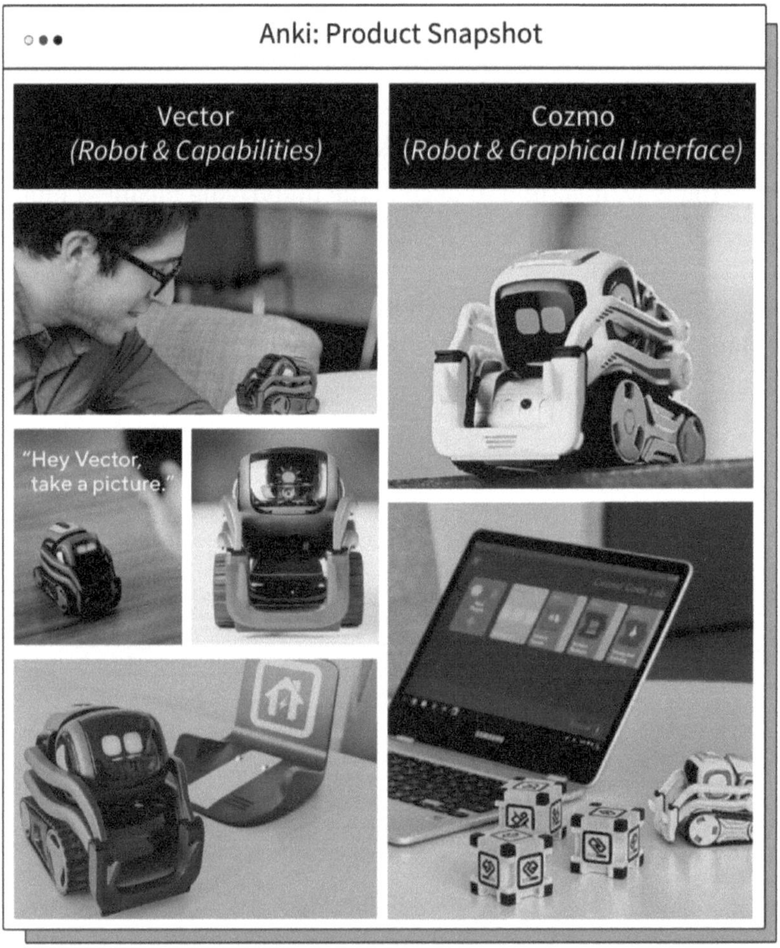

Figure 15.1 — Anki: Vector & Cozmo

Illustration by the author. Robot images from the Anki product page.

Anki's Business Model

Anki's Business Model

- **Products:** Cozmo & Vector, education-focused and AI-powered companion robots.
- **Model:** VC-backed consumer robotics platform scaled through crowdfunding channels.
- **Funding:** Raised approximately $182.5M across multiple venture rounds. Notable investors included Andreessen Horowitz, Index Ventures, Two Sigma, and JP Morgan.
- **Crowdfunding Channel (Post-Anki Era):** Following Anki's shutdown, its assets were acquired by Digital Dream Labs, which launched a Kickstarter campaign to finance the relaunch efforts of Vector. The *Vector* Kickstarter campaign raised approximately $1,877,719 pledged by 8,293 backers in 2020, making it one of the most funded consumer robotics projects on the platform.
- **GTM:** Direct-to-consumer sales combined with retail distribution channels. The brand was positioned around educational robotics and AI-powered social companionship, anchored by the Cozmo and Vector product lines.
- **Growth lever:** Community- and developer-ecosystem-led expansion. Anki released official SDKs and documentation, enabling developers to build custom programs and applications for Cozmo and Vector. Post-acquisition, Digital Dream Labs further advanced this strategy by open-sourcing elements of Vector's platform and leveraging crowdfunding channels to sustain adoption and offset ongoing cloud infrastructure costs.
- **Revenue Model:**
  - **Hardware:** Unit sales of Cozmo and Vector robots.
  - **Software:** Recurring subscription revenue from advanced AI features (Vector AI), offered at $11.99 per month or $99.99 per year.
- **Exit:** Anki ceased operations in 2019; its assets were subsequently acquired by Digital Dream Labs, which assumed responsibility for continuing the Vector platform.

Figure 15.2 — Anki's Business Model

Illustration by the author.

Jibo Product Snapshot — Prototype & Capabilities

Figure 15.3 — Jibo: Prototype & Capabilities

Illustration by the author. Robot images from the Jibo product page.

Jibo's Business Model

Jibo's Business Model

- **Products:** Jibo Social Robot (consumer "family robot" / home companion)
- **Model:** Crowdfunded and VC-backed consumer social robotics platform, with an ecosystem strategy built around "Skills" (apps) and a strong developer program.
- **Funding:** Raised ~$70M+ in venture funding across multiple rounds.
- **Crowdfunding Channel:** A crowdfunding campaign was launched on Indiegogo as an early demand-validation and preorder channel. The campaign closed at approximately $3.6M raised from 6,000+ backers.
- **GTM:** Preorder-led launch followed by direct sales; publicly marketed as the "world's first social/family robot."
- **Growth lever:** Platform and developer-ecosystem narrative ("Skills") designed to expand capabilities beyond the core product roadmap. In simple terms, beyond the robot's built-in social features, the company launched a developer program that gave developers worldwide access to its SDK (JavaScript API) to build Skills (apps). These Skills could be offered to consumers through a recurring subscription model, enabling strong alignment between consumers, the company, and the developers—ultimately establishing a robust revenue flywheel.
- **Revenue Model:**
  - **Hardware:** Unit sales of the Jibo robot
  - **Software:** Subscription-based access to Skills (apps.) and premium capabilities delivered through the developer ecosystem
- **Price Point:** Positioned at a premium tier, with preorders priced at $899 in 2017.
- **Exit:** Jibo shut down its operations in 2018, as demand for its product was significantly impacted by the rapid adoption of smart speaker platforms such as Google Home and Amazon Echo launched by major technology companies during that period. Its IP and assets were acquired by SQN Venture Partners the same year and later by NTT Disruption, announced in 2020, with the business repositioned toward B2B healthcare and education use cases.

Figure 15.4 — Jibo's Business Model

Illustration by the author.

Mayfield Robotics Product Snapshot — Prototype & Capabilities

Figure 15.5 — Mayfield Robotics: Prototype & Capabilities

Illustration by the author. Robot images from the Bosch and Mayfield Robotics product pages.

Mayfield Robotics' Business Model

Mayfield Robotics' Business Model

- **Products:** Kuri Home Robot (consumer home companion robot)
- **Model:** Corporate-backed consumer robotics platform incubated within Bosch's startup accelerator arm, primarily focused on enabling social interaction within family environments.
- **Funding:** Mayfield Robotics was incubated by Bosch, which reportedly committed approximately $100M to support product development and commercialization efforts.
- **Crowdfunding Channel:** None
- **GTM:** Preorder-led launch followed by direct-to-consumer sales. Kuri was publicly positioned as a warm, friendly and socially expressive robot designed predominantly to serve as a trusted companion for the elderly.
- **Growth lever:** Pure hardware-led adoption strategy centered on a minimal yet warm design profile, friendly personality, and deep emotional engagement, supported by a robust technology stack featuring autonomous navigation, voice assistant integration, and built-in video calling features, all positioned as a unified system designed exclusively to help strengthen bonds among family members.
- **Revenue Model:** Centered on unit sales of the Kuri robot.
- **Price Point:** Positioned at a premium tier, with preorders priced at $699 in 2017.
- **Exit:** In 2018, Mayfield Robotics ceased operations and canceled Kuri shipments, citing lower-than-expected consumer demand. In hindsight, Kuri was a product designed on par with the craftsmanship of Japanese masters; however, heavy reliance on a single funding source was possibly the single biggest driver behind this unplanned exit the company had to make. Bosch subsequently discontinued its consumer home robotics initiative, and the Kuri product line was shut down.

Figure 15.6 — Mayfield Robotics' Business Model

Illustration by the author.

Amazon: Product Snapshot — Prototype & Capabilities

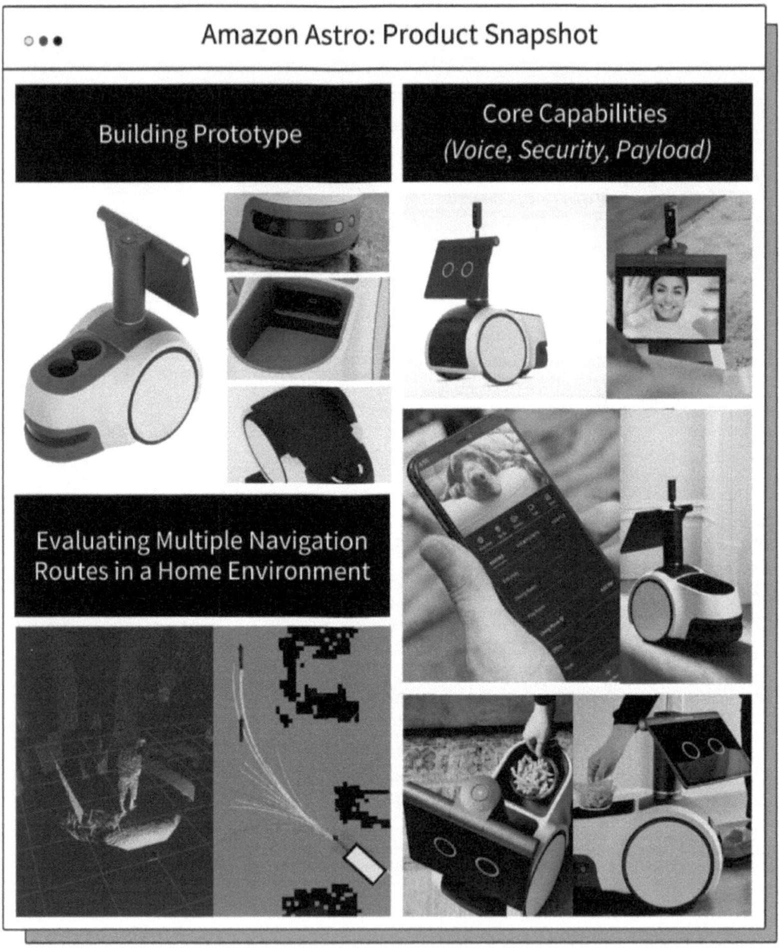

Figure 15.7 — Astro: Prototype & Capabilities

Illustration by the author. Robot images from the Amazon Astro product page.

Amazon's Business Model

Amazon's Business Model (Astro)

- **Product:** Astro, an autonomous mobile home security robot developed by Amazon and positioned as a mobile extension of the Amazon Alexa voice-powered home assistant ecosystem.
- **Funding:** Developed within Amazon's Devices and Services division, leveraging Amazon's multi-billion-dollar annual R&D budget.
- **GTM:** Invite-only launch in 2021 followed by a limited U.S. direct-to-consumer rollout, positioned primarily as a mobile home monitoring device integrated with Amazon's smart home services, including Ring.
- **Product Specifications:** Weighs approximately 20 pounds, features a 10-inch display, and includes a 5-megapixel periscope camera extendable to approximately 42 inches, designed exclusively for home monitoring.
- **Growth lever:** Ecosystem-led adoption strategy anchored in Amazon's broader smart home ecosystem. Astro integrates a wide range of capabilities, including autonomous navigation, telepresence, Alexa-enabled voice assistance, and direct integration with the Ring home security system, all unified within a mobile home monitoring platform.
- **Pricing Point:** Launched at an introductory price of $999.99 in 2021, later transitioning to a standard retail price point of $1,449.99.
- **Business variant:** Astro for Business was introduced at $2,349.99 and was discontinued around September 25, 2024.
- **Revenue Model:** Hardware plus subscription. Device revenue driven by unit sales of Astro, supplemented by recurring subscription revenue tied to home monitoring features enabled through integration with Amazon's Ring security ecosystem.
- **Exit:** The program was discontinued amid limited consumer traction and growing privacy concerns surrounding a mobile, AI-enabled device capable of continuously monitoring one of the most personal and cherished spaces for any family, their home and its members.

Figure 15.8 — Amazon's Business Model

Illustration by the author.

Chapter 16: Delivery Robots — Core Vectors Driving Adoption

Delivery robots such as those designed by Starship Technologies, Serve Robotics, Kiwibot, Coco, and Vayu Robotics have been around since 2014. Over the last seven years, their demand has skyrocketed in our society, driven by three core vectors — booming e-commerce, mounting last-mile delivery costs, and a shrinking labor workforce. This demand is organic and reflected in the rate at which cities around the world are transitioning to a robot-first infrastructure model.

From Los Angeles to London, and Helsinki to Tokyo, delivery robots are being equally embraced by major metros of the world. Clear regulatory guidelines for pilots and large-scale deployments, along with strong demand for rapid doorstep delivery service, are hallmarks of cities that have transitioned to this model. This demand is also reflected in the number of deliveries completed by these robots to date.

Starship alone has completed over 9 million deliveries, Coco 500,000+, Kiwibot 200,000+, and Serve 100,000+.[212-215] These numbers are surging exponentially each quarter, with companies like Vayu building robots designed exclusively to operate on-road, not just sidewalks to scale these deliveries.

With demand surging across verticals ranging from delivery pioneers such as Uber, DoorDash, Grubhub, and Grab, to logistics giants like Swiss Post, Fedex, and DHL, this sector is primed for exponential growth in the coming years. Given the demand and the scale of adoption, it is imperative to study the trajectory of leading startups in this arena, decode the challenges they are confronting, understand how they are addressing those challenges, and identify the corresponding opportunities that lie ahead for the next generation of companies and engineers in this space.

To better understand the trajectory of these leading delivery robot startups and the lessons they offer, it helps to first understand the three core vectors leading this shift in our economy.

Delivery Robots — Three Core Vectors Driving Adoption

U.S. Census data shows that e-commerce sales in the country are surging at roughly a 5.1% annual rate in 2025, and when you put that in perspective, that's pretty significant growth. Amazon alone ships 17.3 million packages each day.[216] DoorDash does 3 million deliveries per day.[217] That's 20.3 million deliveries per day between just Amazon and DoorDash, and these numbers are rising at an expansive rate each year as consumers continue to favor e-commerce channels over physical channels. While the demand for e-commerce services is on the rise, the corresponding workforce needed to sustain that demand is not maturing at the same pace, and the numbers shared by the logistics industry serve as a strong proof point here.

The U.S. trucking industry has reported that it is in dire need of at least 80,000 additional truck drivers.[218] That number is only expected to climb as demand for e-commerce services continues to rise. To put that shortfall in context, it helps to understand that Amazon alone ships more than 6 billion packages each year, and each truck can carry only a few thousand at a time. With an 80,000-driver shortage in the U.S. in 2025, combined with accelerating delivery volumes, labor is emerging as one of the biggest constraints in logistics and the last-mile delivery sector. This has direct implications for the shipping and delivery costs that both companies and consumers ultimately have to bear.

The World Economic Forum reports that last-mile delivery costs rose from 41% of total shipping cost in 2018 to 53% in 2023, meaning that over half of the total shipping cost for every item we purchase online sits in the final leg of its journey from warehouse to our doorstep. Accenture also affirms that last-mile delivery accounts for 53% of total shipping cost, reinforcing that this is the "most expensive leg" of the supply chain network that powers today's e-commerce brands.[219] Optimizing these costs offers the biggest ROI for startups working in the delivery robot space.

Those three core drivers indeed raise an important question: if delivery robots are truly the most optimal solution to address these persistent labor shortages in the last-mile delivery vertical, then why are we not seeing them zipping across our cities on each street and in every neighborhood, much like how we see Uber and Lyft today? Why haven't they attained that level of adoption yet?

That's the question we will try to address in this chapter by examining the path of robotics startups leading this arena, specifically Starship Technologies, Serve Robotics, Kiwibot, and Coco Robotics. To begin with, it helps to first understand what these robots are capable of doing and the underlying challenges these startups share.

Delivery Robots: Capabilities & Challenges

A delivery robot is essentially an autonomous mobile platform capable of navigating public spaces to transport goods such as food, groceries, or parcels to consumers. These robots, as with any autonomous robots, are configured with sensors and core AI models to enable the standard autonomy stack — perception, localization, navigation, and motion control.

There are three key factors that differentiate these robots from traditional autonomous robots.

First, they have to operate in public places such as cities or university campuses, and typically use sidewalks, or in some cases roads, to enable these deliveries. Second, today these robots are largely scoped only to address last-mile delivery needs, and they operate with Level 4 autonomy, unlike autonomous long-haul trucks that can carry freight across the country. Level 4 autonomy implies they can operate autonomously with very minimal human supervision, but only within geofenced, well-defined restricted zones inside cities.

Third is payload capacity. Given their core function is to enable delivery, they are often equipped with storage configurations that allow vendors to deliver dry, frozen, and perishable goods to end users with ease. While those three vectors summarize the capabilities of these delivery robots, their scale of adoption today is bottlenecked by three core factors: urban navigation, community trust, and regulatory variance.

Enabling autonomous navigation in unstructured environments such as cities is still a reasonably hard challenge, since each city comes with its own set of variables, including geographic layout, pedestrian density, sidewalk configuration, demographics, and more. Even when these startups get the autonomous navigation stack right and address that bottleneck, the next major challenge is building trust in the community. It takes time for humans to get used to seeing and interacting with robots. Now, even before these startups can address those two challenges, they must pass the regulatory test. Each country, and even each city, has a diverse set of regulations that govern the ability of these startups to deploy these robots on public streets.

Despite those three core bottlenecks, last-mile delivery startups have garnered significant support from communities around the world by establishing strong partnerships with food delivery companies such as DoorDash and Uber Eats to drive adoption of these services at scale. Put simply, if as a startup you can prove the demand is real and unmet, regulatory bodies will often find a way to work with you to address that demand in their metro. This demand-supply dynamic has led to the success of several major last-mile delivery robot startups today. We will now cover each of these companies one by one and decode their trajectories, along with the corresponding business model and lessons, in the following chapters.

Before we proceed to the next chapter, it is important to distill the shared business model parameters that apply across all of them. In the following chapter, we will examine the elements of their business models that are unique to each company.

Delivery Robotics Startup Business Model — Shared Core Parameters

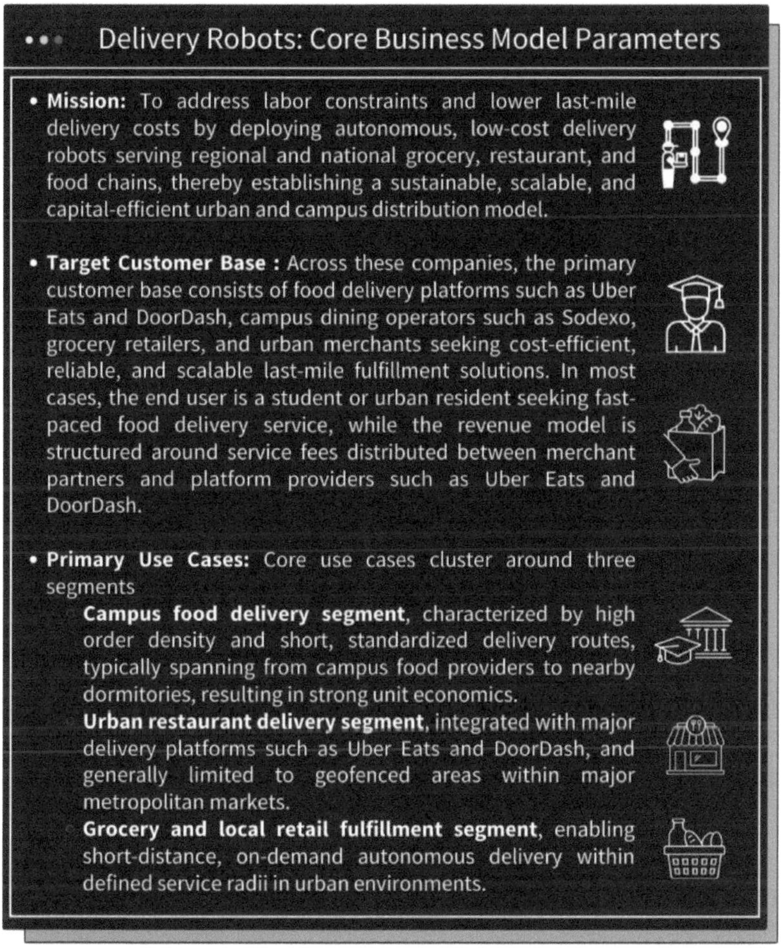

Figure 16.1 — Delivery Robotics Startup Business Model: Shared Core Parameters

Illustration by the author.

Chapter 17: Starship — Estonia to 60+ U.S. Campuses with Sodexo

Starship Technologies was incubated in June 2014 in Tallinn, Estonia, by Skype co-founders Ahti Heinla and Janus Friis with the goal of automating last-mile delivery services using autonomous robots. Heinla's interest in robotics traces back to his participation with Friis in the NASA Centennial Challenge, where they designed autonomous robotic systems for planetary applications. While they didn't win that competition at the time, they recognized the potential of their solution and decided to apply it to address the challenges inherent in the last-mile delivery space. Following a brief meeting in London later that month, they started working on this mission and launched Starship Technologies in July 2014. The company built its first prototype, an autonomous sidewalk delivery robot, later that year and kickstarted pilot programs in 2016 in cities such as London and Redwood City, California, in the United States.[220]

Pilot Programs & Commercial Rollout (2016–2019)

These early-stage pilots were supported by industry and logistics partners, including Just Eat in the UK, Swiss Post and Hermes in European postal and delivery trials, and Pronto, a London-based food delivery service. During this time, Starship also worked closely with city governments in Redwood City and Washington, D.C., securing regulatory approval to operate autonomous delivery robots safely on public sidewalks.[221-222]

Given the lessons this team learned from these early pilots, Starship now had a standardized playbook to unblock bottlenecks and scale its adoption base. In 2017, it ramped up its commercialization efforts and launched robot delivery services across multiple cities in the United States and the United Kingdom. The company became one of the first startups to deploy autonomous delivery robots in public spaces in these two countries.

Continuing this momentum, in April 2018, Starship inked a major deal with Co-op and Tesco to enable a city-wide rollout of its autonomous delivery service in Milton Keynes, UK.[223] This deployment helped the company not only extend its financial runway but also accelerate its adoption footprint. By late 2018, Starship had tested and deployed fleets across more than 100 cities and 20 countries, logging tens of thousands of miles even in these early commercial runs. Given this success in major metros, the company decided to deepen its roots within these cities by expanding its services onto university campuses.

University Programs & Partnerships (2019–2022)

On the majority of these university campuses, food service and facilities management providers like Sodexo play a central role in students' daily lives. For example, Sodexo is one of the largest food service companies in the world and manages dining operations at hundreds of colleges and universities across North America, where it delivers resident dining, retail food programs, and even delivery options tailored to the needs of students studying on these campuses.

In January 2019, Starship partnered with Sodexo to launch robot food delivery at George Mason University in Virginia with a fleet of 25 robots, then the largest campus rollout of its kind. It later expanded to six additional U.S. universities, including Purdue, the University of Pittsburgh, and the University of Wisconsin–Madison. Continuing this momentum in 2020, Starship expanded its university campus delivery footprint by adding services at the University of Mississippi and Bowling Green State University. By 2023, Starship robots were already delivering food and groceries on more than 60 U.S. university campuses. This strong footprint, established during its formative years across both university campuses and major metros, meant Starship was now scaling at galactic speed.[224-226]

Operational Scale and Performance Metrics (2023–2025)

By October 2024, Starship's fleet had grown to 2,000 robots globally, completing more than 7 million autonomous deliveries and traveling over 8.5 million autonomous miles. Given this scale, the team had exhausted its financial runway, and that meant it was time to raise additional capital. In 2025, Starship announced a $50 million Series C funding round, bringing its total funding to approximately $280 million.[212] This investment supported continued global expansion, with over 2,700 robots operating across 270+ geographic zones spanning seven countries. By this stage, the company had surpassed 9 million deliveries, and its fleet of robots had autonomously traveled 12 million miles. Starship has also stated plans to scale its fleet to roughly 12,000 robots by 2027.

Given this scale and pace, a natural question that may arise here is what was the exact business and revenue model this company adopted, and who were the key partners that helped build this adoption curve. To answer that question, we should first take a step back and understand how orders are placed into this system by vendors and fulfilled by these robots.

Starship's Business Model

Starship's robots deliver groceries, meals, and small packages for retail partners such as Co-op and Tesco, as well as delivery platforms worldwide including Grubhub, Wolt, Bolt, and Uber Eats. Customers place orders either through the Starship app or through partner platforms such as Uber Eats. These orders are processed by the partner first, after which an alert is sent to Starship's robots. The robot then routes to the partner's site, collects the package, and heads to the customer's location to complete the delivery.

These robots travel at speeds of about 6 km/h, carry no more than 20 pounds (9 kg) of payload, and operate under a Level 4 autonomy model within designated operational zones in each city or campus. Today, Starship robots are already integrated with major delivery platforms, retailers, and university campuses across the U.S. and UK. This operational scale naturally leads to an important question: what business and pricing model enables Starship to sustain these services?

Starship operates a hybrid B2C and B2B last-mile delivery model. On the consumer side, it charges users a flat fee of roughly $1.99–$2 per delivery for food and groceries in the U.S., and around £1 in the U.K.[227] On the business side, Starship also partners with retailers, campuses, and food service providers to integrate robot delivery into their operations through custom revenue models tailored to each partner. Together, this hybrid model allows Starship to scale its autonomous, on-demand delivery service across cities and universities.

While the revenue model is now clear, the next question that may arise here is the capital expenditures (capex) model: how much does it cost today to build these robots? While costs vary by robot generation, reports suggest that as of 2026, the cost of building each robot amounts to about $4,000, with the majority of that driven by the sensors and compute power each robot requires.[228] With this capex structure, strong partnerships, and its hybrid revenue model, Starship has established a strong foothold across the U.S. and Europe, completing millions of deliveries and logging over 11 million autonomous miles.

Given this trajectory, what are some of the core lessons we can decode for the next generation of founders and engineers? If we zoom out and analyze Starship's journey, there are three core pointers that are highly specific to startups and worth understanding here.

Core Lessons from Starship's Journey

Starship's success is governed as much by its underlying technology as by the strategic decisions made by its founders. The decision to operate in geofenced zones, whether in cities or on university campuses, was especially critical. This approach greatly limited the edge cases the company had to address early on and helped minimize operational burnout. Second was its robust chain of distribution partners. From Uber Eats to Grubhub, Bolt, and Sodexo, Starship built an extensive network of partners across both cities and campuses, enabling the adoption of its service at scale.

Third was the company's ability to establish a durable layer of trust with regulatory bodies and local communities from Day 01. Starship did so by proactively working with city governments and regulators, from London to Washington, D.C., ensuring that its solution was deployed in a way that not only aligned with local laws but also factored in the expectations of the community in each of these metros. This win-win approach is what ultimately positioned Starship as a last-mile delivery pioneer across North America and the UK.

Starship's trajectory and adoption footprint inspired technologists around the world to address emerging challenges in the last-mile delivery space, leading to the launch of another delivery robotics startup in 2017, which we will cover in the next chapter.

Now that we have distilled the core lessons embedded in Starship's trajectory, from its origins through its evolution and the pivotal decisions the team made along the way, it is time to decode the underlying business model. In the final section of this chapter, we will break down the business model and illustrate how the product portfolio maps to select primary use cases.

Starship: Product Snapshot — System Overview & Market Traction

Figure 17.1 — Starship Technologies: System Overview & Market Traction

Illustration by the author. Robot images from the Starship Tech. product page.

Starship Technologies' Business Model

Starship Technologies' Business Model

- **Products:** Autonomous delivery robots for campus and urban markets.
- **Model:** : VC-backed autonomous delivery network operator, scaled through long-term partnerships with universities and delivery platforms.
- **Funding:** $280M+ raised across multiple rounds led by a series of investors including, but not limited to, Goodyear Ventures, TDK Ventures, Shasta Ventures, Matrix Partners, and the European Investment Bank.
- **GTM:** Partner-led deployments across 60+ U.S. university campuses and 30+ European cities, with continued expansion into U.S. urban markets.
- **Customers & Partners:**
  - **Pilots:** Just Eat in the UK, Swiss Post, and Hermes in Europe during early-stage deployments from 2016 to 2019
  - **University Partnerships:** George Mason University, University of Wisconsin–Madison, University of Pittsburgh, Purdue University, and the University of Mississippi
  - **Campus dining operator:** Sodexo enabling scaled campus deployments
  - **Delivery platforms:** Grubhub (US), Uber Eats (US), Bolt (EU)
- **Market Traction:** 9M+ autonomous deliveries, 2,700+ robots deployed, with plans to scale to 12,000+ robots by 2027.
- **Growth lever:** Concentrated expansion within dense urban and campus clusters, where high order density and short, standardized routes drive superior unit economics, complemented by enterprise partnerships with universities, grocers, and delivery platforms.
- **Revenue Model:** Robot-delivery-as-a-service delivered through B2B and B2B2C contracts with campus dining operators, retailers, and delivery platforms, monetized through per-delivery service fees distributed between merchants and platform providers.
- **Price Point:** ~$1.99–$2.00 per delivery (U.S.) and ~£1 per delivery (UK).
- **Status:** Active and scaling as of 2026.

Figure 17.2 — Starship Technologies' Business Model

Illustration by the author.

Chapter 18: Serve Robotics — Postmates Lab to NASDAQ with Uber Eats

In 2011, a group of three young engineers in San Francisco decided to revisit the mission of Webvan, a delivery startup from the dot-com era that had failed to gain traction in the early 2000s. Webvan had attempted to address the last-mile delivery challenge for consumers, relying on warehouses as central inventory points for groceries and other household items, supported by a fleet of trucks responsible for enabling deliveries to customers.

While Webvan failed to gain traction and may have been a bit ahead of its time, these young entrepreneurs in 2011 decided to solve the same challenge and launched what is now known as Postmates. Postmates offers a last-mile delivery service similar to Grubhub and DoorDash, enabling consumers to order a wide range of products—from groceries to home care items—through the company's app. Orders are then processed and delivered through its network of delivery partners.

The operating model is similar to that of Uber, except here the partners deliver goods instead of ferrying customers from one end of the city to another. Postmates, much like Google, also established Postmates X, an internal R&D division, to address emerging challenges in the sector through cutting-edge technologies such as robotics and artificial intelligence. Around 2017, one of the delivery robotics projects developed within these labs matured to the point where the company decided to spin it off as an independent subsidiary of Postmates. It is this subsidiary that is now known as Serve Robotics.[229]

Serve is on a mission to offer autonomous delivery services to consumers in the United States, powered by its fleet of robots that are deeply integrated with major on-demand delivery platforms. Around 2020, while Postmates was acquired by Uber, Serve Robotics branched out from Postmates and started operating as an independent entity.[230]

A few months later, in March 2021, the company raised funds to build its fleet of delivery robots in a seed round led by Neo, with participation from strategic investors such as Uber, among others. This capital runway proved vital for the company's next phase, enabling commercial rollout.

Commercial Deployments & Partnerships (2021)

In 2021, Serve established a strategic partnership with Uber Eats, enabling the rollout of its delivery service powered by autonomous robots in cities such as Los Angeles, Miami, Atlanta, and Chicago. These early deployments positioned Serve as a strong delivery partner across major metros for Uber, driving increased demand for its service.[231] To meet this demand, the company needed to scale its fleet of robots and raise additional capital. The key question was whether to pursue a traditional Nasdaq IPO or raise another private round. Ultimately, Serve chose a more unconventional yet strategic approach.

De-Risking the Path to Public Markets: From OTCQB IPO to Nasdaq Uplisting

While software companies typically progress through Series A, B, and C funding rounds before eventually accessing major public capital markets such as Nasdaq, robotics startups—and hardware companies more broadly—face a different set of risk vectors that often demand a more strategic and risk-averse route. Hardware-intensive businesses operate under significantly higher capital expenditure (capex) and operating cost (opex) structures than their software counterparts. As a result, pursuing a direct Nasdaq IPO can become a far more complex and demanding task for these young startups.

A traditional Nasdaq listing requires companies to meet stringent regulatory standards ranging from governance and audit readiness to underwriting commitments, pricing thresholds, and investor relations. In addition, the process is highly capital-intensive, as startups must partner with investment banking firms to enable the transition to an IPO and incorporate SEC-compliant reporting protocols, including SOX requirements, into their financial and operational infrastructure. This process can erode the very financial runway the company intends to build. Moreover, entering major public markets at an early stage also leads to heightened exposure to market volatility. For early-stage robotics startups, these requirements can impose significant financial and operational pressure, potentially diverting the focus of these teams away from the company's core mission.

By contrast, markets such as the OTCQB Venture Market provide a lower-threshold entry point into public markets, offering a more gradual and agile pathway for emerging technology firms. This staged approach enables companies to build credibility, establish a public trading history, and strengthen internal controls before transitioning to a major exchange. Serve Robotics followed this de-risked route by first raising capital through an OTCQB debut and subsequently uplisting to Nasdaq once it had achieved greater operational maturity and market traction.[232]

By October 2025, Serve reported over 100,000 deliveries conducted through its robotic fleet of around 2,000+ units in partnership with more than 2,500 merchants across the U.S., an important milestone underscoring its progress toward scaled commercial deployment. Today, it operates one of the largest sidewalk delivery robot fleets in the country and is now focused on scaling its partnerships beyond Uber.

Partnerships & Platform Evolution

In October 2025, Serve announced a strategic deployment partnership with DoorDash, enabling its robots to fulfill DoorDash orders across additional U.S. markets. This expansion allows Serve to scale its presence well beyond its prime deployment cities, including Los Angeles, Miami, Atlanta, and Chicago.[233]

Today, Serve continues to advance its robotics platform, specifically focusing on increasing payload capacity, extending range, and strengthening its autonomy stack. Its third-generation robot model can carry larger items such as flat pizzas, travel at speeds aligned with the needs of today's fast-paced metros, and operate with extended battery life, enabling 24/7 operations.

These upgrades enhance Serve's ability to support more merchants, handle larger orders, and serve longer delivery routes across dense urban environments.

Given this trajectory, what are the core lessons we can decode for the next generation of founders and engineers? If we zoom out and analyze Serve's journey, there are at least two core lessons that every founder should take seriously.

Core Lessons from Serve Robotics' Journey

Serve Robotics' evolution from an internal robotics initiative at Postmates to one of the largest autonomous sidewalk delivery robot operators in the U.S. offers founders and engineers at least two high-impact lessons on building and scaling deep tech startups.

Lesson 01: Adopt a De-Risked IPO Strategy — From OTCQB to Nasdaq

Serve understood the importance of raising capital in a proactive manner, given that it is a deep-tech robotics company, one of the most capital- and hardware-intensive categories on the planet. Instead of waiting for years to march toward an IPO, the company realized that, at some point, it would need to raise funds through public markets, since securing sustained private investment in the long run for a startup that is both capex and opex heavy can be a daunting task.

To minimize the friction that most startups face when heading toward an IPO, and to address bottlenecks along the way with a more proactive approach, Serve, rather than heading straight to Nasdaq or mainstream markets, first launched on the OTCQB

Venture Market in March 2024. It then uplisted to the Nasdaq Capital Market in April 2024 under the ticker SERV, with a $40 million public offering.

This was an important decision, as it not only helped Serve manage the operational and regulatory risks inherent on this path for a company at this stage entering public markets, but it also placed the company in front of consumers early, allowing them to become champions and investors. This win-win approach ensured that as Serve began launching its robots in new geographies, it had already started building a community of supporters in each of these regions post-IPO.

Lesson 02: Partnerships Are the Ultimate Distribution Engine

Partnerships were central to Serve's accelerated growth. From the outset Serve integrated directly with major delivery platforms, most notably Uber Eats, enabling real commercial deployment across multiple U.S. cities. In October 2025, Serve expanded beyond that ecosystem through a strategic deployment partnership with DoorDash, allowing its robots to fulfill orders across additional markets. This partnership-first model enabled Serve to scale faster than competitors focused on implementing small-scale pilots or isolated geographic rollouts, as demand and merchant networks were already embedded within the platforms.

Today, Serve Robotics represents one of the most successful U.S. autonomous delivery robotics startups scaling across American sidewalks, establishing deep partnerships with major delivery platforms such as Uber Eats and DoorDash, alongside a strong network of more than 2,500 merchants across the country. While

Serve's journey offers important lessons on capital structure and partnerships, another delivery robotics startup that emerged around 2017 provides equally strong lessons, particularly on go-to-market strategy.

Before we examine the trajectory of that startup in detail, now that we have distilled the core lessons embedded in Serve Robotics' trajectory, from its origins through its evolution and the pivotal decisions the team made along the way, it is time to decode the underlying business model.

In the final section of this chapter, we will break down the business model and illustrate how the product portfolio maps to select primary use cases.

Serve Robotics: Product Snapshot — System Overview & Market Traction

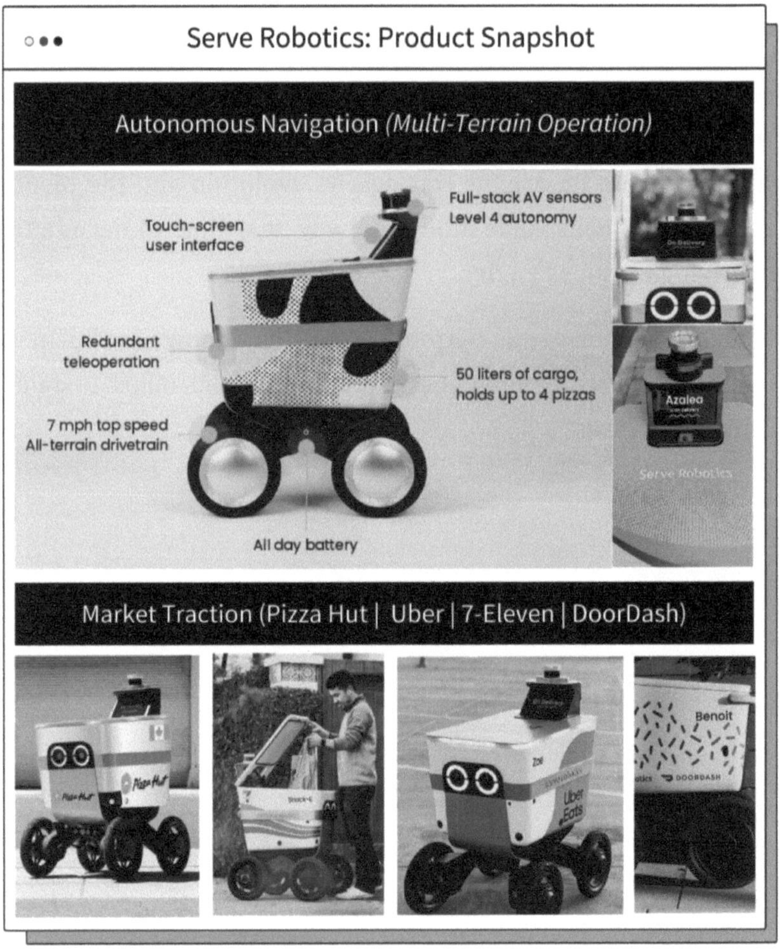

Figure 18.1 — Serve Robotics: System Overview & Market Traction

Illustration by the author. Robot images from the Serve Robotics product page.

Serve Robotics' Business Model

Serve Robotics' Business Model

- **Products:** Autonomous sidewalk delivery robots for restaurants and convenience stores
- **Model:** Publicly listed autonomous delivery network operator, scaled primarily through long-term platform partnerships, most notably with Uber Eats and DoorDash
- **Funding:** The company raised approximately $220 million following its spin-out from Uber in 2021. This was followed by a reverse merger that led to a public listing valuing the company at approximately $40M on the OTCQB Venture Market, and subsequently an uplisting to NASDAQ.
- **GTM:** Platform-led rollout across Los Angeles, Miami, Atlanta, and Chicago, with expansion plans tied to major metropolitan markets served by Uber Eats and DoorDash
- **Customers & Partners:**
  Delivery Platform: Uber Eats & DoorDash as primary distribution partners under multi-year agreements
  Restaurant Partners: Integrated with 2,500+ restaurant merchants operating on the Uber Eats platform
  Convenience Store Partner: 7-Eleven partnership announced in November 2025 to integrate the company's delivery robot fleet across 13,000 stores spanning the U.S. and Canada.
- **Market Traction:** Over 2,000 robots deployed across the U.S.
- **Growth lever:** Two core platform partnerships, Uber Eats and DoorDash, unlocking access to thousands of merchants nationwide at scale
- **Revenue Model:** Robot-delivery-as-a-service delivered through multi-year contracts with platform partners, monetized via a per-delivery service fee model.
- **Price Point:** Targeting approximately $1 per mile per delivery under an emerging customized pricing model developed for B2B partners
- **Status:** Active and scaling as of 2026

Figure 18.2 — Serve Robotics' Business Model

Illustration by the author.

Chapter 19: Kiwibot — Bogotá to UC Berkeley with SkyDeck

Kiwi Campus, widely recognized for its autonomous delivery robots (i.e., Kiwibots), started as an experiment about 3,800 miles from Silicon Valley. The company was incubated in 2015 by founder Felipe Chávez in his native city of Bogotá, Colombia, to scale last-mile deliveries for vendors in the city who were struggling with the surging demand for their goods and rising labor costs.[234] Felipe wanted to bridge this gap and started by offering them a courier service led by a team of trained delivery workers at an affordable pricing point. While that approach served as a foundational solution for these vendors, it was not scalable as labor costs were on the rise.

Even though Felipe and his team tried to offset these costs by bundling orders within specific zip codes and assigning them to delivery workers dedicated to each zone to reduce delivery time and expense, the model ultimately remained constrained by the same underlying challenge: high labor-driven operating costs.

Transition to Autonomous Delivery Robots (2017)

To scale this model in a cost-efficient manner, Felipe decided to augment this delivery team with robots. To enable this transition and raise the corresponding funds needed to build a fleet of robots, Felipe, in January 2017, joined UC Berkeley's SkyDeck accelerator program and launched the first fleet on the UC Berkeley campus.[234] This transition proved vital, as the campus served as fertile ground to validate Felipe's hypothesis of offering autonomous delivery service in designated zones, rather than in open city environments marked by an array of navigation bottlenecks—pedestrian and vehicle flow, regulations, construction zones, and more.

Another added benefit of this approach was the fact that students are more open to exploring and adopting new technology, and in Kiwibot's case, they helped build an organic layer of trust at scale in the company's mission from Day 01. With this strategy, the company was able to launch its robot-led delivery service across campuses in the U.S. in the coming years. These robots are powered by the open-source Robot Operating System (ROS) and equipped with NVIDIA's Jetson Xavier chip, providing them with the compute horsepower and intelligence needed to navigate these campus environments.[235]

That raises an important question: what does Kiwibot's end-to-end operational model look like? Campuses are also filled with edge cases that these robots must address—so how does the team handle these anomalies?

Campus-First Approach — A Three-Layered Operations Model

Today, the operational model adopted by the company is notably hybrid and involves three core layers: the vendor collection layer, where robots are staged near centralized parking lots by delivery workers to pick up orders from vendors within each geographic zone; the last-mile delivery layer, where robots are repositioned at a parking lot closer to customer destinations to complete deliveries within that zone; and the hub reset layer, where robots are returned to Kiwibot's central distribution hub to restart this cycle each hour.

Here is how it works in practice.

The first layer is the vendor collection layer, where a team of delivery workers bundles the orders received in each geographic zone and maps them to the corresponding robots needed to deliver those orders to customers in that zone. Each member of this team manually takes a fleet of robots and deploys them near a centrally located parking lot within the zone, ensuring access to all major vendors associated with the orders received from customers at that point in time. The robots then navigate autonomously to these vendors, collect the orders, and return back to this parking lot.

From there, that same delivery worker relocates the robots closer to another central point that provides the best access to the delivery destinations within the zone. This is where the last-mile delivery layer kicks in. The robots then head to their respective drop-off locations to complete the last-mile deliveries and return back to this vantage point, i.e., the parking lot. Once the cycle is complete, they are picked up and brought back to Kiwi's central distribution hub by the same delivery worker operating in that zone. This process repeats every hour.

Today, these robots operate in a semi-autonomous mode and are actively monitored by a remote supervision team based in Bogotá, nearly 3,800 miles away from the U.S. campuses they serve.[236] With only a two-hour time difference between Bogotá and major North American delivery markets, this structure allows the company to maintain continuous operational oversight during active campus hours while offering customers a safe and cost-efficient delivery service.

A key advantage of this model is its ability to dramatically reduce labor costs—by nearly 70%. The minimum wage in Colombia is roughly $2, significantly lower than in the United States, and because each remote operator can supervise about three robots simultaneously, the system becomes both scalable and economically sustainable.[236] At the same time, it provides meaningful opportunities for these operators, enabling them to familiarize themselves with robotic systems and their underlying principles, while also fostering a passion for STEM careers—all while being paid for their work.

Scaling Beyond Berkeley — B2B Growth and City Pilots (2018–2021)

By 2018, Kiwibot had already demonstrated measurable traction at UC Berkeley. In May of that year, the university reported that the company had successfully completed its 10,000th delivery. Students were able to order from both national chains and local merchants, including Jamba Juice, Chipotle, La Burrita, and Cheese n' Stuff, with a delivery fee of around $3.80 and an average delivery time of approximately 27 minutes.[237]

This early-stage validation of its operating model, and the value it created for students, universities, and corresponding vendors, implied that the company was now ready to scale its adoption footprint. Demand for its service had been validated, and it was time to expand beyond a single campus into other universities across North America and eventually into cities. These early partnerships with national chains helped Kiwibot build trust and credibility, setting the stage for the next phase of its journey.

With this mission in mind, the company started pursuing B2B partnerships at scale with a range of fast-food partners across cities, including but not limited to Jamba Juice, Chipotle, La Burrita, Chick-fil-A, as well as platform partners like Shopify and Ordermark. By April 2021, Kiwibot had built roughly 400 robots and completed approximately 150,000 deliveries.[240] San Jose emerged as a key urban expansion market and was described as the first city integrated at scale into Kiwibot's delivery model. During this period, Kiwibot also matured its manufacturing ecosystem, naming Segway as a production partner for its Kiwibot 4.0 generation.[238]

Now, with its manufacturing engine ready to scale and its operations model matured, it was time for the company to expand across university campuses in North America. To do so at scale, in 2021, the company partnered with Sodexo, a food service provider responsible for delivering dining services across universities on the continent.

Institutional Scaling Through Sodexo & Pricing Model (2021–2022)

This collaboration with Sodexo marked a major step in Kiwibot's journey, enabling a series of campus-wide deployments across multiple universities, including New Mexico State University, Loyola Marymount University, and Gonzaga University. As part of this collaboration model, students on these campuses were able to order and pay for food using campus meal plans through Sodexo's Bite+ app, minimizing the need for them to download Kiwi's own app or other custom platforms. This also offered a relatively straightforward pricing structure composed of a flat fee per order plus a small percentage of the order value. This model generated strong revenue for the company within the first few months of launch, and within a year, in February 2022, the company reported a $20 million contract expansion with Sodexo, with ambitions to deploy more than 1,200 robots across 50 U.S. campuses.[239]

Partnerships and Campus Ecosystem Growth (2022–2023)

Building on this collaboration model, Kiwibot strengthened its campus distribution engine further through partnerships with major food delivery platforms. In December 2022, the Colombian robotics startup announced a strategic partnership with Grubhub, enabling students to order robot deliveries directly through the app they were already using, thus minimizing adoption friction at scale. By then, Kiwibot had completed over 200,000 deliveries. With Grubhub and Sodexo onboard, adoption exploded across campuses.

In August 2023, it secured a long-term contract with the University of Southern Indiana to launch a fleet of 15 robots on that campus alone, in partnership with Sodexo dining services and at a flat $3 delivery fee per order.[241] While this partnership-led model helped the company scale its footprint, it also highlighted a key constraint: the current revenue model was heavily dependent on extremely thin margins per campus. Extending its runway would require new monetization levers. The team decided to tackle this challenge head-on by launching its second revenue arm—advertisements.

Advertising as a Second Revenue Stream (2024)

In 2024, Kiwibot expanded beyond enabling autonomous delivery operations into an advertising-first model, adding a second revenue channel to complement its logistics business. That year, TechCrunch reported that the company acquired ad-tech firm Nickelytics for $25 million, effectively transforming its robots into mobile billboards operating across these campus environments. Nickelytics' customer base included major brands such as AWS, Coinbase, Nationwide, and DirecTV, giving these high-profile advertisers direct access to more than 1,000+ mobile Kiwibots. As part of the acquisition, Nickelytics CEO Judah Longgrear joined Kiwibot as a co-founder.[242] This move positioned Kiwibot as one of the few delivery robotics companies pursuing a strategic and sustainable dual-engine revenue model, one rarely explored by competitors operating in this space.

Core Lessons from Kiwibot's Journey

Today, Kiwi robots are ferrying food and raising brand awareness for major national partners across university campuses in North

America, supported by a growing fleet built and operated by a mission-driven team. The path the company has traveled—from its roots in Bogotá to major markets across the United States—along with the relationships it has built with campus partners such as Sodexo, food delivery giants like Grubhub, platform partners like Shopify and Ordermark, and advertising partners such as AWS, Coinbase, DirecTV, and more, reflects the dedication and planning that have gone into scaling this business. Given the novel approach this team took to organically scale its footprint and revenue streams in a sustainable manner, there are eye-opening lessons embedded in this journey for the next generation of robotics founders and engineers that cannot be overlooked. We will now cover each of them one by one.

Lesson 01: Dual Revenue Streams Enable Sustainable Runway

One of the most important lessons from Kiwibot's journey is that last-mile delivery startups cannot rely on delivery margins alone. Campus delivery is an operationally resource-intensive service and is often constrained by thin per-order profit margins. This economic reality makes it difficult to scale and sustain the business using this logistics revenue channel as the sole revenue arm. Kiwibot addressed this challenge by building a dual-engine revenue model, pairing delivery operations with an advertisement-based monetization channel through its acquisition of the ad-tech company Nickelytics. Thus, by enabling its robots to serve as both mobile billboards and delivery channels, the company unlocked an additional revenue stream that strengthened its financial runway and minimized its dependency on delivery fees alone.

Lesson 02: Partnerships Across the Stack Unlock Scale

Kiwibot scaled by embedding its robots directly into systems that were already adopted by consumers, instead of trying to build an end-to-end logistics platform from scratch. This scale was enabled through the cross-vertical partnerships the company landed strategically from the very beginning. Sodexo helped them roll out across campuses, Grubhub and other platforms put robot delivery inside apps students were already using, and Segway helped them manufacture at scale. These partnerships became the real growth engine, allowing Kiwibot to expand its footprint much faster than it could have in isolation.

Lesson 03: Cross-Border Operations Enable Scale—and a Win-Win Workforce Model

Kiwibot scaled its operations in a sustainable manner by deploying robots on U.S. campuses while enabling remote supervision from Bogotá. This cross-border operations model ensured the company could offer a safe and reliable service to consumers while dramatically reducing labor costs. Thus, it not only allowed Kiwibot to retain sustainable margins and provide affordable delivery services, but also created meaningful opportunities for remote operators to gain hands-on exposure to robotics systems and consider STEM-oriented career pathways. In this way, Kiwibot built an operations engine that was both economically scalable and socially responsible—a rare win-win in this sector.

With Kiwibot proving the model could work at scale, the opportunity for emerging startups in this space became both clear and compelling. One of the next major entrants was Coco Robotics. We will cover Coco's journey in the next chapter.

Now that we have distilled the core lessons embedded in Kiwibot's trajectory, from its origins through its evolution and the pivotal decisions the team made along the way, it is time to decode the underlying business model.

In the final section of this chapter, we will break down the business model and illustrate how the product portfolio maps to select primary use cases.

Kiwibot: Product Snapshot — System Overview & Partnerships

Figure 19.1 — Kiwibot: System Overview & Partnerships

Illustration by the author. Robot images from the Kiwibot product page.

Kiwibot's Business Model — Part 1 of 2

Kiwibot's Business Model

- **Products:** Autonomous sidewalk delivery robots enabling campus food delivery service
- **Model:** Campus-first autonomous delivery operator, scaled through strategic partnerships with institutional foodservice providers and delivery-platforms.
- **Funding:** As of 2026, the company has raised approximately $24M across multiple funding rounds from investors such as Sodexo, Headline, and Berkeley SkyDeck to support multi-campus deployments across the U.S.
- **GTM:** University-centric deployments executed through large campus dining operators such as Sodexo
- **Customers & Partners:**
  - **Campus Dining Operator:** Sodexo partnership to scale deployments across U.S. universities
  - **University Deployments:** Multi-campus rollout model targeting 50+ U.S. campuses
- **Market Traction:** Announced plans with Sodexo to deploy 1,200+ robots across 50 U.S. campuses. Its fleet of 500+ robots has completed over 250,000 cumulative deliveries till date.
- **Growth lever:** High order density within geo-fenced, standardized campus environments, combined with institutional foodservice partnerships that embed delivery robots directly into campus dining operations, establishing recurring revenue streams while improving unit economics.
- **Revenue Model:** Robot-delivery-as-a-service delivered through strategically secured long-term contracts with campus dining operators and institutional partners, leading to sustained revenue streams.

Figure 19.2 — Kiwibot's Business Model: Part 1

Illustration by the author.

Kiwibot's Business Model — Part 2 of 2

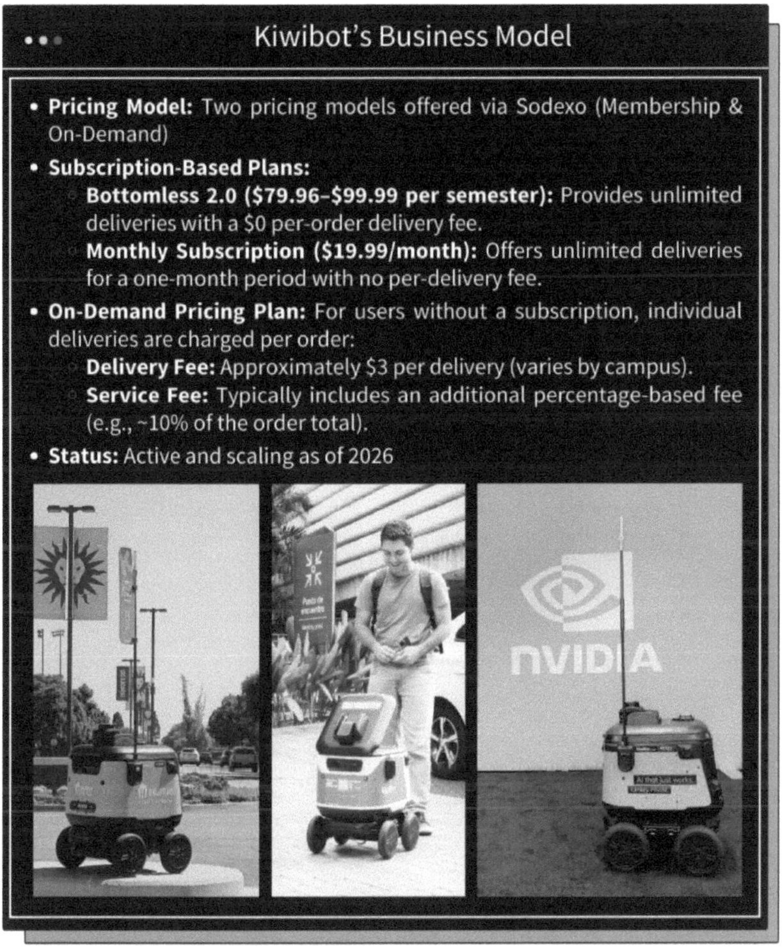

Figure 19.3 — Kiwibot's Business Model: Part 2

Illustration by the author.

Chapter 20: Coco Robotics — UCLA Dorms to Helsinki with DoorDash

Coco Robotics was incubated in early 2020 in Santa Monica, California, by UCLA CS graduate students Zach Rash and Brad Squicciarini, who, post-graduation, started exploring applications of robots in the logistics space. Last-mile delivery turned out to be the obvious choice, given that Santa Monica was becoming a booming tourist destination, with a series of local merchants offering a wide range of cuisines and artwork around the iconic pier located on the banks of the Pacific Ocean in Southern California.[243]

Given the immediate value these robots could offer to local merchants near this pier, the team went full throttle on delivery robots. Like many early-stage robotics ventures, Coco faced skepticism from investors and had to work hard to secure seed capital. After pitching to a series of investors, the team didn't give up and endured, finally raising a $36 million Series A round in August 2021, led by angel investors such as Sam Altman, Ellen Chen, and Mario Del Pero, as well as institutions such as Silicon Valley Bank.[244]

Deployments & Strategic Partnerships — Wolt, Uber Eats, DoorDash (2020–2025)

From there on, the team moved decisively, launching a pilot around the Santa Monica area with both local merchants and securing a strategic digital kitchen partner, C3, to offer robot-led deliveries within roughly two miles of several Southern California locations, including Santa Monica and West Hollywood. These early pilots, seeded in 2021, continued for the next four years and helped the company refine its operations model and the underlying technology stack to better align with traffic footprints, regulatory frameworks, and the taste buds of the local residents.[245]

By 2025, the company had established a strong operating model, a refined tech stack, and a proven partnership-driven revenue channel. Given this maturity, in April of that year, the founders made a decisive move and inked a long-term partnership with DoorDash to offer their autonomous delivery service in select U.S. cities such as Los Angeles and Chicago. This move proved vital, as it allowed local residents in these cities to order food directly from 600+ merchants through the DoorDash app, slashing delivery fees for both merchants and end users by around 70%. That year, the UCLA-born robotics startup officially completed over 14,000 deliveries using its fleet of robots for DoorDash customers. This early traction proved that demand was real and that Coco was capable of addressing it at scale. That proof served as a catalyst for its entry into international markets.[246]

Building on strong traction in the United States, DoorDash made a decisive move to scale this partnership internationally, bringing the service to residents of Helsinki, Finland, through its subsidiary Wolt. Wolt is a food delivery platform the company acquired in June 2022 to support international expansion. It offers services similar to DoorDash, but tailored to residents of Helsinki, a city best known for the Northern Lights and Nokia phones.

By late 2025, Coco robots had completed more than 500,000 deliveries, a figure that marked the company's first major adoption milestone. Today, it operates hundreds of robots in cities like Los Angeles, Chicago, Miami, and Helsinki. In 2026, the company announced plans to expand to 10,000 robots across these markets.

Coco's Funding and Engineering Model

To fund these plans, Coco secured $80 million in strategic financing—a Series B led by investors including Sam and Max Altman, SNR.vc, Pelion Venture Partners, Offline Ventures, DeepWater Asset Management, and others. The funding was intended to scale Coco's operations model, robot production capacity, and key partnerships. That same year, Coco also announced a strategic partnership with OpenAI, enabling Coco to leverage OpenAI's technology while OpenAI gains access to real-world data collected by the delivery robots to help improve its models. Together, these developments brought Coco's total funding to over $110 million since inception, fueling plans to expand its fleet toward 10,000 robots by 2026.[247]

Today, the engineering model adopted by Coco to design these robots follows more or less a similar path as that followed by its competitors. Each robot is suitcase-sized, travels around 5 mph in urban environments, and is powered by a suite of AI models, sensors, and cameras that enable it to perceive its surroundings, plan its route, and navigate to its destination. Each of these robots, on the backend, is integrated with APIs offered by leading food delivery operators such as DoorDash and Uber Eats.

While the engineering approach the team took to its robots might be similar to that of its competitors, what separates Coco from others is the pace at which the company has scaled its adoption footprint. From its inception around 2020 to less than five years in operation, the company has already managed to complete over 500,000 deliveries across two continents, in partnership with food-delivery giants such as DoorDash and Uber Eats, along with local merchants, and has secured a strong capital runway to sustain this trajectory. This trajectory not only symbolizes the novel model the company devised to scale in urban environments at a rapid pace, but also underscores the importance of making decisive moves each quarter, as a founder, to hit that escape velocity of the adoption orbit for your startup—and make an impact in the truest sense.

Now that we have distilled the core lessons embedded in Coco's trajectory, from its origins through its evolution and the pivotal decisions the team made along the way, it is time to decode the underlying business model.

In the final section of this chapter, we will break down the business model and illustrate how the product portfolio maps to select primary use cases.

Coco: Product Snapshot — System Overview & Partnerships

Figure 20.1 — Coco: System Overview & Partnerships

Illustration by the author. Robot images from the Coco Robotics product page.

Coco Robotics' Business Model — Part 1 of 2

Figure 20.2 — Coco Robotics' Business Model: Part 1

Illustration by the author.

Coco Robotics' Business Model — Part 2 of 2

Figure 20.3 — Coco Robotics' Business Model: Part 2

Illustration by the author.

Chapter 21: Dawn of Humanoid Robots

The humanoid robotics market sits under a few billion dollars today, but it is about to experience a period of serious growth. Barclays Research estimates that the market could reach nearly $200B by 2035, given the current trajectory of adoption across sectors.[248] That number should not come as a surprise. Declining birth rates, rising labor shortages, and an aging workforce are increasingly becoming binding constraints on global economic growth. Demand across critical industries continues to rise, but the supply of labor does not.

Healthcare provides one of the clearest examples of what this gap implies for our economy. The United States, in particular, is undergoing a major demographic shift: by 2036, the 65+ cohort will grow by 34%, while the 75+ cohort will grow by 55%, according to the Association of Mature American Citizens (AMAC). Yet as this population continues to grow, the capacity of our healthcare system, measured in terms of workforce strength, including the number of physicians and medical staff per hospital, is declining at an equally rapid pace. In 2025, the U.S. faced a shortage of 57,000 physicians, and that gap is projected to reach 81,000 by 2035.[249] Compounding this shortage, the existing workforce itself is aging rapidly. Over 30% of doctors are already 60+ today. More than 80,000 physicians are nearing retirement, and with training costs approaching $300,000 per doctor, this curve will not ease anytime soon.[250]

Healthcare is running into the same wall that manufacturing has been hitting for years: there simply are not enough workers to meet rising demand. As of 2026, more than 400,000 manufacturing jobs remained unfilled in the United States, and the sector will require 3.8 million new employees by 2033.[251] Across healthcare, hospitality, and manufacturing, the vacancy rate, defined in terms of jobs left unfilled per sector, now hovers around 8% and continues to rise year after year.

This need is already evident across the industries leading adoption: automotive manufacturing, chip production, healthcare, and space exploration. Automotive, in particular, appears to be at the leading edge of this adoption curve. From BYD to Tesla, BMW to Audi, and Hyundai to Mercedes, automotive plants are increasingly being swarmed by humanoid robots.

That raises an important question: why is the automotive industry so aggressively pursuing humanoid robots? What are the underlying forces driving this adoption? To answer that question, it helps to take a step back and examine the core pain points automakers face today.

If you've ever visited an automotive plant, you'll notice that the workforce requirements for these factories distill down to three core principles: mobility, efficiency, and reliability. You need workers who can navigate across factory floors often laden with parts and machines. You need them to carry heavy components and transfer them from one area of the factory to another. And, most importantly, this entire workflow, whether it involves manufacturing parts or transporting them from machines to conveyors or the inventory area, must remain both time- and resource-efficient.

The challenge is that this workforce is shrinking, and more importantly, this work itself is extremely straining for humans. That means the next generation is unlikely to enter this sector unless the work profile undergoes a step change. To meet this demand, we need a workforce that can deliver the level of precision, reliability, efficiency, and payload capacity that cannot reasonably be expected of people at scale.

These are precisely the qualities that make humanoids such an ideal fit.

You may ask: How?

If we take a step back and try to understand the anatomy of a humanoid, you'll notice that at a high level, it is equipped with two legs that allow it to navigate diverse indoor terrain profiles: staircases, uneven surfaces, and a wide range of obstacles, resulting in an extremely high level of terrain mobility. Then it is equipped with two highly dexterous arms, or what we refer to in the robotics world as spatial manipulators. They enable spatial agility. This means our humanoid can use these robotic arms to grab an apple or thread a needle through fabric with extremely high precision. These two features, combined with the ability to teach humanoids to perform a wide range of repetitive tasks using modern AI models, add up to remarkable efficiency. Hence, humanoids are becoming an increasingly compelling solution across automotive and the broader manufacturing domain.

That naturally leads to this question: Which of the manufacturing giants are adopting humanoids, and how?

Global Adoption of Humanoids

From chip manufacturers to automakers and beyond, humanoids are being adopted across nearly every major industry vertical. In fact, it is no surprise that NVIDIA is partnering with Foxconn to deploy humanoids in its chip manufacturing plants.[252] After all, who doesn't want to automate the task of fabricating about 3.6 million enterprise-grade GPUs per year? Beyond NVIDIA, nine automakers around the world are already embracing humanoids on their production lines. This includes BMW, Mercedes-Benz, Tesla, Hyundai, Audi, Volkswagen, Nio, and BYD.

Need proof? Try visiting BMW's Spartanburg plant in South Carolina, here in the U.S., where you will witness Figure AI's humanoids working on the production lines without taking a day off for months in a row. Over a period of just 11 months in 2025, Figure AI's humanoids contributed to the production of more than 30,000 BMW X3 vehicles.[253] Beyond that, the team behind the Optimus bot has deployed humanoids on the factory floors of Tesla. Optimus is currently being trained to automate assembly lines.

Similarly, Hyundai has adopted the Atlas humanoid developed by Boston Dynamics to automate its auto manufacturing plants.[254] In China, a local humanoid manufacturer, UBTECH, has secured partnerships to provide humanoid assistants to an array of automotive manufacturers including, but not limited to, BYD, Geely, FAW-Volkswagen, BAIC, and Foxconn.[255]

In Europe, Neura Robotics has launched its own humanoid to automate factories of Porsche and Mercedes.[256] Beyond automotive, the warehouse and logistics industry is also seeing a surge in the adoption of humanoids. Agility Robotics' Digit humanoid is being trained by Amazon to aid with logistics and last-mile deliveries. Trials are underway at Amazon's San Francisco training campus.[257]

Beyond manufacturing and warehouse teams, humanoids are equally being favored by household users as well. *Neo*, a home humanoid assistant being developed by the 1X team, is gaining popularity among those who dislike doing their laundry. For those concerned about affordability, Unitree has pushed the market forward by launching a humanoid priced under $5,000.

Given this adoption curve, some of you might ask: What about the production capacity? Isn't it extremely difficult to manufacture these systems at scale?

Production Capacity & Pricing

The manufacturing scale of humanoid robots is currently constrained primarily by the cost of one of their most expensive components: actuators. Nearly 50% of humanoid production costs today can be attributed to actuators, namely the motor-and-tendon systems that enable payload capacity and dexterity. While actuator costs remain a major driver of overall manufacturing expense, the adoption curve suggests that supply chains are maturing at a reasonable pace, and we should see a gradual decline in actuator pricing over time.

Thus, while concerns around the production capacity of humanoids are legitimate, this bottleneck is now being addressed at scale across the industry. Tesla, for instance, has confirmed a target production plan of approximately 100,000 units for the coming year. BYD, a Chinese humanoid manufacturer that started with a plan of about 1,500 units in 2025, is now targeting an annual production capacity of 20,000 units.[258-259]

While these targets may seem ambitious, make no mistake, they are well within reach. Need proof? There's no better example of production scale than Agility Robotics, a company that has launched a dedicated production facility for humanoids targeting an output of 10,000 units in its first year.[260]

This surge in production is also rapidly driving down the pricing of these robots. For instance, the Tesla Optimus, currently priced around $30,000, is expected to see lower pricing as the company continues to scale its manufacturing capacity.[261] And for those still skeptical, look at China, where the production cost for a high-end, industrial-grade humanoid already hovers around $5,000 to $10,000.[262] That should give us a clear indication of where the market is headed.

Given this scale of production, the ongoing decline in pricing, and the exponential adoption curve now unfolding across industries, it is worth examining the trajectory of humanoid startups and the underlying strategies driving this adoption. We will explore these strategies in the coming chapters.

Chapter 22: Figure AI — Illinois Farms to BMW Factories

Humanoid robot startups can be distilled down to two core groups: pioneers and challengers. Pioneers are essentially the first-generation humanoid startups that seeded the belief in the market that humanoids are the ultimate scale vector we need to address emerging bottlenecks in the global economy — from aging population and shrinking workforce to rising labor costs and beyond. Pioneers believed that humanoids could help us scale productivity levels in our society and, ultimately, GDP. These are companies like Figure AI, 1X, and Agility Robotics.

Challengers are the next generation of humanoid startups. After this belief in humanoids was seeded by the pioneers, challengers pushed the pedal even harder, accelerating adoption of these humanoid robots across a wide range of industry verticals to an entirely new level. These include UBTECH, Apptronik, AgiBot, Neura, and more.

In this chapter, we will focus on the pioneers, starting with Figure AI. In the following chapters, we will cover the challengers.

Figure AI Origins

Incubated in 2022, secured a $70M Series A in 2023, launched its first production-grade humanoid the same year in less than 12 months, secured its first paid customer, i.e., BMW, in Jan 2024—less than 24 months since the company was founded—helped produce 30,000+ BMW X3 vehicles in less than 9 months, with dozens of its humanoids running a 10-hour daily shift straight for that time in a live factory environment, received applications from 170,000 engineers, but decided to operate with an extraordinarily lean footprint of just 450+ team members.[263-264]

If there is one humanoid startup on this planet that defies the thresholds of speed, scale, precision, and impact, in the startup cosmos, that's Figure AI.

Within just a span of four years from its inception, the company has already shipped ultra-dexterous humanoids that are, in the truest sense, addressing real pain points for this German automaker, i.e., BMW, that leads the global luxury automotive segment. In 2025, its 8 million-square-foot manufacturing facility located in Spartanburg, U.S., boasted the highest yield among its factories anywhere in the world—supported in part by these humanoids operating on the production floor. Given this trajectory, Figure AI merits deeper examination.

We will first start with the story of its founder, and then study the strategies the company has pursued that have led to this impact, along with the lessons we can learn from it.

Founder DNA and the Road to Figure AI

Figure AI was incubated in 2022 by Brett Adcock, a serial entrepreneur best known for consistently hitting the escape velocity of the startup orbit—taking young companies from seed stage to full-blown commercialization at a pace that would rival the finest engines of Ferrari and Lamborghini. If the startup world had its own Olympics with the sole criteria of speed and impact, Brett would be competing for gold. The sheer determination to make an impact on society, and the grit to defy the odds along the way, are reflected just as clearly in the early chapters of his life.

Born on April 6, 1986, Adcock grew up on his family's third-generation farm outside Moweaqua, Illinois. By the age of sixteen, he was already building web-based businesses, including an e-commerce site focused on outdoor electronics and a content platform called *Street of Walls*. He later attended the University of Florida, where he initially studied industrial engineering before graduating in 2008 with a Bachelor of Science in Business Administration. With an entrepreneurial appetite ignited at such a young age and an emerging pedigree in building technology-driven ventures, Brett was now ready to address frontier challenges in society.[265]

He started with Vettery, a talent marketplace he co-founded in 2013, designed to connect technology professionals with employers and help bridge the talent gap in society. This experience helped the young founder develop a deep sense of awareness around the wide range of technical skills nurtured in our society, skills that are the ultimate determinants of the scale of innovation we experience in our economy.[266]

With this foundation, he then leaned into the urban aircraft segment with Archer Aviation, a company he co-founded to enable a sustainable air transit system for modern cities by accelerating the development, manufacturing, and operation of electric vertical takeoff and landing (eVTOL) aircraft. Archer was incubated on the

University of Florida campus and went public in 2021 at a $2.1B valuation. This combination of software and hardware focus gave Brett a strong appetite for taking complex, hardware-centric companies from concept to full-scale commercialization stage, driven by a sheer desire to make a strong impact in society. It is this journey, and the deeper purpose underlying it, that ultimately led to the incubation of Figure AI, a company well known today in the humanoid world, but in reality was first conceived in a cramped phone booth in Palo Alto, California.[267]

Adcock started Figure with a clear mission: to build general-purpose humanoid robots that could help address labor shortages in our economy while automating physically demanding jobs, allowing future generations to live happier, more purposeful lives. The goal has always been to improve the quality of life for both workers and citizens in our society by augmenting frontline and industrial workforce with humanoids to minimize strain and maximize productivity, thereby contributing to GDP growth.

With $100M in funding from his own capital war chest, Brett incubated the company and hired about 40 extraordinary professionals from Boston Dynamics, Tesla, Apple SPG, IHMC, Cruise, and Alphabet X to bring this mission to life, launching its first prototype in March 2023, i.e., Figure 01. Later that year, the company raised a $70M Series A led by Parkway Venture Capital, with participation from Aliya Capital, Bold Capital Partners, Tamarack Global, FJ Labs, and KUKA Robotics CEO Till Reuter, aimed at accelerating commercialization of its first humanoid platform. This funding proved vital, as during this phase the team was in talks with automotive giant BMW to launch a pilot program at one of its U.S. manufacturing facilities.[263,268]

Within a span of mere 12 months since Series A, in January 2024, the company inked a deal to pilot its first prototype, Figure 01, at the Spartanburg, South Carolina plant to automate manufacturing operations, especially the production of BMW X3s. With this deal secured, by February 2024, Figure became one of the most heavily capitalized humanoid startups, raising a Series B of $675M at a $2.6B valuation, with strategic backers including Microsoft, NVIDIA, Jeff Bezos, Intel Capital, and OpenAI Startup Fund.[269]

Within 60 days, the company kickstarted pilots at BMW's Spartanburg plant to automate the production of the BMW X3. In August of that year, Figure launched its next prototype, Figure 02, which was subsequently integrated into the pilot program. The 11-month-long pilot, which started in early 2024 and ended in November 2025, involved multiple Figure humanoids operating Monday through Friday on 10-hour shifts. Together, they contributed to the production of a staggering 30,000 BMW X3s and enabled the automated handling of more than 90,000 parts.[264] Given this impact, BMW decided to scale its adoption of humanoids across its factories, and Figure AI, in turn, needed to extend its capital runway.

In September 2025, the company raised a $1B Series C, led by Parkway Venture Capital, with participation from NVIDIA, Intel Capital, LG Technology Ventures, and others, valuing Figure AI at $39 billion. Since then, it has signed another major U.S. company as its second customer. While details have not yet been disclosed, the company has also held conversations with UPS to automate ground operations for the logistics giant, which operates in over 200 countries, processes more than 22.4 million packages each day, and runs one of the largest airline fleets in the world.[270-271]

Given this trajectory, and the rate at which Figure AI continues to make an impact across society and the economy, there are lessons we must now decode—lessons that will serve as a guiding light for the next generation of founders and engineers, equally determined to walk this path with grace and a deep sense of urgency.

Strategic Lessons: The Figure AI Operating Model

If we analyze the trajectory of Figure AI, essentially the underlying operating model that led to this impact boils down to seven core principles. Principles that are unique to this company and its founder, and that span the full spectrum of the startup cosmos, from positioning to fundraising, go-to-market, and scale. We will now cover each of these one by one.

Lesson 01: Positioning — "Robot Labor" From Day One (B2B)

Figure AI's strategy was industrial-first from inception, grounded in a B2B "robot labor" thesis, not in the consumer humanoid spectrum. This level of focus from Day 01 played a key role in ensuring the company was building a general-purpose humanoid for society, but with a surgical emphasis on industrial use cases. The company positioned its robots as labor substitutes for factories and warehouses, not as social companions.

That level of clarity from Day 01 shaped every downstream decision vector, from its choice of early adopters to the execution of ROI-driven pilots in enterprise environments, while steering away from demo marathons and the endless consumer-facing R&D cycles that dominate much of the robotics world. Even the company's fundraising language reinforced this direction. The Series A announcement was explicitly tied to commercializing Figure 01 humanoid, signaling adoption-centric intent rather than open-ended robotics research.

Lesson 02: Fundraising — The Best Pitch Is a Paid Customer

Each round of capital raised by Figure AI was justified quantitatively: the company could point either to a paid customer deployment or to one already underway. In each round, Figure AI ensured that its robots were either operating on factory floors and generating revenue, or backed by signed agreements from real industrial customers prepared to pay for them—and, more importantly, who believed in the value they had to offer.

Securing a pilot with BMW at one of its largest U.S. production facilities so early on is a clear example of this principle in action. Thus, the strongest fundraising pitch is proof of demand. A paying customer signals credibility in a way no slide deck ever can. With that level of validation in the market, investors place far greater trust in both your mission and your team than they would without it. At that point, a single slide that conveys your mission, the caliber of your team, and undeniable execution can speak louder than any elaborate presentation. All you need is one slide with those three pillars, demonstrating proof of execution that does not violate the principles of time, speed, and impact that define the startup cosmos.

Lesson 03: GTM — Anchor Customer as Market Validator (BMW)

Figure landed an anchor customer unusually early for a humanoid robotics startup: BMW announced a commercial agreement on January 18, 2024, to deploy Figure humanoids to automate its automotive manufacturing plants. More importantly, BMW validated the robots in an actual production environment, not a staged pilot, confirming on September 11, 2024, that Figure 02 was being tested inside Plant Spartanburg.

This is important to note, as gone are the days when startups worked with industrial giants and carved out a separate pilot area or an explicit test setup to conduct deployments. Those test environments do not generate organic proof points and do not help a company truly evaluate its product or address the real bottlenecks that emerge on production-grade sites.

So, as a founder, if you truly want to understand the full capability of your product to address customer pain points, you must secure an anchor customer as you build your first prototypes and put them to the test in a live production environment. As long as you can mitigate risks on both the technical and operational fronts during these early production-grade deployments, by putting human supervisors or robust supervision systems in place, this approach alone will yield far better results than 100 demos ever will.

The GTM lesson here is clear: one marquee industrial buyer who is willing to test your product in live, production-grade environments is far more valuable than 100 pilots or demos conducted in non-production settings. That single customer becomes a powerful trust vector for your mission and team across the entire industry. And that is what ultimately drives inbound opportunities that are truly ROI-positive for your startup.

Lesson 04: Credibility — ROI-driven Pilots, Not Demos

Now, once you have landed a customer who is willing to offer their production site to test your prototypes, you want to be surgically precise about the problems you are trying to solve for them in this deployment, and you want to quantify outcomes ahead of time. Your goal should be to establish clear quantitative thresholds for the output you guarantee before the pilot even begins. If you cannot have that level of belief in your product and your team, then you cannot expect the customer to place that trust in you as a founder. This principle is evident in Figure AI's trajectory.

Brett eliminated the robotics "demo trap" by turning each deployment into a quantified execution scoreboard. The company reported an 11-month trial in a production environment: a full-scale deployment by month 10 on an active assembly line, running 10-hour shifts Monday through Friday. Quantified outputs included 90,000+ parts loaded and contribution to the production of 30,000 BMW X3 vehicles by November 19, 2025. The most important lesson here is that, as a founder, credibility is one of your strongest trust vectors in the industry, and you must continue to earn it by consistently demonstrating quantified outcomes, i.e., ROI, to both investors and customers. Ultimately, it is consistency in outcomes that drives credibility.

Lesson 05: Capital — Constraint-Mapped Strategic Backers (NVIDIA, Microsoft, OpenAI)

Figure AI's fundraising strategy was centered around two key vectors: commercialization and strategic partnerships. While the former is well understood, the latter is what most founders do not spend enough time thinking about. Each investor you bring onboard brings with them not just capital, but also strategic capacity.

In the case of Figure AI, the company understood its core choke points from Day 01. These constraints were compute power, i.e., chips; AI infrastructure, i.e., servers; and access to a strong LLM model that it could leverage to build its custom humanoid intelligence stack. To address each of these bottlenecks, Figure AI strategically partnered with NVIDIA to unblock the chip runway, Microsoft and Amazon to secure the cloud runway, and OpenAI to clear the LLM runway, raising $675M at a $2.6B valuation on February 29, 2024. The lesson is clear: the best investors are often strategic enablers, and when mapped surgically to your A+ constraints, they can unblock your path in a way no amount of capital alone ever will.

Lesson 06: Moat — Own the Autonomy Layer (Helix)

As a startup, you need to build a strong moat from day one, even if you're venturing into a blue-ocean market. History is proof that no market stays blue-ocean, i.e., uncongested, forever. Over time, competitors saturate the space and compress margins, making it extremely hard for startups to compete.

We've seen this before: iRobot once defined the early consumer robotics market, but eventually low-cost entrants commoditized the category. The same dynamic is now unfolding in humanoid robotics, where dozens of startups are rapidly gaining momentum worldwide. Thus, no matter your starting point or the sector you are entering, you should retain a strong moat, either on the technology front or the execution front. Ideally both.

In Figure AI's case, the founder understood that simply building a humanoid robot was not going to help the company thrive, especially with the advent of dozens of humanoid startups around the world. Instead, building a strong custom-trained autonomy stack tailored to industrial use cases could serve as a powerful defense vector, helping advance the company's mission without sacrificing margins or getting drowned in the startup graveyard. Figure AI's moat is not the humanoid form factor, it's the autonomy layer.

The company introduced Helix (Vision-Language-Action) on February 20, 2025, as a generalist AI model to train its humanoids. Today, it serves as the brain of Figure humanoids. Helix helps the team train the robot to perceive its environment, understand instructions conveyed through audio or language inputs, translate them into appropriate actions, and execute real industrial workflows. Over time, the robot trains itself using the operational data captured across Figure AI's fleet of robots deployed in live production environments, i.e., on factory floors. The second version of Helix, Helix 02, extended these capabilities toward full-body autonomy, bringing Figure closer to a truly general-purpose humanoid robot designed to automate next-generation factories end to end.

Thus, by owning this autonomy layer and continuously training it using real-world field data, Figure AI is building a strong defensive moat beyond the hardware itself, significantly minimizing dependence on commoditized external LLM providers. While this moat enables Figure humanoids to stand out on the autonomy front in an increasingly saturated market, the company didn't stop there. To further strengthen its defense war chest, Figure AI also launched its own production facility, BotQ, minimizing dependence on third-party manufacturing partners and building a cost-efficient engine to scale its production capacity.

Lesson 07: Scale — Robots Don't Scale Like Software (BotQ)

To manufacture robots at scale and in a cost-efficient manner, the company announced the launch of BotQ in 2025, a dedicated high-volume humanoid manufacturing facility, with stated capacity to produce up to 12,000 robots per year. This is important to note here because, unlike software startups, if you are a hardware startup, production and shipping costs can add up in no time, eroding your financial runway at a pace you may not have anticipated yet. To address this challenge, Figure AI launched BotQ, cutting reliance on third-party manufacturers and reducing exposure to supply chain bottlenecks that often accelerate the burn rate of young, hardware-centric startups.

Now that we have distilled the core lessons embedded in Figure AI's trajectory, from its origins through its evolution and the pivotal decisions the team made along the way, it is time to decode the underlying business model.

In the final section of this chapter, we will break down the business model and illustrate how the product portfolio maps to select primary use cases.

Also, now that we have covered the lessons on building high-performance teams and products to enable operational endurance for your robotics startup, the next question that naturally follows is: how do you communicate the exact value that your product or services offers to your customers and build trust? In simple terms, what are the core lessons you must keep in mind as you develop your sales and marketing strategy? We will cover these in the next chapter.

Figure AI: Product Snapshot — BMW Production Plant

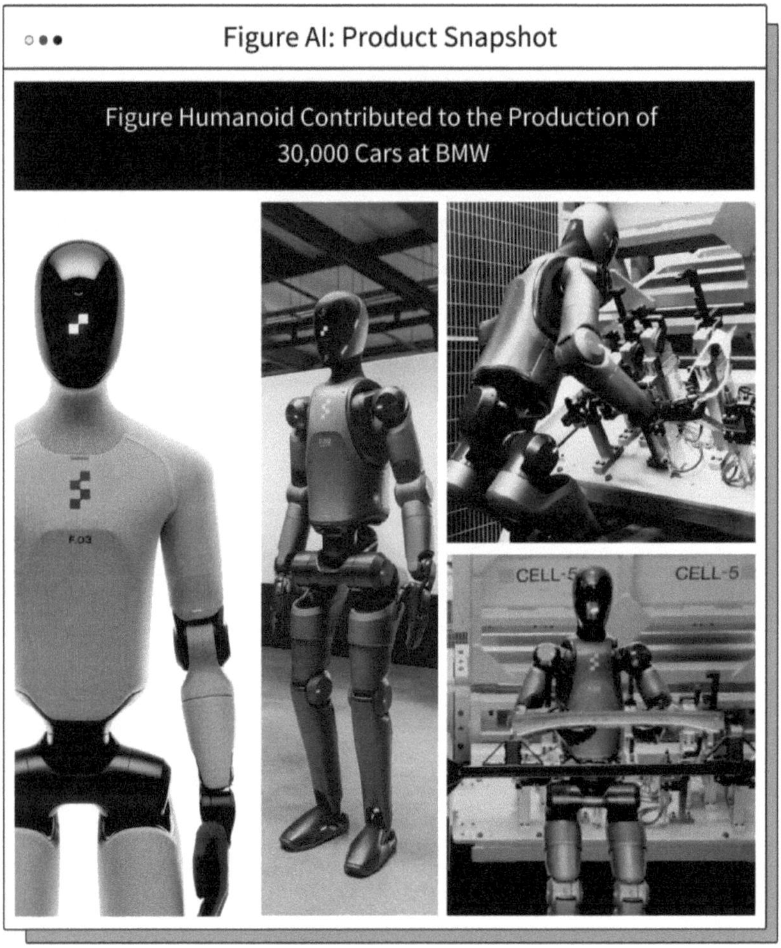

Figure 22.1 — Figure AI: BMW Production Plant

Illustration by the author. Robot images from Figure AI's BMW production announcement page.

Figure AI's Business Model — Part 1 of 2

Figure AI's Business Model

- **Mission:** To build the world's first commercially viable, general-purpose humanoid robot designed to augment the human workforce and address labor shortages across industries.
- **Target Customer Base:** Enterprise operators in logistics and manufacturing verticals seeking cost-effective automation solutions to streamline labor-intensive workflows and scale production capacity.
- **Products:** Figure general-purpose humanoid robot.
- **Model:** A venture capital-backed humanoid robotics company scaling through strategic enterprise partnerships with automotive leaders such as BMW, supported by purpose-built, in-house manufacturing infrastructure built on the principle of "robots building robots."
- **Funding :** Exceeded $1B in committed Series C capital at a $39 billion post-money valuation in 2025, with notable investors including Microsoft, OpenAI, NVIDIA, Intel Capital, ARK Invest, Align Ventures, and others.
- **GTM:** Commercial-first deployment strategy targeting labor augmentation in structured environments, such as factories and warehouses, with a defined pathway toward future consumer-centric applications.
- **Anchor Customer – BMW, Spartanburg Plant, USA:** A fleet of a dozen Figure humanoids supported the production of more than 30,000 BMW X3 vehicles in less than nine months, each operating 10-hour daily shifts within an 8 million square foot production-grade manufacturing facility in Spartanburg, United States.

Figure 22.2 — Figure AI's Business Model: Part 1

Illustration by the author.

Figure AI's Business Model — Part 2 of 2

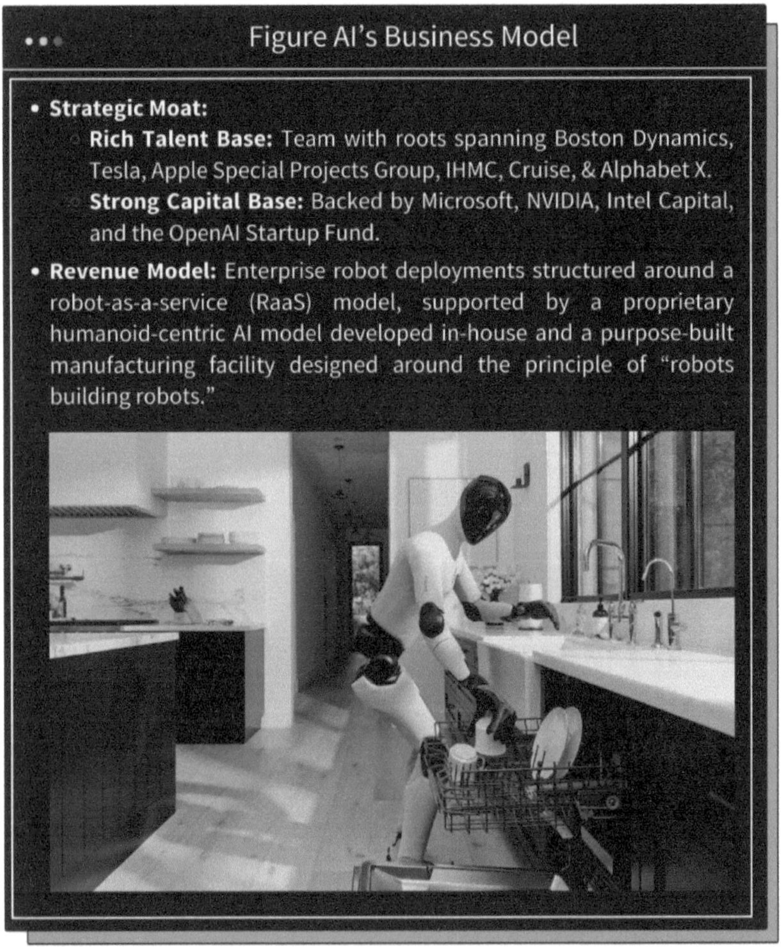

Figure 22.3 — Figure AI's Business Model: Part 2

Illustration by the author.

Chapter 23: 1X — Oslo Labs to Norway's Humanoid Moonshot

Incubated in 2014 in Norway under its original name, Halodi Robotics developed an industrial humanoid prototype during the early days named EVE, sold 140 of them to ADT, the company that provides security workforce services to major enterprises across North America, made a pivot into consumer humanoids and rebranded itself in 2022, secured a $23.5M Series A2 led by the OpenAI Startup Fund in March 2023, raised a $100M Series B led by EQT Ventures in January 2024, unveiled its first consumer-grade humanoid prototype in August 2024, launched public preorders for it in October 2025, and signed a strategic partnership with EQT, a Swedish multi-asset investment firm with deep roots in private equity, healthcare, and infrastructure verticals, in December 2025 to deploy up to 10,000 humanoid robots across EQT's portfolio companies between 2026 and 2030.[272-274]

If there is one humanoid startup that was not only amongst the first to secure a large-scale contract in the early 2020s, but also to make one of the boldest bets ever on consumer humanoids designed exclusively to offer a helping hand to those in society who are in dire need, i.e., the elderly and people with special needs, within just two years of securing its Series A, it is 1X Technologies.

Today, these humanoids are offered to consumers under two pricing models: an upfront purchase price of about $20,000, or a leasing option starting at roughly $499 per month.[275]

While most humanoid startups today are heavily focused on the industrial wedge, 1X has taken a more contrarian route: building humanoids designed to coexist with humans in society, specifically in domestic environments, our homes, which are arguably among the most high-variant deployment sites for these systems. No two homes around the world share the same physical arrangement, emotional profile, traffic footprint, or demographic makeup.

From age groups to emotional sensitivity, and from household goods to behavioral patterns, the variables that emerge in these domestic settings are nearly infinite across the globe. This, in turn, represents one of the greatest opportunities to collect a wide range of real-world and anonymized data through these humanoids, enabling them to be trained to serve society in a safe and reliable manner. Hence, 1X has chosen this frontier.

In doing so, the company has positioned itself at the edge of what may ultimately become the most consequential frontier for humanoid adoption: everyday life.

Given this trajectory, 1X merits deeper examination. We will start with the story of its founder, and then decode the strategies the company has pursued, along with the lessons the next generation of robotics founders and researchers can draw from it.

Founder DNA and the Road to 1X

1X Technologies was founded by Bernt Øivind Børnich, a Norwegian entrepreneur who developed a deep passion for disassembling electronic appliances at an early age. As a child, Bernt would often take things apart at home, ranging from kitchen appliances all the way to the family-owned car with his father, welding metal parts and fixing underlying electric components. This passion over the years transpired into his foray into the robotics cosmos, guiding Bernt to pursue his formal education in this arena. So he did.

After graduating from the University of Oslo with a bachelor's degree in Robotics, this entrepreneur, with a few years of industry experience under his belt at a local software firm, decided to launch a humanoid robotics startup called Halodi Robotics in 2014 with a co-founder, a fellow engineer from the same institution, and seed investment from a local partner.

Halodi specifically designed a humanoid robot called EVE with a mission to offer robotic assistants across select verticals in our society, including healthcare, security, and logistics. After years of research and development that continued from 2014 to 2021, the company managed to ink a strategic contract with ADT to deploy about 140+ of these industrial humanoids across enterprise sites in North America to augment the on-site security workforce for its Fortune 500 clients. A contract of that scale and caliber at the time proved vital for Bernt and his team, as it not only provided the financial runway the company needed, but also became a strong and organic trust vector, one that was of utmost importance for this young entrepreneur to establish strategic partnerships in the industry and build the capital war chest needed to take this mission forward.[276-278]

This was also the time when we witnessed major breakthroughs in the AI orbit, enabling the launch of large language models (LLMs) and vision-language models (VLMs), which made it possible to accelerate the training process for these robots, thereby guiding the company to build a domestic variant of EVE, one that would have to work with 1,000+ variables and navigate millions of edge cases inherent in our society. That led to the birth of NEO, the consumer humanoid robot.

To bring this humanoid to life, in 2022, and to align its mission, Halodi Robotics officially rebranded itself as 1X Technologies, marking a strategic pivot toward general-purpose humanoids built explicitly for domestic environments. Marching toward this mission, 1X raised a $23.5M Series A2 in March 2023, led by the OpenAI Startup Fund, followed by a $100M Series B in January 2024, led by EQT Ventures, with participation from Samsung NEXT and other strategic backers, thereby unblocking its path to both frontier AI models to train its humanoids and commercialization through EQT's global portfolio.[279-280]

By August 2024, 1X unveiled NEO Beta, its first prototype, launching preorders for it in October 2025, and positioning it as one of the first humanoids designed exclusively to enable affordable domestic assistance services in society for those that need it most. Continuing this momentum, the company was able to hit the escape velocity of its prototype orbit in December 2025, when it inked a strategic partnership with EQT to enable the rollout of 10,000 humanoids across EQT's portfolio companies from 2026 to 2030, thus enabling enterprise-scale deployments.[274]

Today, the mission of 1X is clear: to build humanoids that can live alongside us, serve as a personalized home assistant, reduce the karmic fatigue that comes from mundane and strenuous physical tasks, and elevate our quality of life. Given its mission, and the velocity at which this young startup has transformed itself from a Norwegian robotics lab into one of the most strategically backed humanoid ventures in the world, there are lessons here that we must decode. Lessons that will help guide the next generation of builders and researchers working at the edge of this emerging human-robot frontier in our society.

Strategic Lessons: The 1X Operating Model

If we analyze the trajectory of 1X, essentially the underlying operating model that led to this impact boils down to nine core principles. Principles that are unique to this company and its founder, and that span the full spectrum of the startup cosmos, from positioning to fundraising, go-to-market, and scale. We will now cover each of these one by one.

Lesson 01: Model — Dual-Wedge from Day One (B2B & B2C)

1X's strategy from day one has always been clear: build humanoid assistants for society, and that meant a dual-wedge approach targeting both factories (B2B) and homes (B2C) from the very beginning. The company initially focused on the industrial edge first, launching EVE and securing a large-scale contract for it with ADT, proving early traction in factory environments. At the same time, 1X was already kickstarting its consumer engine, which led to the launch of NEO, one of its first consumer-facing humanoid robots designed for domestic environments. In summary, 1X seeded its journey with a singular mission to bring friendly, resourceful humanoid assistants into our society, starting at the industrial edge and gradually scaling its footprint toward the consumer edge, with this dual-wedge model eliminating single points of dependency and defining a resilient business strategy from the get-go.

Lesson 02: Positioning — Humanoid Assistants for Society Eve & NEO (B2B & B2C)

Bernt and his team have been extremely consistent with the positioning of their product from day one: humanoid assistants for society. In the first half of their journey, the company launched EVE, a humanoid assistant for industrial environments, and then NEO, a humanoid assistant for domestic settings. While the scope of these products may have been unique across the two market segments, the purpose and positioning as a whole has always been the same: to truly add a helping hand in society for those who need it most, whether that is a worker operating on the factory floor or an elderly person at home.

As a founder, it is extremely important to retain surgical clarity on your mission and drive your product positioning with that level of consistency and endurance from your starting point to the end of your journey. This combination of clarity, consistency, and endurance is what helps build a strong layer of sustained trust in your brand, and that trust ultimately drives all downstream vectors such as investment, adoption, scale, and more.

Lesson 03: GTM — Build for One, Scale with Partners (ADT & EQT)

1X has always adopted a dual-pronged model to enable commercialization from day one. The first prong is optimizing the robot around one surgically studied use case, and the second is establishing strategic partnerships to accelerate adoption. This approach is evident in how the company launched its two humanoids and scaled their deployment.

On the industrial front, with EVE, 1X landed a strategic partnership with ADT and was able to sell 140+ humanoid robots at a time when most robotics startups across the world were still struggling to get a robot to pick up an object with a 90% accuracy rate, let alone ship a production-grade unit. The underlying lesson is that the company was not trying to address every edge case on day one, but instead optimized its humanoid specifically for security and healthcare verticals early on, which is why the ADT contract worked. EVE was designed to solve one problem exceptionally well, not to be a general-purpose robot from day one. Similarly, when it launched its consumer variant, NEO, 1X not only launched a waitlist to gauge market demand, but also ensured a more proactive path to adoption by inking a deal with EQT to deploy 10,000 robots across EQT's portfolio companies. In summary, 1X has consistently positioned its humanoids around a few high-conviction use cases and then leveraged strategic partnerships to scale distribution and drive real-world adoption.

Lesson 04: Pricing Strategy: Meet Customers Where They Are (Buy or Lease)

Bernt and his team understood the importance of deeply assessing the purchasing power, i.e., the financial capacity, of the two types of customers they intended to serve in society. While industrial partners often come with strong capital chests, consumers demand a more well-thought-out and affordable solution. To address these two adoption vectors with precision, 1X offered its products through two pricing models: a one-time purchase option of around $20,000, which was more industry-friendly, and a more accessible monthly lease model priced at $499 per month. This approach ensured the company was not only aware of the pain points of consumers, but also their financial capacity to address those pain points, and tailored its offering accordingly.

Lesson 05: Capital — Secure Strategic Backers (ADT, OpenAI, EQT)

The 1X team was clear about its mission from the get-go: to build ultra-dexterous humanoids that could serve as a helping hand in society. That implied it understood that these robots would need compute-heavy resources to train and portfolio-heavy companies to scale adoption. Knowing that training and adoption were going to be its two most important vectors to unlock scale, the company strategically partnered with OpenAI, ADT, and EQT. OpenAI unlocked the compute runway, and ADT and EQT unlocked the adoption runway, thus ensuring a win-win approach with these three strategic investors.

Lesson 06: Adoption — Phased Adoption Over Perfection

The 1X team didn't strive for a fully autonomous humanoid from day one to drive adoption. In fact, it launched its semi-autonomous humanoids in both consumer and industrial markets with the clause that remote operators would monitor the robot and intervene as needed to ensure safe and reliable operation. Thus, the company was transparent with its customers that while it would take years to enable full autonomy, in their current state these humanoids were still capable of delivering immense value through this remote supervision model, value that could help scale productivity levels in both society and industrial settings.

Lesson 07: Moat — Data Flywheel Anchored in Unstructured Environments (Homes)

While putting a humanoid in a home environment can be as daunting for a robotics team as it is for us humans to adopt one, given both the trust and engineering vectors at play, the company still decided to develop a consumer humanoid for domestic settings. The underlying principle here is that until you deploy thousands of these in real homes, you will not be able to collect the data needed to train these humanoids to address the very edge cases that make such deployments so difficult. To tackle this challenge, the company approached domestic environments with a more solution-driven and long-term mindset, with the goal of collecting data in a transparent manner from consumers, anonymizing it to ensure compliance with privacy standards, and then using it to train the navigation and spatial manipulation mechanisms for these humanoids.

The core lesson that emerges from this approach is that sites that offer infinite edge cases, if studied with depth and clarity, can serve as a strong training ground, helping you build the data flywheel you will need to make your product 100x better in the long run.

Lesson 08: Talent — Dual-Forts: U.S. and Norway Talent Base

The core lesson on talent from 1X is that if you are building a frontier robotics company, you are going to need strong expertise across multiple verticals: engineering, manufacturing, capital, and advisory. To address these requirements, 1X adopted a dual-fort approach, anchoring its roots across two countries. Today, the company is headquartered in Palo Alto, California, U.S., and operates with a primary manufacturing hub in Hayward, a secondary manufacturing hub in Moss, Norway, and an evolving talent base increasingly absorbed from the local giants in Silicon Valley and Oslo. This dual-fort approach has enabled 1X to tap into cutting-edge AI and robotics talent in the U.S., while maintaining a strong operational and production backbone in Norway. In summary, the company approached its talent base as a strategic wedge, marrying Silicon Valley's innovation-centric engine with Norway's industrial execution strength to build an operationally resilient organization.

Now that we have distilled the core lessons embedded in 1X's trajectory, from its origins through its evolution and the pivotal decisions the team made along the way, it is time to decode the underlying business model.

In the final section of this chapter, we will break down the business model and illustrate how the product portfolio maps to select primary use cases.

1X: Product Snapshot — NEO Humanoid

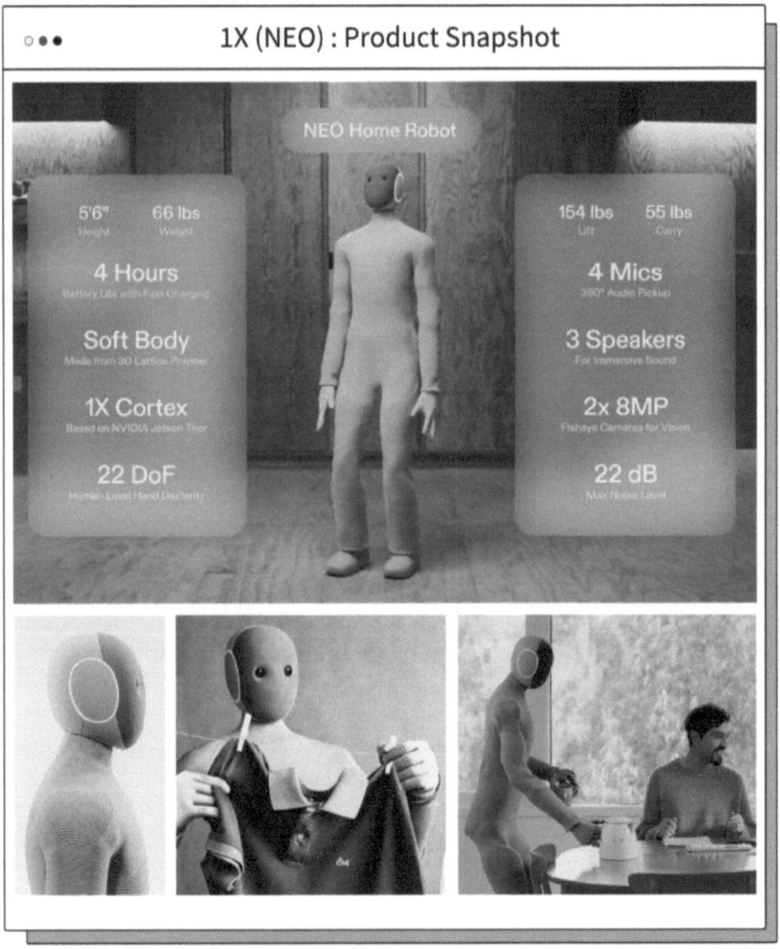

Figure 23.1 — 1X: NEO Humanoid

Illustration by the author. Robot images from 1X's product specifications page.

1X's Business Model — Part 1 of 2

1X's Business Model

- **Mission:** To build safe, general-purpose humanoid robot assistants designed to integrate seamlessly into both home and industrial settings.
- **Target Customer Base:** Consumers and industrial operators.
- **Primary Use Case:** AI-powered humanoids designed to serve as trusted, fully capable assistants for humans at home and at work.
- **Product 01 — NEO:** A next-generation bipedal humanoid robot engineered to serve as a trusted assistant and companion in domestic environments.
- **Product 02 —EVE:** A wheeled humanoid security robot engineered to serve as a capable, task-optimized assistant in commercial settings.
- **Model:** Dual-pronged strategy, with EVE targeting industrial markets and NEO targeting consumer markets.
- **Funding:** Backed by strategic and institutional investors, including OpenAI, Samsung NEXT, and Tiger Global, with cumulative funding exceeding ~$125 million across multiple rounds.
- **Go-To-Market Strategy — EVE:** Enterprise-first deployment model for EVE, anchored in a large-scale, early-stage contract inked with ADT to supply 140 humanoids to augment security guard services for its enterprise clients across North America.
- **Go-To-Market Strategy — NEO:** NEO launched for consumers via pre-orders, targeting a $20,000 capex-centric direct-to-purchase model alongside a $499 per month leasing model focused on automating medium-intensity workflows such as laundry and home cleaning in domestic settings.

Figure 23.2 — 1X's Business Model: Part 1

Illustration by the author.

1X's Business Model — Part 2 of 2

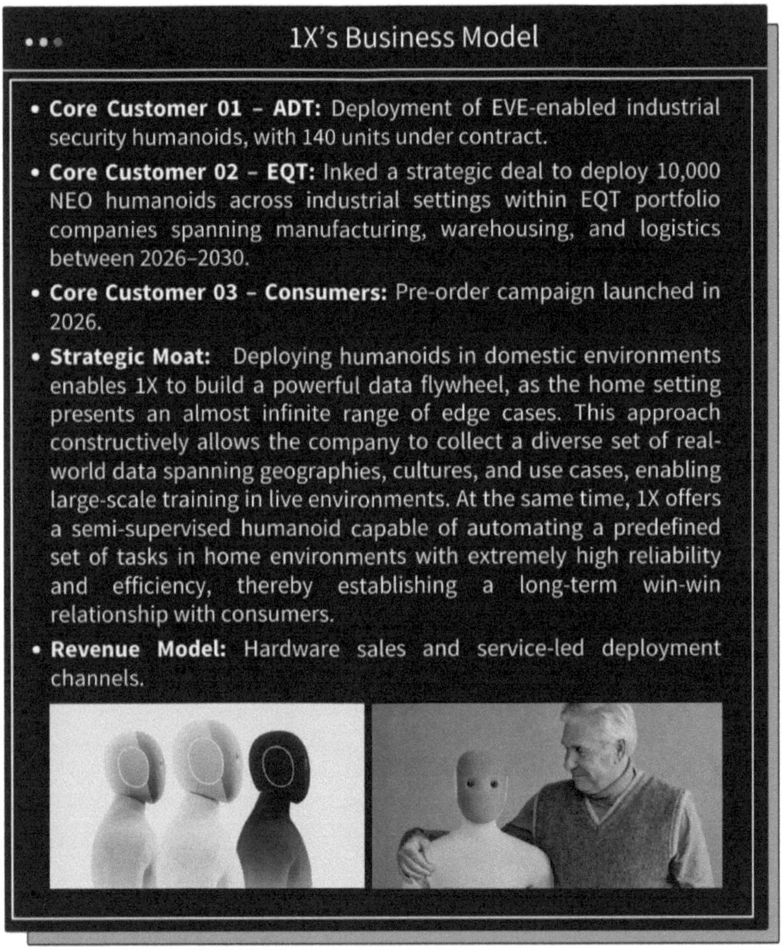

Figure 23.3 — 1X's Business Model: Part 2

Illustration by the author.

Chapter 24: Agility Robotics — Oregon State Labs to Amazon Scale

Incubated originally in the Dynamic Robotics Lab at Oregon State University in the early 2010s, officially forked from this research nest and established itself as an independent commercial entity in 2015, with the mission to bring purpose-built humanoids to warehouses, this young startup developed its first prototype, Cassie, in 2016, positioning it as a licensed research platform to generate a sustained capital runway from day one. It then released its industrial variant in 2017, i.e., Digit, establishing strategic partnerships and selling these units directly to major enterprises including, but not limited to, Ford Motor Company, Amazon, and GXO Logistics, to automate both indoor, i.e., warehouse operations, and outdoor, i.e., the last-mile delivery frontier. Along the way, it raised $400M in capital from strategic investors and launched the world's first full-stack humanoid robot factory, capable of rolling out 10,000+ humanoids a year, in 2023 in Oregon. If there is one humanoid startup that got all three vectors right, capital, adoption, and scale, in the logistics and warehouse vertical from day one, that is Agility Robotics.[281-283]

Given this trajectory, this warehouse humanoid pioneer merits deeper examination. We will first begin with the journey of its founder, and then decode the strategies the company has pursued, along with the corresponding lessons that the next generation of builders and researchers in this space can learn from it.

Founder DNA and the Road to Agility Robotics

Agility Robotics was founded by Jonathan Hurst, Damion Shelton, and Mikhail Jones, all of whom brought deep expertise in legged robotics grounded in academic research spanning both Carnegie Mellon University (CMU) and Oregon State University (OSU). Hurst established the Dynamic Robotics Lab at OSU in the early 2000s, where his research team developed a bipedal robotic platform named ATRIAS with funding from DARPA. He later partnered with CMU alumnus Damion Shelton and Mikhail Jones, an OSU master's student, to nurture this system and mature it into a semi-humanoid platform with no perception system, i.e., no "eyes," called Cassie. Cassie would go on to set the Guinness World Record for the fastest 100-meter run by a bipedal robot and ultimately shape the company's commercial humanoid program.[284]

The team started licensing this platform to academic research labs, funding their trajectory to evolve Cassie into a full-legged humanoid, ultimately launching Digit in 2017. The launch of Digit humanoid caught the eye of the North American automotive giant Ford Motor Company, which was evaluating robot-led solutions to modernize logistics workflows across its production plants, and inked a deal with Agility to secure several dozen of these humanoids.[285]

Given this adoption by Ford, the company was now on the radar of other logistics, manufacturing, and warehouse operators across the continent. A major commercialization milestone arrived on June 5, 2024, when Agility announced that GXO had signed a contract and deployed Digit full-time at its logistics facility near Atlanta, Georgia, under an industry-first, multi-year Robots-as-a-Service (RaaS) agreement. This marked one of the first commercial deployments of a humanoid robot contributing to a production-grade logistics facility.[286]

Subsequently, the company landed a similar partnership with Amazon to automate both warehouse operations and enable the last-mile delivery frontier. Today, Amazon is actively training Digit and exploring its integration into Rivian electric delivery vans to offer a more cost-efficient last-mile service to consumers. To fund this path, the company raised an $8M Series A led by Playground Global, followed by subsequent rounds of investment from strategic partners such as DCVC, Amazon's Industrial Innovation Fund, and others. By 2025, Agility had raised around $400M at a multibillion-dollar valuation to advance its mission.[287-288]

As the team continued to unlock the capital and adoption bottlenecks through these partnerships, the next major challenge was manufacturing these humanoids at scale. Given the sheer number of parts and sensors needed to build these robots, the team understood that a cross-Atlantic production model would not serve this purpose, as it would significantly extend iteration cycles and impede execution speed. To address this bottleneck, the founders decisively launched RoboFab in 2023, a dedicated manufacturing facility in Salem, Oregon, designed to ramp up production of Digit to match the industrial throughput expected by its customers. With Digit now positioned as a warehouse-ready humanoid platform, and GXO and Amazon as anchor deployment partners, Agility represents one of the clearest examples of a logistics-first humanoid commercialization model.[260]

Given this trajectory, the company merits deeper examination. We will now decode its model and the underlying lessons, insights that will help guide the next generation of builders and researchers working at the edge of this emerging human-robot frontier in our economy.

Strategic Lessons: The Agility Robotics Operating Model

If we analyze the trajectory of Agility Robotics, essentially the underlying operating model that led to this impact boils down to 9 core principles. Principles that are unique to this company and its founders, and that span the full spectrum of the startup cosmos, from positioning to fundraising, go-to-market, and scale. We will now cover each of these one by one.

Lesson 01: Model — Logistics-First Humanoids

Agility Robotics' strategy from day one has always been clear: build humanoids to automate logistics flows in factories and warehouses, explicitly targeting industrial customers. The founders made a strategic decision to steer away from their initial research-focused humanoid, Cassie, into a purpose-built industrial humanoid with the launch of Digit in 2017, and later partnered with key organizations in this domain, i.e., GXO and Amazon, to scale adoption of these humanoids. A core lesson here is that, as a founder, you must be clear about both your near-term adoption wedge and the long-term vertical where you want to build a lasting foothold. In many cases, these will align, but in some cases, as with Agility and 1X, they may differ. The path often begins with the segment that is easiest to gain traction in, before expanding into the vertical that ultimately generates sustained revenue at scale for your team.

Lesson 02: Capital — Revenue First, Perfection Second (Academia & GXO)

What's probably one of the hallmark lessons in this trajectory is that the company launched and sold a bipedal robot without any top half, i.e., no eyes and no perception system, called Cassie, to academic research labs, and used that early-stage revenue to fund its trajectory. It was this early-stage commercialization pathway that enabled Agility to build its next platform, Digit, a full-scale humanoid. Digit, in turn, was adopted by Ford, Amazon, and GXO while still not being a fully autonomous humanoid, and this second revenue stream helped the company scale its product iteration cycles toward full autonomy.

The core lesson here is that, as a founder, you do not have to wait until your product is perfect before launching it into the market. That reflects an enterprise-grade product development mindset. You are a startup, and you must surgically define the right transition point at each phase of your journey to move your product from the lab into the real world, just as Agility did with Cassie and Digit, without pouring capital into decade-long R&D cycles.

Lesson 03: Moat & Scale — Own One Wedge, Deliver at Scale (RoboFab)

A third core lesson from Agility's trajectory is that the company did not try to boil the ocean. Instead, it anchored itself around one clear adoption wedge: warehouse and logistics sites. By focusing relentlessly on a single industrial vertical, Agility was able to ink deals with enterprise partners, generate revenue, and build a defensible position in the market from the very beginning.

But enterprise adoption alone is not the moat. The core challenge for humanoid startups is scale. Once demand emerged, Agility recognized that manufacturing, and not talent density, would become the binding constraint. That is why the company made the decisive move to establish RoboFab in 2023, a dedicated humanoid manufacturing facility designed to ramp Digit production toward industrial throughput, with the capacity to roll out 10,000+ robots per year.

The core lesson here is that in humanoids, the winners will likely be the ones that own one solid wedge and then build a powerful manufacturing engine around it to meet industrial demand at scale. Agility understood early that scale is the moat, and RoboFab was the infrastructure bet it had to make to turn the enterprise traction it experienced into a long-term defensive shield.

Lesson 04: Adoption — Sustained and Exponential Adoption Curve (Amazon, Mercado Libre)

Last but one of the most important lessons here is that this startup remained laser-focused on ensuring that each year it was able to secure at least one to two high-profile customers at scale for Digit since its launch. First, it was GXO's logistics sites near Atlanta. Then it was Amazon, actively training Digit and exploring its potential integration across both Rivian delivery vans and its network of 200+ warehouses. And now, Mercado Libre, deploying Digit at its San Antonio, Texas site to automate parts of the warehouse operations that power its e-commerce business.

Each of these adoption vectors translated into quantified outcomes for both the customer and future prospects. At the Texas warehouse site, the fleet of Digit robots was able to handle and process roughly 100,000 totes during these early deployments. Totes are essential baskets used to carry items per order and transport them from the inventory area to packaging and ultimately to the shipping zone within the warehouse. In summary, the founders remained laser-focused on landing at least one large-scale deployment each year, and more importantly, retaining those enterprise partners over time.

That wraps our segment on pioneers, i.e., humanoid startups leading this frontier across both the consumer and industrial fronts. Now, in the following chapters, we will cover the challengers, companies that are building the next generation of humanoids and competing with the pioneers head-on, starting with UBTECH.

Before we proceed to the next chapter, now that we have distilled the core lessons embedded in Agility Robotics' trajectory, from its origins through its evolution and the pivotal decisions the team made along the way, it is time to decode the underlying business model.

In the final section of this chapter, we will break down the business model and illustrate how the product portfolio maps to select primary use cases.

Agility Robotics: Product Snapshot — Digit Humanoid at Amazon & GXO Facilities

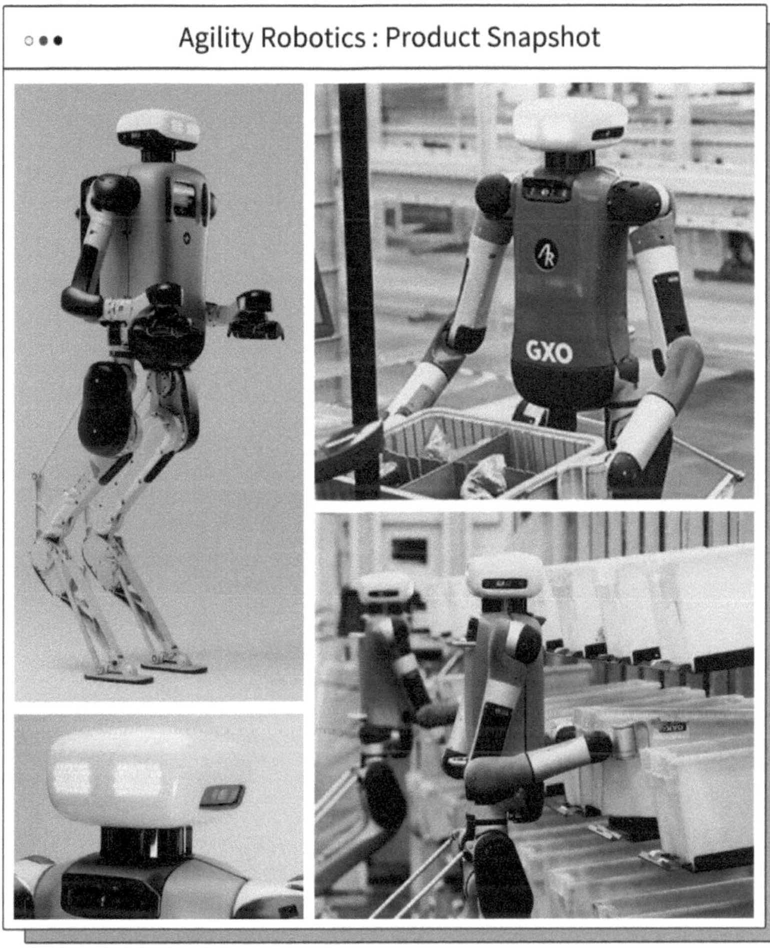

Figure 24.1 — Agility Robotics: Digit humanoid at Amazon and GXO Facilities

Illustration by the author. Robot images from Agility Robotics resources page.

Agility Robotics' Business Model — Part 1

Agility Robotics' Business Model

- **Mission:** To build humanoid robots capable of augmenting the industrial workforce, enabling safer working conditions, and addressing persistent labor shortages, thereby scaling throughput in a sustainable manner for operators across sectors.
- **Target Customer Base:** Companies in the logistics, manufacturing, and e-commerce sectors.
- **Primary Use Case:** Custom-trained humanoids designed to automate industrial workflows, with an explicit focus on highly repetitive tasks such as tote handling and pick-and-place-enabled sorting flows that are dominant in warehouse and distribution center environments.
- **Products:** Digit humanoid robot, a bipedal robot designed to automate warehouse and logistics workflows.
- **Model:** Commercial humanoid robots engineered to automate labor-intensive tasks and scale production capacity for industrial operators.
- **Funding:** Backed by strategic and institutional investors, including Amazon and DCVC, with cumulative funding exceeding $641 million across multiple rounds, resulting in a valuation of roughly $2.12 billion in 2025.
- **Go-To-Market Strategy:** Enterprise deployment model anchored in commercial pilots, supported by Amazon teams testing Digit in fulfillment and last-mile delivery operations.
- **Core Customers:** Amazon, GXO Logistics, Toyota Motor Manufacturing Canada, Mercado Libre, and Ricoh USA.

Figure 24.2 — Agility Robotics' Business Model: Part 1

Illustration by the author.

Agility Robotics' Business Model — Part 2

Figure 24.3 — Agility Robotics' Business Model: Part 2

Illustration by the author.

Chapter 25: UBTECH — Shenzhen Roots to Airbus and Audi Fleets

Incubated in 2012, in the Silicon Valley of China, Shenzhen, this humanoid startup made its entry into the robotics cosmos with a rather unusual yet strategic approach. Instead of building full-scale humanoids from the get-go, this team started with two small-scale robots: one designed to serve as a STEM education platform for students, and the other as a personal companion for families. The STEM platform resembled WALL·E from the iconic Pixar movie and was named Jimu Astrobot, while the 16-inch humanoid assistant was introduced as Alpha 1P. Both debuted on the global stage at IFA Berlin in 2017.[289]

The company strategically chose a small-scale form factor for these two projects to maximize iteration velocity, minimize burn, and build a sustained revenue trajectory by exclusively targeting academia and consumers from day one. Building on this momentum, in 2018, Alpha 1P evolved into Lynx, a more mature home humanoid integrated with voice assistant ecosystems such as Amazon Alexa, and positioned as a wellness companion capable of helping users learn yoga. By 2020, Lynx matured into a production-grade humanoid system called Alpha 1S and was deployed in healthcare facilities across China. This deployment significantly elevated the demand for its humanoids, and naturally the capital it needed to fund their production. To deliver on this demand and unblock its financial runway, the company raised funds from strategic investors including Tencent and ICBC. With this capital secured, from 2021 to 2022, the team accelerated its commercialization efforts.[289-290]

During those years, first, it inked a major deal with preschools in Seoul, South Korea, delivering 300+ of these STEM robotic kits, and in parallel, launched mobile robots equipped with UV-enabled disinfection systems across hospitals in China and the United States.[289] These early-stage commercial deployments helped the team build a strong revenue pipeline for the company, one that it desperately needed to sustain its long-term runway and continue its march toward the mass production of these humanoids.

To further fuel its commercialization efforts and scale its product iteration velocity, in December 2023, the company went public on the Hong Kong Stock Exchange, raising approximately HK$1 billion in its IPO. With that freshly injected capital, it launched its production-grade humanoid platform, the Walker S Series, including Walker S1 and S2, in October 2024.[290-291]

Then, within the span of just fifteen months, from late 2024 through January 2026, it inked major enterprise deals to supply humanoid systems to industrial giants across multiple sectors, including but not limited to Airbus, Texas Instruments, Foxconn, BYD, ZEEKR, and Audi.[292-297]

This is the story of UBTECH Robotics, a company that has been on a mission to build and launch fully capable humanoids, robots designed to add a true helping hand across society, for more than a decade. Incubated by Zhou Jian, it is driven by the vision of bringing intelligent, human-centric robots into every home, school, and business, making daily life more convenient, efficient, and intelligent.[298] Given how challenging it is to build a sustainable robotics company, let alone a humanoid company, this team has demonstrated an extraordinary level of perseverance in bringing this mission to life.

Given this trajectory, this humanoid pioneer merits deeper examination. We will now analyze the strategies the company has pursued, along with the corresponding lessons that the next generation of builders and researchers in this space can learn from it.

Strategic Lessons: The UBTECH Operating Model

If we analyze the trajectory of UBTECH Robotics, essentially the underlying operating model that led to this impact boils down to 7 core principles. Principles that are unique to this company and its founder, and that span the full spectrum of the startup cosmos, from positioning to fundraising, go-to-market, and scale. We will now cover each of these one by one.

Lesson 01: Model — Humanoids Anchored in the Manufacturing Epicenter of the World (Shenzhen)

UBTECH's strategy has always been structurally different from the university spinout archetype that defines many Western humanoid companies. Instead of emerging from an industrial research lab similar to DARPA or an academic robotics department, UBTECH was incubated in Shenzhen, China's hardware and manufacturing epicenter, and built its humanoid roadmap inside this city that is essentially the production engine for nearly all major enterprise giants around the world. From BYD cars to DJI drones, Apple iPhones to TP-Link routers, to TCL displays, the city pretty much hosts factories for all major electronics and automotive giants around the world. It is here that the company both understands the challenges large-scale manufacturers face and builds humanoids to surgically address them. The core lesson here is that if you have the opportunity to build your product in close proximity to your customers, it helps you develop a surgical understanding of their true needs and address them with precision, thereby accelerating your product iteration cycles and, in parallel, adoption.

Lesson 02: Product Wedge — Start Small, Then Scale to Full-Scale Humanoids (Jimu and Alpha)

Rather than attempting to build full-scale industrial humanoids from day one, UBTECH started with two small-scale platforms, Jimu Astrobot and Alpha 1P, targeting academia and consumers looking for personal assistants. This was not a detour, but a strategic wedge. Smaller robots enabled faster iteration cycles, lower burn rate, and rapid market entry. This strategy enabled the company to generate sustained revenue from day one after launching these products into the market. The adoption of 300+ of these early-stage production units by schools in South Korea is a strong proof point here.

The core lesson is that founders do not need to start with the final form factor, especially when building large-scale systems such as humanoids or autonomous vehicles. The path to scale in both cases often begins with smaller, simpler platforms that enable rapid, cost-efficient product iteration and launch cycles.

Lesson 03: Revenue — Proactively Build Sustainable Cash-Flow Bridges

UBTECH's early-stage education and service robots helped the company build a sustained revenue trajectory long before Walker, its production-grade humanoid, entered factories. From STEM robots deployed in South Korea to mobile robots supporting hospital operations in the U.S. and China, UBTECH monetized intermediate product layers while working toward a more mature humanoid stack.

The lesson here is clear: in robotics, product development cycles are unforgivingly long and capital-intensive, so the only way through is to proactively build revenue bridges at each stage, not wait until you have a perfect, production-ready system for scaled deployment. If you wait too long, this phase can erode your financial runway well before your product matures enough to meet industrial demand.

Lesson 04: Moat — 24/7 Factory Uptime Enabled by Battery Swap

What's truly novel about UBTECH's approach is that, given its presence in the manufacturing epicenter of the world, Shenzhen, the team understood that factories operate nearly 24/7. Any robot expected to work in these mission-critical environments would need to align with that operating model. Recognizing this, UBTECH introduced the Walker S2 series with an automated battery-swap feature from day one, ensuring industrial customers could integrate these robots into their manufacturing operations and run them perennially, significantly minimizing downtime for their production engine.

The core lesson here is clear: if you are building humanoids for factories, you must design them to match the factory operating model, one that demands 24/7, high-throughput operations. This approach not only minimizes adoption friction but also helps you, as a robotics startup, build a strong moat by enabling customers to achieve extraordinary uptime, ultimately maximizing revenue and establishing a true win-win partnership with them.

Lesson 05: Data Flywheel — Factories as Fleet Training Grounds (Zeekr)

After its initial launch, the company left no stone unturned to deploy Walker at scale at Zeekr's automotive production plant, even though it came at a price that, on paper, may not justify the returns for a young startup. The underlying principle was that this deployment was never primarily about short-term revenue. It was strategically designed to enable UBTECH to collect high-quality data at scale from real, live production environments.

Deploying dozens of humanoids operating in parallel as a fleet gave the company a massive advantage. It exposed the robots to a wide range of edge cases, helped the company rapidly iterate its product and address these challenges, throttled down failure rates, and ultimately maximized system reliability and throughput at scale. This approach, securing one strong large-scale deployment that serves as a true training ground, allowed

UBTECH to build the high-speed data flywheel it needed to iterate at galactic speed.

In summary, if you are building a new class of robot such as a humanoid, your first few customers may not be optimized for revenue, but they should absolutely be optimized toward product maturity. Your goal is to deliver value to your customers while ensuring your product matures at an exponential rate. UBTECH did this by creating a data flywheel that enabled rapid product iteration, ultimately producing a matured and mission-critical version of these humanoids that are now being deployed across sectors at scale. Thus, while this early deployment may not generate immediate financial return on paper, the compounding effect it creates sets the stage for all future returns that will pay for this investment in the long run.

Lesson 06: GTM — Build a Cross-Vertical Anchor Customer Base (Airbus, Foxconn, BYD, Audi, Texas Instruments)

UBTECH has adopted a multi-vector GTM strategy to build its anchor customer base across sectors spanning automotive, aviation, electronics manufacturing, and chip manufacturing. Strategic deployments involving Foxconn, BYD, Zeekr, Texas Instruments, Audi, and Airbus reflect a calibrated and structured go-to-market strategy anchored in the enterprise validation flywheel spanning an increasingly complex and equally diverse set of industrial environments.

Each anchor customer serves as both a proof point and a wedge into adjacent sectors, helping the company build organic credibility, grow its total addressable market base, accelerate adoption velocity, and minimize concentration risks that often emerge when a startup is overly exposed to a single narrow vertical.

The lesson here is that, as a humanoid startup, your commercialization strategy should not be anchored in one vertical, but ideally should be anchored across a portfolio of enterprise customers spanning multiple verticals. This array of diverse, high-value anchors enables you to compound your adoption footprint over time while minimizing single points of dependency.

Lesson 07: Stack — BrainNet AI Model to Enable Fleet-Scale Training & Deployments

The team building the Walker robot understood that, in order to truly succeed as a humanoid startup, especially one focused on bringing this at scale to industrial environments, it had to develop the core software stack, i.e., the underlying AI model it would need to train these robots at scale. Fleet-scale training was the real moat here, and one it had to build to hit the escape velocity of the capability orbit for these humanoids. Recognizing this need, UBTECH developed BrainNet, an AI model that allows humanoids to learn, coordinate, and improve as a fleet in live production environments.

At Zeekr's smart factory in Zhejiang, BrainNet powered the world's first multi-robot, multi-task, multi-scenario collaborative training deployment for the company, with dozens of Walker S1 units operating across the assembly line of this automotive giant. It is this multi-robot deployment, enabled by BrainNet, that later allowed the company to iterate its humanoids at a rapid pace and secure deals with industry giants across verticals ranging from Airbus to BYD, Foxconn, and Texas Instruments.

The core lesson is that when you are building robots for industrial environments, you must figure out a way to scale the training process in a time- and resource-efficient manner. Sequential training is not an option, since the market will get saturated by the time you reach your adoption milestones. To accelerate your product maturity cycle, you need to turn each deployment into a training powerhouse for your fleet. UBTECH did that with BrainNet, ensuring every humanoid deployed in the real world becomes a training vector for the rest of the fleet, thereby unlocking galactic levels of speed, scale, and impact.

Now that we have distilled the core lessons embedded in UBTECH's trajectory, from its origins through its evolution and the pivotal decisions the team made along the way, it is time to decode the underlying business model.

In the final section of this chapter, we will break down the business model and illustrate how the product portfolio maps to select primary use cases.

UBTECH: Product Snapshot — Walker S2 Humanoid

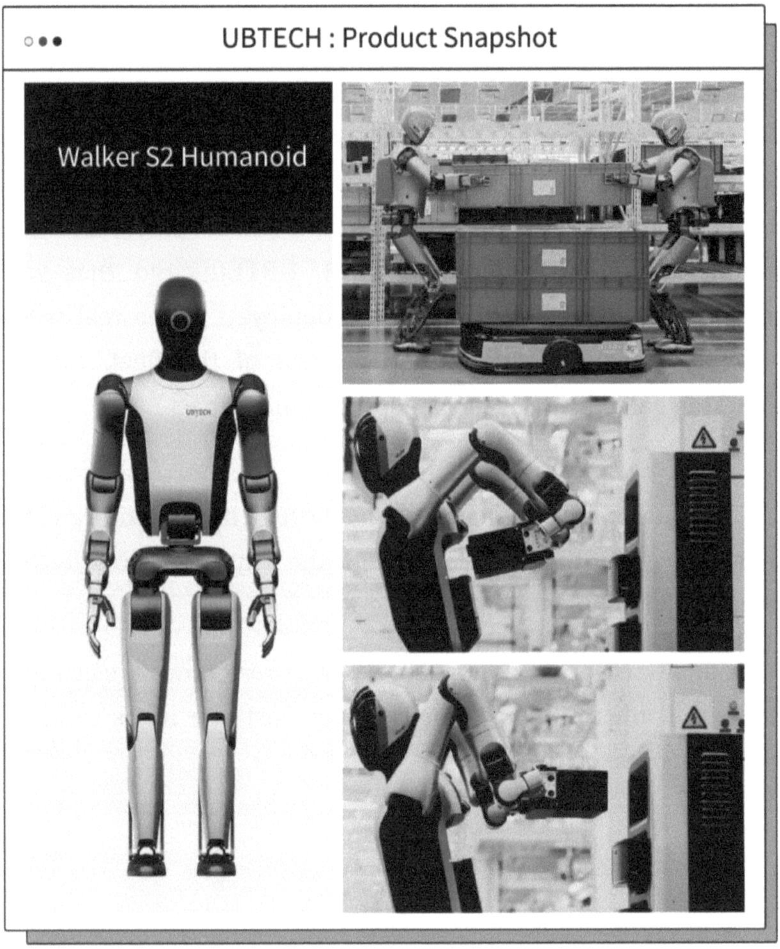

Figure 25.1 — UBTECH: Walker S2 Humanoid

Illustration by the author. Robot images from UBTECH Robotics' product specifications page.

UBTECH Robotics' Business Model — Part 1

UBTECH Robotics' Business Model

- **Mission:** To bring intelligent humanoid robots into every family, amplifying convenience and productivity levels in everyday life, while fostering a rich, human-robot-centric ecosystem.
- **Target Customer Base:** Consumers, educational institutions, research organizations, and industrial operators
- **Primary Use Case:** Humanoid robots engineered to deliver long-term value across key economic sectors, from education and hospitality to industrial automation and logistics.
- **Product 01 — Consumer & Educational Robots:** Compact humanoid robots such as Alpha Mini, along with STEM kits including the Jimu and uKit series.
- **Product 02 — Industrial Humanoid Robots:** The Walker S humanoid robot series, engineered exclusively to automate industrial workflows spanning manufacturing, assembly, inspections, and logistics operations.
- **Model & Strategic Moat — Humanoids Anchored in the Manufacturing Epicenter of the World (Shenzhen):** UBTECH was incubated in Shenzhen, China's hardware and manufacturing epicenter, and developed its humanoid roadmap within this ecosystem, which serves as the production engine for many major enterprise leaders around the world. From BYD vehicles and DJI drones to Apple iPhones, TP-Link routers, and TCL displays, Shenzhen hosts manufacturing plants for major global electronics and automotive brands. This positioning enables UBTECH to conduct pilots within these dense manufacturing clusters and optimize its humanoid systems at an exponential pace, thus providing it with a strong edge to compete head-on at a global scale.

Figure 25.2 — UBTECH Robotics' Business Model: Part 1

Illustration by the author.

UBTECH Robotics' Business Model — Part 2

UBTECH Robotics' Business Model

- **Funding:** Publicly listed on the Hong Kong Stock Exchange, with a market capitalization of approximately HKD 60 to 70 billion, equivalent to roughly USD 8 to 9 billion, as of 2026.
- **Go-To-Market Strategy:** A product-led commercialization model powered by direct sales and distributor-led partnership channels.
- **Core Customer Segment 01 — Consumers and Educational Institutions:** Adoption of the Alpha and Jimu series robots.
- **Core Customer Segment 02 — Industrial Operators:** Integration of the Walker S series humanoid robots into factory floors, assembly lines, and logistics operations to scale throughput and address labor constraints.
- **Current Customer Base:** Airbus, BYD, Foxconn, BAIC, Texas Instruments, and Audi.
- **Price Point — Consumer Robot:** Alpha Pro priced at approximately $675.
- **Price Point —Industrial Humanoid:** Base model launched at approximately $41,000, with fully capable commercial variants positioned at around $112,000. In the coming years, UBTECH is expected to lower the price of these humanoids as it continues to scale its production capacity to serve rising demand across sectors.
- **Revenue Model:** Product sales of educational and humanoid robots.

TOYOTA AIRBUS BYD

Figure 25.3 — UBTECH Robotics' Business Model: Part 2

Illustration by the author.

Chapter 26: Apptronik — NASA Labs to Mercedes-Benz Factories

From legendary BBQ dishes served at the iconic Franklin Barbecue restaurant, to serving as fertile ground for the Sequoias of Silicon Valley, there are only a handful of cities on this planet that know how to weave together the threads of music, food, nature, and technology as well as Austin, Texas.

The city has become a major tech hub, hosting companies such as Tesla, Apple, and Google, and serves as a launchpad for an array of deep-tech startups pushing the frontier of innovation. It is also home to SXSW, perhaps the only festival in the world where a global technology conference and world-class live musical performances converge in near-nirvanic harmony, transcending young minds into a state of heightened awareness and igniting a perennial flame of creativity and joy in their soul.

At the heart of this city lies The University of Texas at Austin, one of the most powerful academic engines of innovation in the United States. In 2023, the university ranked No. 3 among U.S. institutions for the number of innovations patented across a wide range of verticals, trailing only the University of California system and MIT. It was on this campus, in early 2015, that a group of robotics researchers from the university's Human-Centered Robotics Lab partnered with NASA to represent the university at the DARPA

Robotics Challenge. Their work directly contributed to the development of NASA's first mission-critical humanoid systems: Valkyrie, developed at the Lyndon B. Johnson Space Center. The primary objective of this humanoid was to support future NASA missions by helping establish infrastructure on outer planets in advance of the arrival of astronauts, ultimately serving as a robotic assistant to these teams during deep-space expeditions.[299]

Post the launch of this first prototype, Valkyrie, this core team continued their research and decided to build its equivalent for our planet. Around 2016, they established what is now one of the frontier U.S. ventures focused on bringing general-purpose humanoid robots into factories worldwide. From 2016 to 2019, the company pursued this humanoid research vision with funding from private institutions and government grants, ultimately developing the skeletal and foundational architecture for these robots.[300]

This work culminated in the launch of a fully electric humanoid platform, the company's flagship production-grade humanoid system: Apollo, launched in 2023. Apollo is a human-scale humanoid robot engineered specifically for industrial deployment.

Over the next two years, the company aggressively raised capital through an ultra-long Series A round that was seeded at $28M and grew to $350M by February 2025, co-led by B Capital, Capital Factory, and Google. It was further extended to $403M through strategic investors including Mercedes-Benz, Japan Post Capital, and RyderVentures, ultimately reaching $935M, with large-scale contributions from AT&T Ventures, John Deere, and the Qatar Investment Authority (QIA), valuing the company at $5B by

February 2026. This capital marathon, spanning 2025 through 2026, was followed by an Ironman of commercialization, pushing the team's endurance to its edge.[301]

During this period, this young startup inked contracts to deploy its humanoids in real production environments: in Mercedes-Benz factories in Germany and Hungary to automate assembly lines, at GXO logistics sites to streamline warehouse operations, and within Jabil facilities to modernize manufacturing workflows. With a lean team of roughly 300 employees, this team has managed to launch a production-ready humanoid platform designed exclusively for industrial operations, while securing a staggering $900M+ in investment at a $5B valuation from strategic partners around the world. All within a span of just three years since it unveiled its first production-ready platform, Apollo, in 2023.[302]

This is the story of Apptronik, and its founders—Jeffrey Cardenas, Nick Paine, Luis Sentis, and Bill Helmsing. It's a journey that reflects the sheer amount of perseverance, discipline, and focus a team needs to sustain over years to bring these modern humanoids to life at a pace that would rival those competing on the F1 tracks of Monza—the legendary Temple of Speed, and one of the fastest circuits in the world. Given this trajectory, its impact, and the pace at which it has delivered that impact, the company merits deeper examination.

We will now analyze the strategies the company has pursued, along with the corresponding lessons that the next generation of builders and researchers in this space can learn from it.

Strategic Lessons: The Apptronik Operating Model

If we analyze the trajectory of Apptronik, essentially the underlying operating model that led to this impact boils down to five core principles. Principles that are unique to this company and its founders, and that span the full spectrum of the startup cosmos, from positioning to fundraising, go-to-market, and scale. We will now cover each of these one by one.

Lesson 01: Model — General-Purpose Humanoid with Industrial-First Focus

The first lesson here is rooted in the positioning strategy the company has adopted since day one. While Apptronik's commercialization efforts are strictly geared toward industrial deployments, its positioning vector from the get-go has always centered on a general-purpose humanoid designed to assist humans, and one built to operate in a world designed by humans.

Both the vision and the priority are clearly defined. The long-term ambition is to build a truly general-purpose humanoid. But the immediate priority is far more focused: to build a system that can serve as a general-purpose worker explicitly within select industrial environments first, as demonstrated by the adoption curve led by Mercedes-Benz, GXO, and Jabil. The core lesson here is that, as a founder, you must know both your vision and your priority, or the scope of that vision each year, to ensure that every downstream decision remains aligned with those two vectors.

Lesson 02: GTM — Multi-Vector Anchor Customer Base (Mercedes-Benz, GXO, Jabil, John Deere, Ryder)

The second strategy the company has adopted that deserves serious attention is on the GTM front. This is not necessarily unique to Apptronik; however, it is important to understand how it has executed this strategy. It is well understood that you need an anchor customer to kickstart commercialization efforts, and here it launched Mercedes-Benz as its first and long-term anchor customer. What is important to note is that it did not stop within that vertical.

After inking a deal with Mercedes to deploy its humanoids at factory sites in Germany and Hungary, the founders immediately started expanding their footprint beyond automotive into logistics, manufacturing, and more importantly, heavy industry. The partnerships and investor signals across GXO, Jabil, John Deere, Ryder, and Mercedes are a strong proof vector here.

The core lesson is that you want to start in one vertical and plant your seed, but you must not stop there. You must continue planting seeds across multiple verticals at the early stage, as long as you have the capital to do so. The number of verticals you can target will be directly proportional to your horsepower. At a minimum, you need 2–3 to minimize market concentration risk once competitors enter the space. Apptronik today is targeting five vectors: manufacturing, automotive, agriculture, logistics, and retail.

In summary, the deeper Apollo gets embedded into diverse industrial verticals, the harder it becomes to displace Apptronik from the market.

Lesson 03: Capital & Moat — Your Cap Table Is Your Most Powerful Moat

The third lesson here is on the approach the company adopted to secure capital, and more importantly, to establish strategic partnerships. To begin with, it helps to understand that your funding rounds are one of the strongest vectors you will need to build a durable layer of defense for your company and your mission. Hence, as a founder, with each round of capital you raise, your job is to evaluate which of the walls you need to build your defense shield are weak, and proactively bring in partners to strengthen those walls and unblock those bottlenecks.

From production to commercialization and beyond, there are a series of challenges that you will inevitably encounter as you scale your company. When you do run into one of these, you will notice that these partnerships prove far more valuable, and can help you address these constraints that no amount of capital ever will. Capital is often just a side effect of executing funding rounds, not the other way around.

If we carefully examine this principle in the context of Apptronik, it doesn't take long to realize that it treated each funding round as an opportunity to bring in partners that mapped directly to its strategic bottlenecks. In this case, Mercedes became both investor and anchor customer. Google brought the AI infrastructure the company needed. Jabil helped the company enter the electronics manufacturing vertical. John Deere signals entry into agriculture and heavy industry. Thus, these strategic partnerships are the real moat, and capital is simply the enabler, or a side effect of that moat, not the other way around.

Lesson 04: Stack — Apollo as a Pre-Trained Industrial Humanoid Platform

The fourth lesson here revolves around the marriage between capability and positioning, one that this team has executed with a surgical precision for its customer base.

On the capability front, both Apollo's design and engineering are tightly aligned with its intended deployment ecosystems. If we observe carefully, Apollo is explicitly designed to operate alongside humans, inside warehouses and factories that were built for humans. Its form factor reflects that alignment: roughly human-scale, standing about 5 feet 8 inches tall, weighing approximately 160 pounds, and capable of lifting up to 55 pounds. These physical specifications are not arbitrary, they are a direct consequence of its positioning.

The same alignment applies to its underlying engineering stack. Apptronik is not building Apollo as a single-purpose robot, nor as a consumer humanoid designed for homes. Instead, it is building Apollo as a standardized, pre-trained humanoid platform, a base system that industrial partners can deploy, adapt, and customize for their workflows, thereby elevating throughput, without needing to rebuild the robotics foundation from scratch. This is indeed a platform-style approach.

The company provides the base humanoid, the hardware, control systems, perception stack, and core task primitives, while partners build the application layer on top. In that sense, Apollo becomes an industrial operating system for next-generation factories and warehouses. Trailer unloading, palletizing, and case picking represent some of the earliest adoption vectors for this humanoid. These are high-frequency tasks, economically painful, and physically taxing on human workers, making them ideal entry points for ROI-driven early humanoid deployments for the company.

Apollo's interface design also reinforces this human-centric deployment model. Its circular eyes are in fact stereoscopic cameras that power the underlying perception system. Its monochromatic E Ink display provides a simple but expressive mouth to connect with factory professionals at an emotional level. The chest features an OLED screen designed to communicate key information to nearby human co-workers. Thus, the team has designed Apollo to connect with us humans at both a contextual and emotional level.

Apollo is indeed an industrial humanoid platform, surgically engineered, packaged, and positioned from day one to integrate into existing industrial workflows at both a social and functional level.

Lesson 05: Scale — Robots Building Robots (The Jabil Partnership)

One of the most strategically important partnerships this humanoid startup has been able to land is with Jabil, one of the world's largest manufacturing services companies. This partnership is critical for Apollo's trajectory, as it helps it unblock two critical risk vectors: scale of production and integration directly with an anchor partner that already brings that same scale to enterprise giants.

In simple words, the partnership with Jabil extends far beyond capital investments. Jabil now aims to help Apptronik both set up its production facilities to mass-produce these robots at scale, and more importantly, it plans to do so by enabling these robots to produce more of them, with minimal intervention from factory professionals. That is the first-order effect of this partnership.

The second is that, in doing so, Apollo will end up getting infused directly into the very production ecosystem that manufactures products for enterprise giants around the world, including, but not limited to, Apple, Amazon, Johnson & Johnson, Ericsson, Cisco Systems, Hewlett-Packard, and Tesla.

Jabil's manufacturing footprint spans the Americas, Europe, and Asia. The company operates across approximately 53.86 million square feet of production facilities, employs more than 200,000 workers, and serves as an industrial manufacturing backbone for Silicon Valley's enterprise ecosystem. Thus, this partnership is not merely about scaling the production of humanoids. It is about infusing Apollo directly into the bloodstream of the global industrial manufacturing engine, accelerating not only production, but also distribution and adoption at scale.

Now that we have distilled the core lessons embedded in Apptronik's trajectory, from its origins through its evolution and the pivotal decisions the team made along the way, it is time to decode the underlying business model.

In the final section of this chapter, we will break down the business model and illustrate how the product portfolio maps to select primary use cases.

Apptronik: Product Snapshot — Apollo

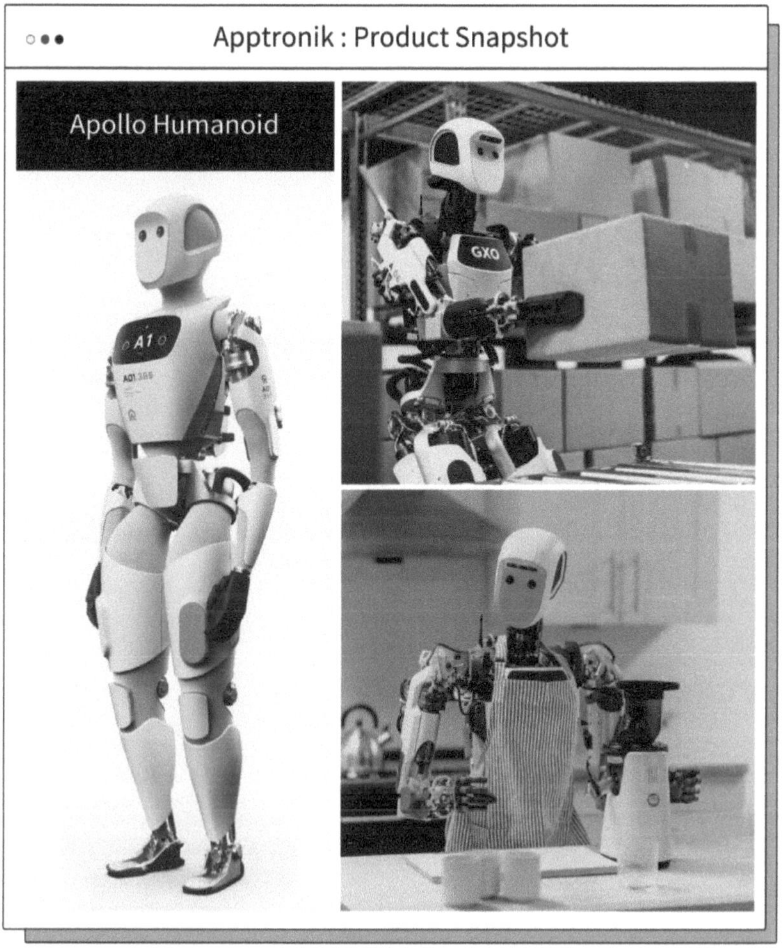

Figure 26.1 — Apptronik: Apollo Humanoid

Illustration by the author. Robot images from Apptronik's product specifications page.

Apptronik's Business Model — Part 1 of 2

Apptronik's Business Model

- **Mission:** To elevate the universal quality of life through humanoid robots capable of performing work that is physically demanding, repetitive, or economically inefficient for humans. The company's mandate is rooted in the belief that humans and machines, working together in synergy, will pave the way for a brighter, more advanced future.
- **Target Customer Base:** Manufacturing and logistics operators.
- **Primary Use Case:** Apollo, a general-purpose humanoid designed to automate workflows in industrial environments.
- **Product — Apollo Humanoid Robot:** A general-purpose industrial humanoid engineered primarily for factory and warehouse environments, designed to integrate seamlessly into existing workflows, scale throughput, and drive operational efficiency for global manufacturing and logistics operators.
- **Model:** A venture-backed humanoid robotics company with ultra-strong capital arteries, nourished by strategic investors and advancing toward large-scale industrial deployments through enterprise partnerships.
- **Funding:** Valued at approximately $5 billion as of 2026 and backed by strategic investors including, but not limited to, Google, Mercedes-Benz, John Deere, the Qatar Investment Authority, and AT&T, positioning Apptronik among the most highly capitalized humanoid robotics startups globally.
- **Go-To-Market Strategy:** Industrial-first commercialization strategy

Google | John Deere | Mercedes-Benz | AT&T

Figure 26.2 — Apptronik's Business Model: Part 1

Illustration by the author.

Apptronik's Business Model — Part 2 of 2

Figure 26.3 — Apptronik's Business Model: Part 2

Illustration by the author.

Chapter 27: AgiBot — Chengdu Origins to Thailand Scale

Jiangxi Province is one of the few regions in China known for its majestic mountain ranges, serene rivers, and ancient relics. On the western bank of the local Gan River lies the city of Ji'An, home to one of the country's youngest and most widely recognized entrepreneurs in robotics.

Born in 1993 in a subtropical environment shaped by harsh monsoon seasons and heavy rainfall, with barely 1,600 hours of sunshine each year, or roughly 60 clear days, it was here that he learned to endure the wrath of both life and nature. Over time, that endurance forged at such a young age would come to shape both his character and ambitions.

During his school years, it was reflected in the early experiments he conducted on electronic circuits at his home in Ji'An. There, he started building and refining these circuits, demonstrating them at regional competitions, continuing to pursue his passion throughout his academic years with a strong sense of discipline and focus. These sustained efforts finally came to fruition in 2013, during his sophomore year, when he competed at the national level and secured first place in an electronic design contest with his homemade smart temperature control system. This was a critical milestone in his journey, as it instilled a deep sense of self-belief and strengthened his commitment to engineering.[7-8]

Soon after, he left home to pursue formal education nearly 1,500 kilometers west of Ji'An, in Chengdu, at the University of Electronic Science and Technology of China, one of the country's oldest academic institutions, with roots dating back to 1956. There, he pursued a degree in biomedical engineering. During his university years, he continued to participate in a wide range of technology competitions, winning more than 40 awards. These experiences became a pivotal catalyst, strengthening both his technical foundation and his resilience, and ultimately guiding him toward a more challenging path — entrepreneurship.

With strong expertise in electronics and circuit design honed at university, and a deep desire to build consumer electronics products reflected in the projects he had nurtured from a young age, he secured a five-million-yuan investment (approximately US$683,600) in seed funding from local investors to support his vision. He kick-started these efforts by building 3D-printed components embedded with electronics circuitry for local businesses. He pursued this venture while continuing his studies and later that year enrolled in a master's programme in robotics.[7-8]

This dual-pronged approach, focused on academic training and field experiments, helped him develop deep expertise in applied robotics and led to the development of an autonomous bipedal humanoid robot that earned him an innovation award at the 2018 World Robot Olympiad.[7] This marked another critical milestone in his journey, guiding him to steer his ship deeper into the electronics arena and prompting him to explore the practical application of humanoid robots to address emerging challenges in the economy.

China's commanding position in the electronics manufacturing vertical, particularly within the cellphone sector, made it an ideal arena to evaluate the potential application of these robots in large-scale production facilities. To facilitate this evaluation-focused initiative, he joined the cellphone manufacturer Oppo that year and, after a two-year tenure, transitioned to the telecommunications giant Huawei, gaining firsthand insight into the operational bottlenecks these industry leaders faced on the manufacturing front. Having closely studied the core pain points of these companies, key contributors to China's GDP, he decided to address these challenges head-on using his deep expertise in robotics.

During his time at Huawei, he developed a prototype version of a robotic arm to help automate core segments of the company's assembly line. That project propelled him onto the global stage when a demonstration video of the robotic arm that he published online garnered 2.5 million views worldwide within 24 hours of its release.[7]

Building on this momentum and on the cumulative experience he gained working hand in hand with two leading telecommunication companies, he developed a deep understanding of both robotics and their strategic application in automating electronic manufacturing lines. It was this experience, together with the resilience he had cultivated at a young age in the monsoon-clad region of Ji'An, that led him to launch AgiBot in 2023, a company now widely recognized as one of the top two humanoid robotics startups in China. Within the first year following the launch of its humanoid robots in 2025, the company sold more than 5,000 units and generated $142 million in revenue, securing clients across a wide range of industry verticals.[303]

From auto parts manufacturing giants such as Fulin Seiko to telecommunications leaders like China Mobile, AgiBot secured large-scale contracts in its inaugural year and continues to expand its footprint at a remarkable pace. This trajectory has been nurtured by a strong cohort of strategic investors who have backed the mission of this young founder with unwavering conviction. This group of industry partners includes, but is not limited to, Tencent, LG, Baidu, Matrix Partners, and the Shanghai Lingang Economic Development Group, a government-owned technology accelerator committed to fostering enterprise innovation and strengthening regional economic development.[304-305]

This is the story of AgiBot, and this young founder is none other than Peng Zhihui. Given his journey and the resilience he cultivated at a very young age, later reflected in the speed at which AgiBot established itself within the industry, we must now turn to the lessons that emerge from this remarkable trajectory, lessons that will help guide the next generation of entrepreneurs and researchers in this field.

Strategic Lessons: The AgiBot Operating Model

If we analyze the trajectory of AgiBot, essentially the underlying operating model that led to this impact boils down to eight core principles. Principles that are unique to this company and its founder, and that span the full spectrum of the startup cosmos, from positioning and fundraising to go-to-market strategy and scale. We will now examine each of these one by one.

Lesson 01: Model — General-Purpose Humanoid with Select-Vertical Focus

The first lesson is rooted in the positioning strategy AgiBot has adopted since day one: a general-purpose humanoid platform, but with an explicit focus on select verticals — electronics manufacturing, tourism, social services, and retail. More importantly, these verticals were carefully chosen and tightly aligned with its Phase 1 Asia-first expansion strategy.

China serves as the primary adoption ground for the manufacturing sector, leveraging its dense electronics ecosystem for rapid deployment and iteration. Meanwhile, Thailand, Malaysia, and Singapore function as high-visibility, high-traffic testbeds for the tourism and hospitality sectors. The company's go-to-market strategy clearly reflects this model.[306-308]

Lesson 02: GTM — Multi-Vertical Deployment Strategy (Electronics | Tourism | Social | Retail)

Instead of developing a concentrated customer base in a single vertical, which most humanoid startups do at an early stage, AgiBot diversified from the outset, seeding its roots across industrial, electronics, tourism, social, and retail segments.

Industrial traction is evidenced by a 100-unit order it received from Fulin Precision, a local automotive parts manufacturer. Electronics validation came through deployment with Longcheer, a company that manufactures smartphones, tablets, and other gadgets for global consumer electronics brands. Similarly, its foray into the tourism and hospitality sector is reflected in its partnership with Thailand's Smart Tourism Board. Beyond the industrial and tourism sectors, its presence in the broader hospitality segment is demonstrated through the launch of its mini-humanoids, specifically designed for concierge roles and social events such as weddings and similar ceremonies.[306-309]

Its presence in the education sector is equally evidenced through LinkCraft, a training portal it launched to scale the training process for its next-generation AI models using videos and other media assets contributed by students and professionals.

This multi-vertical deployment strategy has also produced a second-order effect that far exceeds revenue. The company has now collated rich cross-domain data at scale, spanning multiple geographic regions and industry verticals. It is this data that has become its organic moat and has the potential to strengthen its AI models that power these humanoids and make them truly general-purpose in the truest sense.

Lesson 03: GTM — Global Testbeds to Reduce Geographic Concentration Risk (Asia, E.U., U.S.)

Another key lesson that emerges from the AgiBot journey is that within the first year of launching its humanoid, the team started seeding roots outside the domestic industrial ecosystem. Specifically, it recognized the potential of emerging and mid-sized markets as strong testing grounds for its robots, given its mission to build truly general-purpose robots. These markets offered a wide range of variables and edge cases for the robots to navigate, thus enabling the company to collate real-world data and train its models at a pace that far exceeds its competitors.

This principle is evident in the strategy the company adopted to expand outside China, starting with the launch of production and testing sites in Kazakhstan in early 2025, followed by its debut in Singapore toward the end of that year. This was followed by the launch of its testbeds in Milan, Italy, Malaysia, and Thailand in 2026. The Milan deployment served as a strong European test vector, while the Malaysia and Thailand deployments served as strong Asian test vectors, especially for tourism and hospitality segments.[310-311]

These deployments were later followed by another leg of expansion with its debut in the U.S. at CES 2026. Within a span of 24 months since the launch of its first production-grade humanoid, the company had seeded its roots across international testbeds spanning Asia, Europe, and North America, minimizing geographic concentration risk while simultaneously building one of the most powerful technology and revenue engines in the field.

Lesson 04: Revenue — 5,000 Robots & $142M in Revenue with 300 Employees in 12 Months

As the company pursued the international expansion route, the team remained laser-focused on its revenue targets. Within the first year of launching its first production-grade humanoid, AgiBot deployed 5,000 robots across industrial and social ecosystems, predominantly in Asia, and generated a staggering $142M+ in revenue with a lean team of 300 employees. A significant percentage of that revenue came from China Mobile, world's largest mobile operator by subscribers and revenue, with over 1B subscribers and $111B annual revenue in 2025. That milestone is worth noting, as it reflects the sheer magnitude of impact a lean team can make in this domain within just 12 months post-launch of a production-grade solution in the market.[303-305,312]

Lesson 05: Strategic Partners — Baidu, LG, Tencent, TCL, Lingang New Area Fund

As we analyze the trajectory of this startup, another key lesson that emerges is the role strategic investors play in shaping a company's long-term trajectory. In the case of AgiBot, one such investor was the Lingang New Area Fund.

Lingang New Area Fund functions as the local government's equivalent of Y Combinator, a state-owned entity that provides resources ranging from capital and facilities to regulatory guidance and industrial access, effectively accelerating the growth of young start-ups. Similarly, investments from companies such as Baidu, LG, Tencent, and TCL signal that the company, from the outset, prioritized partnerships with organizations that directly benefited from its solutions and strategically aligned them as investors. The result was a clear win-win framework that reinforced both commercial traction and a resilient financial runway.

Lesson 06: Moat 1 — Production in Emerging & Developed Markets (Kazakhstan & China)

One of the most consequential lessons from the AgiBot journey emerges from the company's decision to geographically diversify production as a long-term moat. Instead of concentrating its manufacturing engine solely in one country, AgiBot anchored its production footprint across both China and Kazakhstan. This was a strategic decision that allowed the company to source cost-efficient components from China while producing these systems in an economy that offers a cost-effective labor workforce, relatively flexible regulatory policies, and a dense export network.

Kazakhstan serves as a strategic Eurasian logistics hub that connects Europe and China. Its industrial positioning, historically strengthened by its leadership in uranium production, has contributed to the development of a robust export pipeline and globally integrated trade routes. As a result, the country offers one of the most cost-effective logistics networks in the world, along with a strong and competitively priced labor workforce that strengthens AgiBot's production moat.[310-311]

The Kazakhstan production and testing site, launched in early 2025, served not only as a manufacturing base but also as a live industrial validation environment outside the domestic ecosystem.[311] This dual-market production strategy significantly minimizes geopolitical concentration risk, helps the company build a resilient supply-chain system, and enables rapid iteration cycles across different regulatory and operational frameworks, thus culminating in a durable manufacturing moat that extends far beyond pure technology and product expertise.

Lesson 07: Moat 2 — Social Platform to Accelerate the Training Process (LinkCraft)

A critical component of AgiBot's competitive moat is the data flywheel it has built alongside its resilient production engine. AgiBot's launch of LinkCraft, its social training portal designed to collect video and task-based inputs from students and professionals, reflects a deliberate effort to crowdsource real-world behavioral data at scale.[313-314]

Instead of relying solely on simulations or factory-generated datasets, the company has developed a distributed social training layer that feeds diverse edge cases directly into its model-training pipeline. This approach significantly accelerates the learning process for these humanoids, proactively exposing them to a wide range of anomalies they might encounter in the field and enables the company to build not just a better humanoid, but a powerful self-learning engine at scale. In summary, AgiBot has managed to build a robust training engine using this data moat, which now enables it to march toward its mission of building a truly general-purpose humanoid platform at a pace that far exceeds that of its competitors.

Lesson 08: Moat 3 — Tourism & Hospitality as a Strategic Adoption Wedge (Thailand, Malaysia, Singapore)

A defining strategic insight from the AgiBot journey is its strategic entry into high-visibility consumer-heavy sectors, which helped it normalize its adoption curve while seeding its brand deep within the core layers of our society.

By entering the tourism and hospitality verticals at an early stage in its journey, particularly through partnerships aligned with Thailand's smart tourism initiatives, AgiBot positioned its humanoids in high-footfall, human-robot-interaction-heavy environments such as concierge services, events, and public-facing venues. These deployments were strategically designed to build a strong social proof engine. In tourism- and hospitality-driven markets like Thailand, Malaysia, and Singapore, robots interact daily with thousands of users from across the globe. This approach builds a dual-layered moat for the company. First, it exposes its humanoids and the underlying conversational models to diverse demographic cohorts that visit these countries each year, enabling accelerated improvements in their performance. Second, it helps the company embed its brand deeply within the diverse demographic fabric of our society. It is these two vectors that allow AgiBot to scale its data engine and amplify trust in its brand across society globally at a galactic pace.[306-309]

Now that we have distilled the core lessons embedded in AgiBot's trajectory, from its origins through its evolution and the pivotal decisions the team made along the way, it is time to decode the underlying business model.

In the final section of this chapter, we will break down the business model and illustrate how the product portfolio maps to select primary use cases.

AgiBot: Product Snapshot — A3 Humanoid

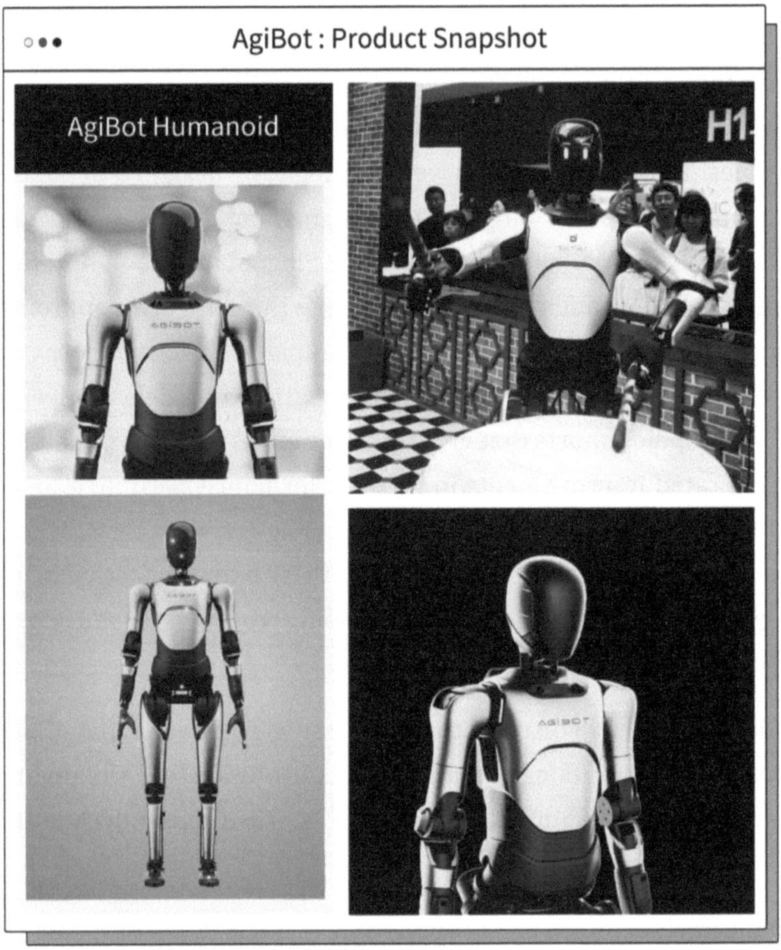

Figure 27.1 — AgiBot: A3 Humanoid

Illustration by the author. Robot images from AgiBot's product specifications page.

AgiBot's Business Model — Part 1 of 2

Figure 27.2 — AgiBot's Business Model: Part 1

Illustration by the author.

AgiBot's Business Model — Part 2 of 2

AgiBot's Business Model

- **Manufacturing Scale:** Agibot rolled out its 5,000th mass-produced humanoid robot in early 2026.
- **Funding:** A private company backed by LG Electronics and Mirae Asset.
- **Go-To-Market Strategy:** A product-led commercialization model strengthened by strategic industrial partnerships spanning the automotive and electronics manufacturing sectors.
- **Core Customer Segment 01:** Automotive & electronics manufacturing firms.
- **Core Customer Segment 02:** Developer and partner ecosystem leveraging APIs.
- **Current Customer Base:** Minth Group , China Mobile, Longcheer Technology, Joyson Electronic, and Fulin Precision.
- **Price Point — Humanoid Robots:** Pricing is configuration-dependent and varies based on model, hardware specifications, and deployment requirements.
- **Market Position:** Shipped more than 5,100 humanoid robots in 2025, capturing approximately 39 percent of global market share.
- **Revenue Model:** Hardware sales across the A-Series, X-Series, and G-Series platforms, complemented by services.

Figure 27.3 — AgiBot's Business Model: Part 2

Illustration by the author.

Chapter 28: Hard-Won Lessons from 25 Robotics Startups

Now that we have examined the trajectories of 25 founders and their startups, we will distill some of the important lessons from this research in this chapter. This is by no means a fully inclusive list of all the core lessons, but as I reflected on each of these trajectories, these are the ones I felt deserved a place in this chapter, as they are unique in their own ways and extremely valuable in the long run for the next generation of entrepreneurs and researchers who are about to kickstart their journey in this arena.

Lesson 01: Boston Dynamics, Market Strategy — Win Small Markets Before You Chase Big Ones

If you notice, in the case of Boston Dynamics with Spot, it was able to penetrate a reasonably niche market by targeting the defense segment first, followed by remote industrial inspection use cases with quadruped robots—but the trade-off was the size of that market and the urgency of the need. Many of these industrial operators didn't trust quadrupeds and were reluctant to adopt them at scale to automate inspection workflows.

While a small market segment like this makes it easier for early-stage robotics startups to gain traction, it will also eventually become your major revenue bottleneck. So, once you have validated your product in this market and established a steady stream of revenue and a strong array of clients to build your roots as a startup, adapt your product to serve adjacent target markets, just as Boston Dynamics did by extending its product portfolio from Spot to the Stretch robot arm, and eventually toward full-scale general-purpose humanoid, pivoting from inspection-only workflows to enabling full-scale automation for industrial operators.

These larger markets provide the financial runway you ultimately need to sustain your company—so know when to pivot.

Lesson 02: Amazon Robotics, Acquisitions — Use Acquisitions to Eliminate Technical Bottlenecks, Not to Rent Users

Amazon acquired Kiva to bring warehouse automation technology in-house, not to rent a user base. Similarly, it acquired Canvas Technology to add a strong set of eyes to its robots and unblock the perception bottleneck. In both cases, Amazon leveraged acquisitions to address a critical technical bottleneck; it did not attempt to rent a user base. A rented user base will not last long. Trust cannot be rented; it can only be earned. Therefore, acquisitions should always be approached with this objective in mind.

Lesson 03: Kiva, Product Positioning — Lead with Outcomes, Not Robots

Kiva never sold robots. It sold outcomes. The message was simple and quantified: triple warehouse throughput per worker. At scale, that outcome mattered deeply to customers. Eventually, this messaging translated into: buy our robots and get acquired by Amazon. That was the reality for many of its customers who were eventually acquired by Amazon—from Diapers.com to Zappos and more.

In robotics, outcomes close deals. The lesson here is to define and quantify the results you are guaranteeing in clear, measurable terms—and commit to them.

Lesson 04: Fetch — Distribution, Not Autonomy, Determines Scale

Partners played a critical role in Fetch's journey. DHL enabled the evaluation and validation of Fetch's solution on international soil, demonstrating its potential for global impact. Honeywell and Zebra helped validate its integrator-led distribution model, enabling Fetch to bring its product to thousands of enterprise customers almost overnight, effectively replacing dozens of legacy labor-heavy workflows with frontier robot-led, resource-efficient workflows—helping these customers scale their throughput.

This hub-and-spoke distribution model demonstrates the importance of evaluating and selecting the right distribution partners. If you do this right, it can significantly accelerate your journey toward the commercialization phase and ultimately help you achieve the scale velocity needed to escape the startup orbit. In doing so, you can meaningfully scale your revenue streams.

Lesson 05: 6 River Systems — Founder DNA Matters More Than Capital

In the case of 6 River Systems, the founders brought deep expertise in warehouse automation and robotics and had previously worked at Kiva, the warehouse automation pioneer that was later acquired by Amazon. Even more importantly, one of the founders had previously worked at Shopify, leading its logistics business segment. In summary, this trio had the perfect DNA for this company to succeed. One brought deep expertise in warehouse automation and robotics, another in technology, and the third a deep bond with Shopify. These three capabilities were critical to its success. Hence, founder DNA always outweighs capital, and 6 River Systems exemplifies this principle like no other.

Lesson 06: Symbotic — Chase Few High-Quality Customers That Offer Scale

Whether you are targeting small businesses or enterprise customers, you are better off chasing a few high-quality customers. Quality here matters far more than quantity. These are customers that either have the potential to grow at scale or are already operating at scale. In the case of Symbotic, all four of its early customers—Target, Walmart, Giant Tiger, and Albertsons—were already operating at scale. This implies that even though it might take a six-month-long pilot with each to secure a small-scale deal in the early stages, it is worth it because, from a long-term perspective, it unlocks scale and operational efficiency for a startup trying to build its financial runway.

In the case of Walmart, the pilot system deployed at the Florida facility, even though it did not immediately result in a contract for Symbotic, allowed the team to iteratively improve the fleet of robots operating at the site over a period of three years. This was critical because, after this phase, Walmart returned and decided to roll out the robots across all 42 distribution centers nationwide. That is revenue for a lifetime. Hence, it is important to chase a few high-quality customers that offer scale.

Lesson 07: Clearpath Robotics — No Matter Your Starting Point, You Can Fund Your First Prototype

The most important lesson here is that the founders of Clearpath, fresh out of the University of Waterloo, were able to come up with creative ways to fund their first prototype. While raising a seed round at an early stage could be an extremely daunting task, and often not the best move when you are still validating your thesis, it goes without saying that without capital resources even this part of the journey becomes extremely challenging.

To address this challenge, the founders strategically cut down their expenses by operating out of a basement space and the accelerator facilities offered by the university, and additionally secured grants from government agencies such as the Ontario Centre of Excellence and the Natural Sciences and Engineering Research Council of Canada. So, no matter your context, even before you raise a seed round, there are ways to fund these early prototypes, and government grants are indeed a strategic channel.

Lesson 08: AutoStore — Solve One A+ Problem That Truly Matters

The importance of this lesson cannot be overstated: focus on solving one core problem — and solve it exceptionally well. The right problem is one you deeply understand and one that is so painful for your customer that, if solved with the correct approach, it should at least double throughput and cut operating costs in half. That is exactly what the Hatteland family did. They encountered this problem firsthand while trying to scale their own warehouse to support a growing electronics business. They wanted to do so without expanding its physical footprint. Their robot-led solution maximised aerial and floor-space utilisation while enabling high-speed inventory handling and order fulfilment workflows.

In effect, what started as an experiment to devise a robot-led solution for their own warehouse became a system powerful enough to serve top-tier customers worldwide, from Medline and the FBI to Whole Foods and H Mart. The takeaway: find one problem that truly matters, and build a world-class solution around it.

Lesson 09: Universal Robots — Early Adopters Are Everywhere

Most founders in robotics tend to chase the same set of enterprise customers during the early stages—hospitals, mall operators, oil refineries, or large-scale manufacturing facilities. While these customers do offer scale and are important for long-term growth, they are often the hardest to penetrate and the slowest to adopt new technologies. In the early stages of a robotics startup, what matters most is not scale but speed of adoption. You need customers who are willing to experiment, deploy your product, and validate that it delivers real value.

These early adopters are everywhere. Think small and mid-sized farms, family-owned metal and recycling businesses, the leather industry, or even the fishing industry. There are countless sub-verticals that remain largely un-roboticised. Just like the founders of Universal Robots landed opportunities with dairy farmers in Iceland, metal product manufacturers in Denmark, and small-scale automotive parts suppliers in Germany, you can do the same and secure your early adopters with ease. In summary, opportunities to validate your product exist everywhere—you just have to look beyond the obvious markets.

Lesson 10: ANYbotics — Own One Most Painful Use Case in the Market

ANYbotics made an intentional decision to position its product around hazardous environments. Across customers ranging from Vale and Equinor to nuclear facilities in Finland and chemical plants operated by BASF, the pattern was consistent: the company focused on environments where the operational risk was highest and the cost of failure was significant.

Although the underlying platform was capable of operating in more general-purpose settings such as warehouses, the company chose not to compete in saturated markets. Instead, it concentrated on a small number of verticals where the need was acute and the value proposition was strongest. This approach made it easier for the right customers to understand the product, for the team to sell that product, and for the company to stand out in an ocean of robot-led solutions. Strong positioning comes from focus. Products that are designed and marketed around a specific, high-stakes use case tend to see faster adoption than those positioned broadly across many markets.

Lesson 11: Exotec — Quantified Milestones Is the Language of Investors

One of the most important lessons we can learn from this team is that each investment round they raised was driven by quantified goals that the team promised and delivered to its customers and investors.

The seed fund of $3.5M in 2015 was structured around the objective of landing one large-scale deal in the industry, and that goal was achieved in 2017 with the CDiscount deployment at the Bordeaux, France site. Similarly, Series B in 2018 was raised for a total of $17.7M, with the goal of selling at least 1,000 robots in the coming year. That goal was achieved, as in 2019 the company secured large-scale contracts with Carrefour, CDiscount Expansion, and UNIQLO.

Likewise, Series C of $90M was raised in 2020 to establish a strong footprint in North America, and by 2021 the company had already secured contracts with Gap, Decathlon, and Geodis. Thus, it is essential to raise capital with quantified milestones, remain transparent about these goals with investors, and deliver ahead of schedule to build enduring trust, and respect in your brand from Day 01. Trust signals your ability to deliver on commitments, while respect comes from doing so in a manner aligned with your values.

Lesson 12: iRobot — Capital Structure Determines Survival

Capital structure often shapes outcomes, especially in the hardware-centric business segments like consumer robotics. Being capital-intensive, if you do not define your funding strategy with clarity from the very beginning, you will struggle to secure a sustainable financial runway.

That is why it is almost always important to define a hybrid capital structure from day one. This means not relying on a single class of investors to fund the company. This is a mistake many founders make, often turning to high-interest debt-based channels at the last minute. Debt buys you time but also adds unreasonable deadlines that are hard to meet, and rightfully so. For this reason, founders should think through and define a hybrid capital structure for their startup from the outset. How do you do that?

You need to adopt a hybrid strategy. Put simply, if you want to build a financially resilient robotics company, you should start with accelerators to establish early-stage credibility in the market. Next, expand into retail through crowdfunding channels to build a community of loyal users who genuinely love and value your product. Take feedback from this community seriously and use it to refine your product. Then engage institutional customers to scale the business. Many startups skip the retail hop, but doing so often negatively impacts your runway in the long run.

Lesson 13: Anki — Demand Was Real, but Unit Economics Weren't

While Anki proved that the demand for consumer robots is indeed real, the economics in this arena are indeed unforgiving. Manufacturing these robots and securing components at a price that would make the venture sustainable was probably the hardest problem for the founders to solve. In addition to that, robots like Cozmo and Vector were not fully consumer-ready and required ongoing software support and continued access to the company's cloud-based servers, which inflated operating expenses. Given that the company adopted a hardware-first sales model, a one-time purchase without any recurring revenue layer, these robots could not generate enough lifetime value to cover the capital expenditures and operating costs required to support them at scale, even in the face of strong consumer demand. Thus, while these robots won the hearts of consumers, the business model failed to win the confidence of investors and suppliers, demonstrating that even strong demand cannot save a business built on weak unit economics.

Lesson 14: Jibo — A Great Product Delivered Too Late Is Indistinguishable from a Bad One

Jibo had a strong mission and organic demand validated at scale, with 6,000+ backers contributing a total of $3 million in the crowdfunding campaign. In this case, the vision, product, and economics were aligned, but the company failed to get the product out to its users in time. The multi-year gap between the announcement of its first prototype in 2014 and commercial shipment in 2017 gave Big Tech time to flood the market with cheaper substitutes.

During those years, big players such as Amazon, Google, and others had started proliferating this space with voice-assisted smart speakers that were introduced at about one-tenth of the price of Jibo and provided a reasonably similar functionality, albeit without the same embodied form factor and social experience. Turns out consumers didn't care much about the experience, but prioritized affordability and convenience. Amazon alone ended up selling about 71 million smart speakers during those years, while Google sold an additional 52 million. This shift in the market posed an existential challenge for Jibo and ultimately led to its shutdown. Thus, timing matters as much as vision and economics in robotics. In simple words, a great product launched too late in the market is often indistinguishable from a bad one and often gets caught in Big Tech's nets.

Lesson 15: Kuri — Capital Structure Can Make or Break a Robotics Startup

A single funding lifeline is a fatal vulnerability, especially if you are a founder running a robotics startup. Kuri, designed by Mayfield Robotics, was an elegant, beloved, and well-engineered social robot, but it had only one financial lifeline, namely Bosch. Once the parent company pulled the plug, the startup had no runway left.

This is especially painful because Kuri was a product of extraordinary calibre, with design and engineering that rivalled the craftsmanship of the finest Japanese masters and the precision of German technologists. And yet, none of that mattered once the funding lifeline disappeared overnight. Thus, consumer robotics startups need resilient capital structures that strategically marry at least four diverse sources in harmony: accelerators, retail investors, VCs, and strategic partners. Fragility here is not an option.

Lesson 16: Astro — Trust Was the Product Constraint

A smart speaker sits in one place. A security camera is fixed to a wall. Astro was different: it was a mobile camera roaming through private living spaces. To enable it to work in your home, you had to grant it full rights to map your home and scan it 24/7. That created immediate privacy-led friction, which throttled mainstream adoption of the robot. Astro's trajectory reinforces a recurring lesson in consumer robotics: even with strong underlying engineering and one of the most trusted consumer brands in the world, home robots do not scale without a must-have use case that significantly outweighs privacy and trust concerns. When the balance tilts the other way, trust becomes the product constraint.

Lesson 17: Starship Technologies — Market Positioning & Regulatory Alignment (University Campuses)

One of the most important lessons from Starship Technologies is the company's laser focus on its primary target segment: university campuses. Since its inception, the company has remained disciplined in serving this market, even as autonomous delivery robots had started gaining traction across broader logistics and retail sectors. Over time, Starship has established partnerships with 65+ universities across the United States, a scale enabled in part through its collaboration with Sodexo, the global food service operator that provides dining services across these campuses. This singular focus on its core segment from day one has allowed Starship to stand out as a robot-led, cost-efficient, and high-speed delivery service provider for students within these environments.

In parallel, the company and its founders prioritized early regulatory alignment. From day one, Starship proactively worked with city governments and regulators—from London to Washington, D.C.—to ensure that its robots were deployed in ways that aligned with local laws and community expectations. The result was a win-win model: students received high-speed, cost-efficient delivery, campuses benefited from elevated operational efficiency, and regulators ensured responsible deployment of autonomous systems.

Lesson 18: Serve Robotics — De-Risking the Path to Public Markets (OTCQB to NASDAQ)

Serve understood the importance of raising capital in a proactive manner, given that it is a deep-tech robotics company, one of the most capital- and hardware-intensive categories on the planet. Instead of waiting for years to march toward an IPO, the company realized that, at some point, it would need to raise funds through public markets, since securing sustained private investment in the long run for a startup that is both capex and opex heavy can be a daunting task.

To minimize the friction that most startups face when heading toward an IPO, and to address bottlenecks along the way with a more proactive approach, Serve, rather than heading straight to Nasdaq or mainstream markets, first launched on the OTCQB Venture Market in March 2024. It then uplisted to the Nasdaq Capital Market in April 2024 under the ticker SERV, with a $40 million public offering. This was an important decision, as it not only helped Serve manage the operational and regulatory risks inherent on this path for a company at this stage entering public markets, but it also placed the company in front of consumers early, allowing them to become champions and investors. This win-win approach ensured that as Serve began launching its robots in new geographies, it had already started building a community of supporters in each of these regions post-IPO.

Lesson 19: Kiwibot — Dual Revenue Streams Enable Sustainable Runway

One of the most important lessons from Kiwibot's journey is that last-mile delivery startups cannot rely on delivery margins alone.

Campus delivery is an operationally resource-intensive service and is often constrained by thin per-order profit margins. This economic reality makes it difficult to scale and sustain the business using this logistics revenue channel as the sole revenue arm. Kiwibot addressed this challenge by building a dual-engine revenue model, pairing delivery operations with an advertisement-based monetization channel through its acquisition of the ad-tech company Nickelytics. Thus, by enabling its robots to serve as both mobile billboards and delivery channels, the company unlocked an additional revenue stream that strengthened its financial runway and minimized its dependency on delivery fees alone.

Lesson 20: Coco — Partnerships Compound Faster Than Capital

From its inception around 2020 to less than five years in operation, Coco has already completed over 500,000 deliveries across two continents, in partnership with food-delivery giants such as DoorDash and Uber Eats. By partnering with these platform providers and working closely with local merchants operating on them, Coco was able to scale its distribution channels in an extremely time- and capital-efficient manner. Building on strong traction in the United States, DoorDash later made a decisive move to extend this partnership internationally, bringing the service to residents of Helsinki, Finland through its subsidiary Wolt—a delivery platform it acquired in 2022 to accelerate global expansion—effectively scaling Coco's robotic delivery network alongside it. The core lesson is that if, as a founder, you establish a few high-quality partnerships early on, you can proactively eliminate many of the distribution bottlenecks startups typically face along this path, while compounding revenue streams and accelerating market traction at a pace critical for early-stage companies to thrive over the long run.

Lesson 21: Figure AI — Fundraising: The Best Pitch Is a Paid Customer

Each round of capital raised by Figure AI was justified quantitatively: the company could point either to a paid customer deployment or to one already underway. In each round, Figure AI ensured that its robots were either operating on factory floors and generating revenue, or backed by signed agreements from real industrial customers prepared to pay for them—and, more importantly, believed in the value they had to offer.

Securing a pilot with BMW at one of its largest U.S. production facilities so early on is a clear example of this principle in action. Thus, the strongest fundraising pitch is proof of demand. A paying customer signals credibility in a way no slide deck ever can. With that level of validation in the market, investors place far greater trust in both your mission and your team than they would without it. At that point, a single slide that conveys your mission, the caliber of your team, and undeniable execution can speak louder than any elaborate presentation. All you need is one slide with those three pillars, demonstrating proof of execution that does not violate the principles of time, speed, and impact that define the startup cosmos.

Lesson 22: 1X — Market Strategy: Dual-Wedge from Day One (B2B & B2C)

1X's strategy from day one has always been clear: build humanoid assistants for society, and that meant a dual-wedge approach targeting both factories (B2B) and homes (B2C) from the very beginning. The company initially focused on the industrial edge first, launching EVE and securing a large-scale contract for it with ADT, proving early traction in factory environments. At the same time, 1X was already kickstarting its consumer engine, which led to the launch of NEO, one of its first consumer-facing humanoid robots designed for domestic environments. In summary, 1X seeded its journey with a singular mission to bring friendly, resourceful humanoid assistants into our society, starting at the industrial edge and gradually scaling its footprint toward the consumer edge, with this dual-wedge model eliminating single points of dependencies and defining a resilient business strategy from the get-go.

Lesson 23: Agility Robotics — Revenue First, Perfection Second (Academia & GXO)

What's probably one of the hallmark lessons in the trajectory of Agility Robotics is that the team launched and sold a bipedal robot without any top half, i.e., no eyes and no perception system, called Cassie, to academic research labs, and used that early-stage revenue to fund its trajectory. It was this early-stage commercialization pathway that enabled Agility to build its next platform, Digit, a full-scale humanoid. Digit, in turn, was adopted by Ford, Amazon, and GXO while still not being a fully autonomous humanoid, and this second revenue stream helped the company scale its product iteration cycles toward full autonomy.

The core lesson here is that, as a founder, you do not have to wait until your product is perfect before launching it into the market. That reflects an enterprise-grade product development mindset. You are a startup, and you must surgically define the right transition point at each phase of your journey to move your product from the lab into the real world, just as Agility did with Cassie and Digit, without pouring capital into decade-long R&D cycles.

Lesson 24: UBTECH — GTM: Build a Cross-Vertical Anchor Customer Base (Airbus, Foxconn, BYD, Audi, Texas Instruments)

UBTECH has adopted a multi-vector GTM strategy to build its anchor customer base across sectors spanning automotive, aviation, electronics manufacturing, and chip manufacturing. Strategic deployments involving Foxconn, BYD, Zeekr, Texas Instruments, Audi, and Airbus reflect a calibrated and structured go-to-market strategy anchored in the enterprise validation flywheel spanning increasingly complex and equally diverse set of industrial environments.

Each anchor customer serves as both a proof point and a wedge into adjacent sectors, helping the company build organic credibility, grow its total addressable market base, accelerate adoption velocity, and minimize concentration risks that often emerge when a startup is overly exposed to a single narrow vertical. The lesson here is that, as a humanoid startup, your commercialization strategy should not be anchored in one vertical, but ideally should be anchored across a portfolio of enterprise customers spanning multiple verticals. This array of diverse, high-value anchors enables you to compound your adoption footprint over time while minimizing single points of dependency.

Lesson 25: Apptronik — Scale: Robots Building Robots (The Jabil Partnership)

One of the most strategically important partnerships Apptronik has been able to land is with Jabil, one of the world's largest manufacturing services companies. This partnership is critical for Apollo's trajectory, as it helps unblock two critical risk vectors: scale of production and integration directly with an anchor partner that already brings that same scale to enterprise giants.

In simple words, the partnership with Jabil extends far beyond capital investments. Jabil now aims to help Apptronik both set up its production facilities to mass-produce these robots at scale, and more importantly, it plans to do so by enabling these robots to produce more of them, with minimal intervention from factory professionals. That is the first-order effect of this partnership. The second is that, in doing so, Apollo will end up getting infused directly into the very production engine owned by Jabil that manufactures products for enterprise giants around the world, including, but not limited to, Apple, Amazon, Johnson & Johnson, Ericsson, Cisco Systems, Hewlett-Packard, and Tesla. In summary, the core lesson is that with a single strategic partnership, Apptronik has been able to simultaneously scale both production and distribution. In doing so, it has effectively created what may be described as a mission-critical partnership—one that unlocks scale across all fronts.

Lesson 26: AgiBot — Tourism & Hospitality as a Strategic Adoption Wedge (Thailand, Malaysia, Singapore)

A defining strategic insight from AgiBot's journey is its deliberate entry into high-visibility, consumer-heavy sectors that help normalize its adoption curve while seeding its brand deep within the core layers of society. By venturing into the tourism and hospitality verticals at an early stage in its journey—particularly through partnerships aligned with Thailand's smart tourism initiatives—AgiBot positioned its humanoids in high-footfall, human-robot-interaction-heavy environments such as concierge services, events, and public-facing venues. These deployments were strategically designed to build a strong social-proof engine. In tourism- and hospitality-driven markets like Thailand, Malaysia, and Singapore, robots interact daily with thousands of users from across the globe. This approach builds a dual-layered moat for the company. First, it exposes its humanoids and the underlying conversational models to diverse demographic cohorts that visit these countries each year, enabling accelerated improvements in their performance. Second, it helps the company embed its brand deeply within the diverse demographic fabric of society. It is these two vectors that allow AgiBot to scale its data engine and amplify trust in its brand globally at a galactic pace.

Chapter 29: How Founders & Engineers Can Apply This Book

This book is organized into seven core segments, each covering the journeys of founders and their startups operating across distinct classes of robots—from industrial mobile robots and warehouse systems to consumer robots, delivery bots, and humanoids. Depending on the arena in which you plan to build your startup, or are already building one, the lessons across these segments—along with the business models they culminate in—are intended to equip you with a series of approaches you can adopt to address the bottlenecks inherent along your path and ultimately bring your mission to life.

My goal here is to ensure that through these case studies you are able to distill the decisions that helped these companies navigate some of the most challenging stages of their journeys and apply those insights to your own.

Below is a simple framework you can use to apply the insights from this book to your startup journey.

01. Map Your Bottlenecks

Take a piece of paper, ideally an A4 sheet, and make a list of your A+ bottlenecks across the full spectrum of your startup's lifecycle: funding, customers, scale, partnerships, manufacturing, product–

market fit, and early adopters. Keep these in mind as you start studying the trajectories of the startups and founders in this book.

02. Start With Your Segment

Based on your primary class of robots, deeply study the trajectories of the startups and their corresponding founders operating in that space that are covered in the respective segments of this book. Segment I is applicable across the board, so start there and then proceed to your primary segment from Segments II–VIII based on the class of robots you are building. Whether you are building humanoids, warehouse robots, quadrupeds, social robots, or delivery robots, start here.

03. Decode Founder Decisions

After reading each chapter, reflect on and document the specific decisions the founders made, along with the specific parameters of their business models that apply to your startup and can help you address the bottlenecks across the areas you listed in Step 01.

04. Learn From Adjacent Segments

Now repeat this process for segments that may not directly apply to you but are still important, since the underlying decisions founders made and the driving factors behind them are often domain-agnostic.

05. Consolidate Insights — Build Your Founder Playbook

Now reflect on the lessons, decisions, and parameters of the business models that you have documented as you advanced through this book, listing solutions to each of the bottlenecks you identified in the areas that matter most to you and your team today. At this point, you will have distilled these insights into a founder playbook that captures the approaches most relevant to your startup and its mission.

06. Unblock Your Bottlenecks & Bring Your Mission to Life

With your founder playbook now in place, start applying these insights to systematically address the most critical bottlenecks in your startup today. Your goal here is to study these founder journeys and translate their decisions into actions that you can execute today—actions that will help you move at a rapid pace, scale in a resource-efficient manner, and ultimately bring your mission to life: one that you started this journey with in the first place and one that your team has put all their hopes, efforts, time, and dreams into.

You're Not Alone On This Path

Building a robotics company—or helping one overcome its bottlenecks and succeed—is equally hard. I know that many of you reading this book are founders, researchers, and engineers working to bring the next generation of robots to life. Some of you may already be building robotics startups, while others are preparing to take that step into entrepreneurship in this space. In either case, this journey is not easy. The winds of distraction and the cliffs of failure can be felt with every step you take along this path. That said, you are not alone.

1. **Contact my team:** Reach out to my team using the chatbox located on https://boringsage.com if you ever have a

question about anything. If not myself, someone from our team will address your question.

2. **Explore our resources:** We regularly share deep dives on career trajectories in this space, startup lessons, and technology breakthroughs on our blog. Based on the problem you are trying to solve, in addition to this book, you will find these resources valuable. You can explore them here: https://boringsage.com/blog

3. **Join our community:** We have a community of robotics professionals who have signed up for our monthly newsletter. We regularly post updates on the resources we share on our blog, upcoming launches of our books, products, and courses. You can sign up for free here: https://www.boringsage.com/pulse

Thank You

At Boring Sage, our mission is to bring deep, field-tested professional expertise to those who don't yet have access to it firsthand. From equipping founders with hard-earned lessons from the field to helping engineers approach their careers with a structured framework, our goal is to enable you to navigate the risks inherent on this path proactively—so you can bring the next generation of robots to life with peace of mind and address emerging challenges in society.

It is my hope that the lessons and frameworks discussed here will prove incredibly valuable in your journey. I wish you the very best as you carry these lessons into the field, and I look forward to seeing your robots in action in the real world.

Image Credits

All diagrams, schematics, and illustrations in this book were created by the author unless otherwise stated.

Portions of select figures include images reproduced, excerpted, or adapted from publicly available materials from the following organizations and sources, used for educational, research, commentary, and illustrative purposes: Agility Robotics; AgiBot; Amazon Robotics / Amazon, Inc.; ANYbotics; Apptronik; AutoStore; Boston Dynamics; Clearpath Robotics; Coco Robotics; Exotec; Fetch Robotics; Figure AI; iRobot; Kiwibot; Kiva Systems; Mayfield Robotics; Serve Robotics; Starship Technologies; Symbotic; UBTECH Robotics; Universal Robots; and 1X Technologies.

Source materials include official company product pages, technical presentations, research publications, engineering blogs, corporate announcements, investor materials, product specification pages, and publicly released media assets, including official company websites, press resources, and YouTube channels.

All third-party trademarks, logos, and images remain the property of their respective owners. Their inclusion in this book does not imply any form of endorsement, sponsorship, or affiliation.

The author has made reasonable efforts to identify and credit original sources where applicable. If any copyright holder believes material has been used in error or without proper attribution, the author and publisher welcome notification so that corrections can be made in future editions.

References

[01] E. Guizzo, "Amazon acquires Kiva Systems for $775 million," IEEE Spectrum, Mar. 19, 2012. [Online]. Available: https://spectrum.ieee.org/amazon-acquires-kiva-systems-for-775-million

[02] E. Ackerman, "Boston Dynamics' founder on the future of robotics," IEEE Spectrum, Jul. 31, 2023. [Online]. Available: https://spectrum.ieee.org/boston-dynamics/marc-raibert

[03] Alert Venture Foundry, "John Lert, Founder and CEO," Alert Venture Foundry, 2025. [Online]. Available: https://alertventure.com/john-lert

[04] A. Feldman, "Rick Cohen's Symbotic expands deal to automate Walmart's warehouses to all 42 regional distribution centers," Forbes, May 23, 2022. [Online]. Available: https://www.forbes.com/sites/amyfeldman/2022/05/23/rick-cohens-symbotic-expands-deal-to-automate-walmarts-warehouses-to-to-all-42-regional-distribution-centers/

[05] A. Feldman, "Meet the billionaire robot overlord reinventing Walmart's warehouses," Forbes, Dec. 13, 2021. [Online]. Available: https://www.forbes.com/sites/amyfeldman/2021/12/13/meet-the-billionaire-robot-overload-reinventing-walmarts-warehouses/

[06] Financial Times, "A remote village, a world-changing invention and the epic legal fight that followed," Financial Times, 2024. [Online]. Available: https://www.ft.com/content/7f12906e-fc6a-403b-aa41-aca03724b76f

[07] B. Lam and P. Ngo, "Meet Peng Zhihui, the 32-year-old mind behind Agibot, China's rising robotics star," VnExpress International, Apr. 14, 2025. [Online]. Available: https://e.vnexpress.net/news/tech/tech-news/meet-peng-zhihui-the-32-year-old-mind-behind-agibot-china-s-rising-robotics-star-4873865.html

[08] X. Shen, "Meet Peng Zhihui, the latest young robotics star to catch the Chinese premier's attention," South China Morning Post, Apr. 10, 2025. [Online]. Available: https://www.scmp.com/tech/tech-trends/article/3305986/meet-peng-zhihui-latest-young-robotics-star-catch-chinese-premiers-attention

[09] B. Børnich, "Bernt Børnich – Founder and CEO at 1X," LinkedIn, 2026. [Online]. Available: https://www.linkedin.com/in/bernt-bornich-032ba38/

[10] Oregon State University, "Jonathan Hurst," Oregon State University College of Engineering, 2026. [Online]. Available: https://engineering.oregonstate.edu/people/jonathan-hurst

[11] Agility Robotics, "Agility Robotics broadens relationship with Amazon," Agility Robotics Press Release, Oct. 24, 2023. [Online]. Available: https://www.agilityrobotics.com/content/agility-robotics-broadens-relationship-with-amazon

[12] A. Keay, "The robots of CES 2023," AMT (Association for Manufacturing Technology), Feb. 21, 2023. [Online]. Available: https://www.amtonline.org/article/the-robots-of-ces-2023

[13] ETH Zurich Robotic Systems Lab, "ANYbotics," ETH Zurich, 2026. [Online]. Available: https://rsl.ethz.ch/partnership/spinoff/anybotics.html

[14] E. Guizzo, "How Boston Dynamics is redefining robot agility," IEEE Spectrum, Nov. 27, 2019. [Online]. Available: https://spectrum.ieee.org/how-boston-dynamics-is-redefining-robot-agility

[15] Boston Dynamics, "Innovation at Boston Dynamics: Our history," Boston Dynamics, 2026. [Online]. Available: https://bostondynamics.com/about/history/

[16] E. Ackerman, "Hyundai buys Boston Dynamics for nearly $1 billion. Now what?," IEEE Spectrum, Dec. 11, 2020. [Online]. Available: https://spectrum.ieee.org/hyundai-buys-boston-dynamics

[17] Boston Dynamics, "Case studies," Boston Dynamics, 2026. [Online]. Available: https://bostondynamics.com/case-studies/

[18] Association for Advancing Automation, "Unimate – The first industrial robot," A3 Robotics (Automate.org), 2026. [Online]. Available: https://www.automate.org/robotics/engelberger/joseph-engelberger-unimate

[19] Defense Advanced Research Projects Agency (DARPA), "DARPA's ATLAS robot unveiled," DARPA, Jul. 11, 2013. [Online]. Available: https://www.darpa.mil/news/2013/atlas-robot-unveiled

[20] J. Markoff, "The umbilical link of man to robot," The New York Times, Oct. 21, 2013. [Online]. Available: https://www.nytimes.com/2013/10/22/science/robot-games-a-challenge-for-the-machines-and-the-controllers.html

[21] J. Markoff, "Google adds to its menagerie of robots," The New York Times, Dec. 14, 2013. [Online]. Available: https://www.nytimes.com/2013/12/14/technology/google-adds-to-its-menagerie-of-robots.html

[22] A. Tilley, "SoftBank acquires Boston Dynamics from Alphabet," Forbes, Jun. 8, 2017. [Online]. Available: https://www.forbes.com/sites/aarontilley/2017/06/08/softbank-acquires-boston-dynamics-from-alphabet/

[23] Boston Dynamics, "Boston Dynamics acquires deep learning based vision company," Boston Dynamics, Apr. 2, 2019. [Online]. Available: https://bostondynamics.com/news/boston-dynamics-acquires-deep-learning-based-vision-company/

[24] Boston Dynamics, "Hyundai Motor Group completes acquisition of Boston Dynamics from SoftBank," Boston Dynamics, Jun. 21, 2021. [Online]. Available: https://bostondynamics.com/news/hyundai-motor-group-completes-acquisition-of-boston-dynamics-from-softbank/

[25] Hyundai Motor Manufacturing Alabama, "The Hyundai team of plants," Hyundai Motor Manufacturing Alabama. [Online]. Available: https://www.hmmausa.com/manufacturing-plant/the-hyundai-team-of-plants/

[26] DHL Group, "DHL to deploy 1,000+ additional robots across UK operations to support business growth in e-commerce and life sciences health care sector," DHL Group, Jul. 9, 2025. [Online]. Available: https://group.dhl.com/en/media-relations/press-releases/2025/dhl-to-deploy-1000-additional-robots-across-uk-operations.html

[27] A. Hancock, "UPS to cut 30,000 jobs as it moves away from Amazon," BBC News, Jan. 27, 2026. [Online]. Available: https://www.bbc.com/news/articles/cdxjpj7j754o

[28] Amazon, "Employees," Amazon Sustainability. [Online]. Available: https://sustainability.aboutamazon.com/employees

[29] S. Dresser, "Amazon launches a new AI foundation model to power its robotic fleet and deploys its 1 millionth robot," Amazon News, Jun. 30, 2025. [Online]. Available: https://www.aboutamazon.com/news/operations/amazon-million-robots-ai-foundation-model

[30] S. Lebowitz, "Amazon is celebrating its 25th birthday. Here's how the e-commerce giant got its name," Business Insider, Jul. 5, 2019. [Online]. Available: https://www.businessinsider.com/amazon-jeff-bezos-chose-company-name-2018-5

[31] Red Stag Fulfillment, "How many Amazon warehouses are there? (2025 data)," Red Stag Fulfillment, 2025. [Online]. Available: https://redstagfulfillment.com/how-many-amazon-warehouses-are-there/

[32] Amazon, "Worldwide operations: Amazon Multi-Channel Fulfillment (MCF)," Amazon Supply Chain. [Online]. Available: https://supplychain.amazon.com/mcf/global

[33] B. Heater, "Amazon launches a cloud-based robotics testing platform," TechCrunch, Nov. 26, 2018. [Online]. Available: https://techcrunch.com/2018/11/26/amazon-launches-a-cloud-based-robotics-testing-platform/

[34] B. Heater, "Amazon acquires autonomous warehouse robotics startup Canvas Technology," TechCrunch, Apr. 10, 2019. [Online]. Available: https://techcrunch.com/2019/04/10/amazon-acquires-autonomous-warehouse-robotics-startup-canvas-technology/

[35] S. Scott, "Meet Scout," Amazon News, Jan. 23, 2019. [Online]. Available: https://www.aboutamazon.com/news/transportation/meet-scout

[36] I. A. Hamilton, "Amazon has agreed to buy Zoox for a reported $1.2 billion," Business Insider, Jun. 26, 2020. [Online]. Available: https://www.businessinsider.com/amazon-reportedly-buys-self-driving-car-startup-zoox-for-one-billion2020-6

[37] C. Tritschler, "Meet Astro, a home robot unlike any other," Amazon News, Sep. 28, 2021. [Online]. Available: https://www.aboutamazon.com/news/devices/meet-astro-a-home-robot-unlike-any-other

[38] T. Greenawalt, "Amazon has more than 1 million robots that sort, lift, and carry packages—see them in action," Amazon News, Oct. 22, 2025. [Online]. Available: https://www.aboutamazon.com/news/operations/amazon-robotics-robots-fulfillment-center

[39] D. Milmo, "Amazon 'testing humanoid robots to deliver packages'," The Guardian, Jun. 5, 2025. [Online]. Available: https://www.theguardian.com/technology/2025/jun/05/amazon-testing-humanoid-robots-to-deliver-packages

[40] J. Weatherbed, "Amazon is reportedly training humanoid robots to deliver packages," The Verge, Jun. 5, 2025. [Online]. Available: https://www.theverge.com/news/680258/amazon-training-package-delivery-humanoid-robots

[41] S. Dresser, "Amazon announces 2 new ways it's using robots to assist employees and deliver for customers," Amazon News, Oct. 18, 2023. [Online]. Available: https://www.aboutamazon.com/news/operations/amazon-introduces-new-robotics-solutions

[42] J. Durham, "Amazon builds first foundation model for multirobot coordination," Amazon Science, Aug. 11, 2025. [Online]. Available: https://www.amazon.science/blog/amazon-builds-first-foundation-model-for-multirobot-coordination

[43] Amazon Staff, "Amazon's delivery glasses: The newest innovation designed to enhance the delivery experience," Amazon News, Oct. 22, 2025. [Online]. Available: https://www.aboutamazon.com/news/transportation/smart-glasses-amazon-delivery-drivers

[44] C. Contreras, "Kiva Systems founders share company's story at Robotics Invest event in Boston," Robotics 24/7, Jun. 9, 2023. [Online]. Available: https://www.robotics247.com/article/kiva_systems_founders_share_companys_story_during_robotics_invest_event_boston/amazon_robotics

[45] E. Guizzo, "Kiva Systems: Three Engineers, Hundreds of Robots, One Warehouse," IEEE Spectrum, Jul. 1, 2008. [Online]. Available: https://spectrum.ieee.org/three-engineers-hundreds-of-robots-one-warehouse

[46] C. Harman, "USA: Webvan pulls the plug on Internet operations," Just Food, Jul. 10, 2001. [Online]. Available: https://www.just-food.com/news/usa-webvan-pulls-the-plug-on-internet-operations/

[47] A. Bindman, "The disastrous rise and fall of a $10 billion Bay Area unicorn," SFGATE, Oct. 15, 2024. [Online]. Available: https://www.sfgate.com/food/article/rise-fall-bay-area-startup-webvan-19829522.php

[48] G. Sandoval, "Webvan defends business model despite woes," CNET, Nov. 13, 2000. [Online]. Available: https://www.cnet.com/tech/tech-industry/webvan-defends-business-model-despite-woes/

[49] National Inventors Hall of Fame, "Peter Wurman," National Inventors Hall of Fame, 2022. [Online]. Available: https://www.invent.org/inductees/peter-wurman

[50] R. Malone, "Staples fastens onto Kiva," Forbes, Dec. 19, 2005. [Online]. Available: https://www.forbes.com/2005/12/19/staples-kiva-robots-cx_rm_1219robots.html

[51] C. Steiner, "Bot-in-time delivery," Forbes, Feb. 26, 2009. [Online]. Available: https://www.forbes.com/forbes/2009/0316/040_bot_time_saves_nine.html

[52] M. Wulfraat, "Is Kiva Systems a good fit for your distribution center? An unbiased distribution consultant evaluation," MWPVL International, Feb. 2012. [Online]. Available: https://www.mwpvl.com/html/kiva_systems.html

[53] R. Wauters, "Amazon closes Zappos deal, ends up paying $1.2 billion," TechCrunch, Nov. 2, 2009. [Online]. Available: https://techcrunch.com/2009/11/02/amazon-closes-zappos-deal-ends-up-paying-1-2-billion/

[54] J. Scanlon, "How Kiva robots help Zappos and Walgreens," Bloomberg Businessweek, Apr. 15, 2009. [Online]. Available: https://www.bloomberg.com/news/articles/2009-04-15/how-kiva-robots-help-zappos-and-walgreens

[55] Fibre2Fashion News Desk, "Gap, Old Navy, Banana Republic, Piperlime, and Athleta adopt Kiva robotic system," Fibre2Fashion, Jan. 2009. [Online]. Available: https://www.fibre2fashion.com/news/textiles-technology-news/newsdetails.aspx?news_id=68342

[56] E. Ackerman, "Zebra Technologies to acquire Fetch Robotics for $305 million," IEEE Spectrum, Jul. 1, 2021. [Online]. Available: https://spectrum.ieee.org/zebra-technologies-acquire-fetch-robotics

[57] J. Pransky, "The Pransky interview: Melonee Wise, CEO, Fetch Robotics," Industrial Robot: An International Journal, vol. 43, no. 3, pp. 253–257, May 16, 2016. [Online]. Available: https://www.emerald.com/ir/article-abstract/43/3/253/183311/The-Pransky-interview-Melonee-Wise-CEO-Fetch

[58] Atoms & Bits, "Melonee Wise, Agility Robotics CTO, on the evolution of robotics, the current 'Cambrian Explosion' of innovation, and the barriers companies face inviting robots into the home," Atoms & Bits, Dec. 4, 2023. [Online]. Available: https://atomsbitsnewsletter.substack.com/p/melonee-wise-agility-robotics-cto

[59] A. Vance, "Robotics research lab Willow Garage shuts down," Bloomberg Businessweek, Feb. 20, 2014. [Online]. Available: https://www.bloomberg.com/news/articles/2014-02-20/robotics-research-lab-willow-garage-shuts-down

[60] A. Nash, "Fetch from startup to exit, with Melonee Wise," Sense Think Act Podcast, Aug. 24, 2021. [Online]. Available: https://www.sensethinkact.com/episodes/1-melonee-wise

[61] D. Terdiman, "How two buddy industrial robots could upend the logistics industry forever," VentureBeat, Apr. 29, 2015. [Online]. Available: https://venturebeat.com/ai/how-two-buddy-industrial-robots-could-upend-the-logistics-industry-forever

[62] RK Logistics Group, "RK Logistics increases productivity with robots," May 30, 2017. [Online]. Available: https://rklogisticsgroup.com/rk-logistics-increases-productivity-with-robots/

[63] Wärtsilä Corporation, "Wärtsilä and DHL deploy cutting-edge mobile robots from Fetch Robotics to streamline warehouse operations," Press release, Jul. 11, 2017. [Online]. Available: https://www.wartsila.com/media/news/11-07-2017-wartsila-and-dhl-deploy-cutting-edge-mobile-robots-from-fetch-robotics-to-streamline-warehouse-operations

[64] Honeywell, "Honeywell partners with Fetch Robotics to deliver autonomous mobile robots to distribution centers," Press release, Oct. 16, 2018. [Online]. Available: https://www.honeywell.com/us/en/press/2018/10/honeywell-partners-with-fetch-robotics-to-deliver-autonomous-mobile-robots-to-distribution-centers

[65] Ryder System, Inc., "Ryder redefines the smart warehouse; deploys innovative mix of startup technologies," Press release, Jan. 17, 2019. [Online]. Available: https://newsroom.ryder.com/news/news-details/2019/Ryder-Redefines-the-Smart-Warehouse-Deploys-Innovative-Mix-of-Startup-Technologies/default.aspx

[66] Fetch Robotics, "Fetch Robotics announces integrated fulfillment solution with Zebra Technologies," Press release, Mar. 9, 2020. [Online]. Available: https://www.prnewswire.com/news-releases/fetch-robotics-announces-integrated-fulfillment-solution-with-zebra-technologies-301019369.html

[67] Zebra Technologies Corporation, "Zebra Technologies makes Newsweek's list of Most Loved Workplaces®," Jun. 13, 2024. [Online]. Available: https://investors.zebra.com/news-and-events/news/news-details/2024/Zebra-Technologies-Makes-Newsweeks-List-of-Most-Loved-Workplaces/default.aspx

[68] M. Murphy, "Congratulations to 6 River Systems on joining Shopify: A clear sign that the robotics market is accelerating!," Menlo Ventures, Sep. 9, 2019. [Online]. Available: https://menlovc.com/perspective/congratulations-to-6-river-systems-on-joining-shopify-a-clear-sign-that-the-robotics-market-is-accelerating/

[69] B. Heater, "6 River Systems co-founder on the state of warehouse robots," TechCrunch, Mar. 24, 2023. [Online]. Available: https://techcrunch.com/2023/03/24/6-river-systems-co-founder-on-the-state-of-warehouse-robots/

[70] R. Hamilton, "Rylan Hamilton – LinkedIn profile," LinkedIn. [Online]. Available: https://www.linkedin.com/in/rylanhamilton/

[71] C. Cacioppo, "Christopher Cacioppo – LinkedIn profile," LinkedIn. [Online]. Available: https://www.linkedin.com/in/cacioppo/

[72] L. Kolodny, "6 River Systems unveils warehouse robots that show workers the way," TechCrunch, Mar. 30, 2017. [Online]. Available: https://techcrunch.com/2017/03/30/6-river-systems-unveils-warehouse-robots-that-show-workers-the-way/

[73] E. Demaitre, "Robotic carts from 6 River Systems maximize productivity with the cloud, human pickers," The Robot Report, Jul. 16, 2018. [Online]. Available: https://www.therobotreport.com/robotic-carts-from-6-river-systems-maximize-productivity-with-the-cloud-human-pickers/

[74] A. Knapp, "This robot startup just raised $25 million to make warehouse fulfillment easier," Forbes, Apr. 4, 2018. [Online]. Available: https://www.forbes.com/sites/alexknapp/2018/04/04/this-robot-startup-just-raised-25-million-to-make-warehouse-fulfillment-easier/

[75] DHL, "Robotics Day 2018 – Sense the future," DHL Innovation Center, Nov. 14, 2018. [Online]. Available: https://www.event-dhl.com/ehome/355710/789299/

[76] DHL Supply Chain, "DHL Supply Chain expands its fleet of robots from 6 River Systems," DHL Press Release, Sep. 1, 2022. [Online]. Available: https://www.dhl.com/cz-en/home/press/press-archive/2022/dhl-supply-chain-expands-its-fleet-of-robots-from-6-river-systems.html

[77] DHL Group, "DHL to deploy 1,000+ additional robots across UK operations to support business growth in e-commerce and life sciences health care sector," DHL Group Press Release, Jul. 9, 2025. [Online]. Available: https://group.dhl.com/en/media-relations/press-releases/2025/dhl-to-deploy-1000-additional-robots-across-uk-operations.html

[78] XPO, Inc., "Corporate profile," XPO Investor Relations. [Online]. Available: https://investors.xpo.com/

[79] S. Gharib, "Shipping company XPO Logistics spends $500 million on tech a year," Fortune, Jul. 6, 2017. [Online]. Available: https://fortune.com/2017/07/06/xpo-logistics-ceo-investment-strategy/

[80] XPO Logistics, "XPO Logistics expands use of intelligent robotics in the UK ahead of holiday shopping season," Press Release, Nov. 19, 2019. [Online]. Available: https://www.xpo.com/newsroom/press-releases/xpo-logistics-expands-use-of-intelligent-robotics-in-the-uk-ahead-of-holiday-shopping-season/

[81] FreightWaves Staff, "XPO's largest robotic deployment to date will bring 5,000 units to its warehouse," FreightWaves, Oct. 3, 2018. [Online]. Available: https://www.freightwaves.com/news/xpos-largest-robotic-deployment-to-date-will-bring-5000-units-to-its-warehouse

[82] Bernard & Company, "Fulfillment automation provider 6 River Systems to unveil new mobile sortation solution at ProMat 2019," Bernard & Company, 2019. [Online]. Available: https://bernardandcompany.com/category/6-river-systems/

[83] M. Oitzman, "6 River Systems unveils new mobile sortation solution at ProMAT 2019," Automated Warehouse, Apr. 7, 2019. [Online]. Available: https://www.automatedwarehouseonline.com/6-river-systems-unveils-new-mobile-sortation-solution-at-promat-2019/

[84] E. Protalinski, "Shopify's 6 River Systems CEOs talk Chuck robots, automation, and the pandemic," VentureBeat, Jul. 27, 2020. [Online]. Available: https://venturebeat.com/ai/shopify-6-river-systems-chuck-robots-automation-pandemic

[85] B. Heater, "Shopify sells 6 River Systems to new owner," TechCrunch, May 8, 2023. [Online]. Available: https://techcrunch.com/2023/05/08/shopify-sells-6-river-systems-to-new-owner/

[86] Ocado Intelligent Automation, "Ocado Mobile Robot System (OMRS)," Ocado Group, 2023. [Online]. Available: https://ocadointelligentautomation.com/systems/omrs-ocado-mobile-robot-system

[87] S. Kirsner, "Symbotic's 25 m.p.h robots," *The Boston Globe*, Feb. 6, 2012. [Online]. Available: https://www.bostonglobe.com/business/2012/02/06/innovation-economy/JM5oWby5Qr7Yc3Jy17A7WP/story.html

[88] K. Zeman, "Albertsons Companies invests in automation technology and makes improvements to its distribution centers," Supply Chain World, Sep. 21, 2018. [Online]. Available: https://www.scw-mag.com/news/1098-albertsons-companies/

[89] C. Powell, "Giant Tiger opens new distribution facility," Canadian Grocer, Sep. 28, 2018. [Online]. Available: https://www.canadiangrocer.com/giant-tiger-opens-new-distribution-facility

[90] Business Wire, "UNFI and Symbotic announce agreement to implement industry-leading warehouse automation systems," Business Wire, Sep. 26, 2022. [Online]. Available: https://www.businesswire.com/news/home/20220926005808/en/UNFI-and-Symbotic-Announce-Agreement-to-Implement-Industry-Leading-Warehouse-Automation-Systems

[91] M. Smith, "High-tech consolidation center set to open in July, adding efficiency to Walmart's supply chain," Walmart Corporate, Jan. 14, 2019. [Online]. Available: https://corporate.walmart.com/news/2019/01/14/high-tech-consolidation-center-set-to-open-in-july-adding-efficiency-to-walmarts-supply-chain

[92] B. Heater, "Symbotic set to take over Walmart's robotics business," TechCrunch, Jan. 16, 2025. [Online]. Available: https://www.techcrunch.com/2025/01/16/symbotic-set-to-take-over-walmarts-robotics-business/

[93] B. Heater, "Walmart is adding Symbotic robots to warehouses across the country," TechCrunch, May 23, 2022. [Online]. Available: https://www.techcrunch.com/2022/05/23/walmart-is-adding-symbotic-robots-to-warehouses-across-the-country

[94] B. Heater, "Walmart will be bringing Symbotic robots to 25 distribution centers," TechCrunch, Jul. 13, 2021. [Online]. Available: https://www.techcrunch.com/2021/07/13/walmart-will-be-bringing-symbotic-robots-to-25-distribution-centers/

[95] B. Heater, "SoftBank and Symbotic team to offer automated warehouses as a service," TechCrunch, Jul. 24, 2023. [Online]. Available: https://www.techcrunch.com/2023/07/24/softbank-and-symbotic-team-to-offer-automated-warehouses-as-a-service/

[96] Symbotic Inc., "Symbotic announces agreement to implement multiple industry-leading warehouse automation systems for Walmex and strategic international expansion in Mexico," Symbotic, Oct. 10, 2024. [Online]. Available: https://www.symbotic.com/about/news-events/news/symbotic-announces-agreement-to-implement-multiple-industry-leading-warehouse-automation-systems-for-walmex-and-strategic-international-expansion-in-mexico/

[97] Clearpath Robotics, "Customer stories," Clearpath Robotics. [Online]. Available: https://www.clearpathrobotics.com/customers-robot-case-stories/

[98] Rockwell Automation, "Rockwell Automation signs agreement to acquire autonomous robotics leader Clearpath Robotics," OTTO Motors by Rockwell Automation, Sep. 5, 2023. [Online]. Available: http://www.ottomotors.com/company/newsroom/press-releases/rockwell-automation-signs-agreement-to-acquire-autonomous-robotics-leader-clearpath-robotics/

[99] S. Silcoff, "Waterloo's Clearpath Robotics sold to Rockwell Automation for US$600-million plus," The Globe and Mail, Sep. 12, 2023. [Online]. Available: https://www.theglobeandmail.com/business/article-waterloos-clearpath-robotics-sold-to-rockwell-automation-for-us600/

[100] RoboR@m, "Clearpath Robotics success story," RoboR@m, Sep. 12, 2010. [Online]. Available: https://www.roboram.wordpress.com/clearpath-robotics-success-story/

[101] H. Siegel, "Coming of age: Clearpath's Ryan Gariepy on growing a robotics startup," Robohub, Sep. 1, 2015. [Online]. Available: https://www.robohub.org/coming-of-age-clearpaths-ryan-gariepy-on-growing-a-robotics-startup/

[102] P. van der Vorst, "Husky goes lunar," Clearpath Robotics, Feb. 3, 2010. [Online]. Available: https://www.clearpathrobotics.com/blog/2010/02/husky-goes-lunar/

[103] Clearpath Robotics, "Coimbra University uses robust, reliable UGV for rugged mine-clearance robot," Clearpath Robotics Success Story. [Online]. Available: https://www.clearpathrobotics.com/wp-content/uploads/2014/05/Success-Story_Coimbra_WEB4.pdf

[104] M. Hennessey, "Clearpath launches small, fast, rugged robot: Jackal UGV," Clearpath Robotics, Sep. 15, 2014. [Online]. Available: https://www.clearpathrobotics.com/blog/2014/09/clearpath-launches-jackal-ugv/

[105] G. De Vynck, "Clearpath Robotics gets funds in push for robot trucks," Bloomberg, Mar. 18, 2015. [Online]. Available: https://www.bloomberg.com/news/articles/2015-03-18/clearpath-robotics-gets-funds-in-push-for-robot-trucks

[106] D. Soltys, "Clearpath Robotics raises $14 million Series A from RRE & iNovia," BetaKit, Mar. 18, 2015. [Online]. Available: https://www.betakit.com/clearpath-raises-14-million-series-a-from-rre-inovia/

[107] Clearpath Robotics, "Clearpath raises $30M to expand indoor self-driving vehicle market," OTTO Motors by Rockwell Automation, Oct. 5, 2016. [Online]. Available: https://www.ottomotors.com/company/newsroom/press-releases/clearpath-raises-30m-to-expand-indoor-self-driving-vehicle-market-otto-motors/

[108] E. Ackerman, "Clearpath Robotics introduces Kingfisher M200," IEEE Spectrum, Oct. 18, 2012. [Online]. Available: spectrum.ieee.org/iros-2012-clearpath-robotics-introduces-kingfisher-m200

[109] Clearpath Robotics, "Customer story: Deployable, autonomous vibration control of bridges using Husky UGV," Robohub, Jan. 5, 2017. [Online]. Available: https://www.robohub.org/customer-story-deployable-autonomous-vibration-control-of-bridges-using-husky-ugv/

[110] D. Nield, "This new robot from MIT can politely navigate through crowds," ScienceAlert, Aug. 31, 2017. [Online]. Available: https://www.sciencealert.com/mit-s-new-robot-can-politely-navigate-through-crowds

[111] Clearpath Robotics, "Husky performs SLAM on stereonets to help predict rock falls and rock bursts," Clearpath Robotics. [Online]. Available: https://www.clearpathrobotics.com/husky-queens-mining-slam-stereonets/

[112] Clearpath Robotics, "University of Virginia maps historic tunnel using Jackal UGV," RoboticsTomorrow, Jan. 16, 2018. [Online]. Available: https://www.roboticstomorrow.com/article/2018/01/university-of-virginia-maps-historic-tunnel-using-jackal-ugv/11439

[113] M. Santora, "Clearpath Robotics offers ROS-enabled cobot arms from Universal Robots," The Robot Report, Aug. 1, 2018. [Online]. Available: https://www.therobotreport.com/clearpath-robotics-universal-robots-partner/

[114] D. Hambling, "The US Army is creating robots that can follow orders," MIT Technology Review, Nov. 6, 2019. [Online]. Available: https://www.technologyreview.com/2019/11/06/132036/the-us-army-is-creating-robots-that-can-follow-ordersand-ask-if-they-dont-understand/

[115] S. Guillen, "Husky UGV turns over a new leaf," RoboticsTomorrow, Aug. 13, 2019. [Online]. Available: roboticstomorrow.com/article/2019/08/husky-ugv-turns-over-a-new-leaf/13980

[116] K. Korosec, "OTTO Motors raises $29M to fill factories with autonomous delivery robots," TechCrunch, Jun. 4, 2020. [Online]. Available: https://www.techcrunch.com/2020/06/04/otto-motors-raises-29m-to-fill-factories-with-autonomous-delivery-robots/

[117] A. Weber, "Rockwell begins production of autonomous mobile robots in Milwaukee," Assembly Magazine, Oct. 31, 2025. [Online]. Available: https://www.assemblymag.com/articles/99635-rockwell-begins-production-of-autonomous-mobile-robots-in-milwaukee

[118] K. Inagaki, "SoftBank invests $2.8bn in robotics group AutoStore," Financial Times, Apr. 6, 2021. [Online]. Available: https://www.ft.com/content/8b923389-da74-447b-ab8d-448c7991aad6

[119] AutoStore, "Customer cases: 1,900+ systems globally and counting," AutoStore, 2025. [Online]. Available: autostoresystem.com/cases

[120] T. Ray, "SoftBank spends $2.8 billion to buy 40% stake in warehouse robotics firm AutoStore," ZDNET, Apr. 5, 2021. [Online]. Available: https://www.zdnet.com/article/softbank-spends-2-8-billion-to-buy-40-stake-in-warehouse-robotics-firm-autostore/

[121] AutoStore, "Elotec first to install AutoStore," AutoStore, 2025. [Online]. Available: https://www.autostoresystem.com/cases/elotec-first-to-install-autostore

[122] S. Banker, "Medline is the largest user of AutoStore robotics in the world," Forbes, Mar. 3, 2024. [Online]. Available: https://www.forbes.com/sites/stevebanker/2024/03/03/medline-is-the-largest-user-of--autostore-robotics-in-the-world/

[123] D. Robitzski, "The FBI is buying 140 robots to handle paper records," Futurism, Nov. 18, 2020. [Online]. Available: https://futurism.com/the-byte/fbi-buying-140-robots-paper-records

[124] The Robot Report Staff, "FBI installs AutoStore records filing and retrieval system for 2B pages of documents," The Robot Report, Nov. 10, 2020. [Online]. Available: https://www.therobotreport.com/gsa-installs-autostore-records-filing-and-retrieval-system-fbi/

[125] AutoStore, "H Mart partners with AutoStore to deploy automated micro-fulfillment centers," Mar. 30, 2021. [Online]. Available: https://autostoresystem.com/news/h-mart-partners-with-autostore-to-deploy-automated-micro-fulfillment-centers

[126] K. Inagaki, "SoftBank invests $2.8bn in robotics group AutoStore," Financial Times, Apr. 6, 2021. [Online]. Available: https://ft.com/content/8b923389-da74-447b-ab8d-448c7991aad6

[127] S. Gopinath and A. Wilen, "AutoStore $1.8 Billion IPO Is One of Norway's Largest," Bloomberg, Oct. 8, 2021. [Online]. Available: https://bloomberg.com/news/articles/2021-10-08/autostore-sets-1-8-billion-ipo-one-of-norway-s-largest-ever

[128] M. Oitzman, "New AutoStore R5+ robot handles larger bins," The Robot Report, Jan. 13, 2022. [Online]. Available: https://therobotreport.com/new-autostore-r5-robot-handles-larger-bins/

[129] StrongPoint ASA, "World's First AutoStore Solution With Frozen Food Zone to Be Installed by StrongPoint," Business Wire, Oct. 24, 2022. Available: https://www.strongpoint.com/news/worlds-first-autostore-installation-handling-frozen-goods-installed-by-strongpoint/

[130] B. Wessling, "AutoStore launches modular cube storage system Pio," The Robot Report, Jan. 12, 2023. Available: https://therobotreport.com/autostore-launches-modular-cube-storage-system-pio

[131] DHL Supply Chain, "DHL Supply Chain and robotics technology company AutoStore announce partnership expansion," Nov. 7, 2023. Available: https://group.dhl.com/en/media-relations/press-releases/2023/dhl-supply-chain-and-robotics-technology-company-autostore-announce-partnership-expansion.html

[132] LogistiekProfs Editorial, "AutoStore robots do grocery shopping Albert Heijn," LogistiekProfs, Dec. 5, 2023. Available: https://www.loginext.nl/en/artikelen/autostore-robots-doen-boodschappen-albert-heijn

[133] DC Velocity Staff, "AutoStore says Thailand factory will double production of AS/RS robots," DC Velocity, Jun. 13, 2024. Available: https://www.dcvelocity.com/articles/61051-autostore-says-thailand-factory-will-double-production-of-as-rs-robots

[134] CDOTrends editors, "AutoStore's Robot Revolution Heads East: New Factory Powers Up Asia Pacific Expansion," CDOTrends, Jun. 24, 2024. Available: https://www.cdotrends.com/story/4074/autostores-robot-revolution-heads-east-new-factory-powers-asia-pacific-expansion

[135] M. Zboraj, "AutoStore Unveils Subzero Robots for Grocery Fulfillment," Progressive Grocer, Sep. 13, 2024. Available: https://progressivegrocer.com/autostore-unveils-subzero-robots-grocery-fulfillment

[136] S. Banker, "Medline Is The Largest User Of AutoStore Robotics In The World," Forbes, Mar. 3, 2024. Available: https://www.forbes.com/sites/stevebanker/2024/03/03/medline-is-the-largest-user-of--autostore-robotics-in-the-world/

[137] MMH Staff, "Kenco goes live with AutoStore solution at DC in Jeffersonville, Indiana," Modern Materials Handling, Aug. 23, 2024. Available: https://mmh.com/article/kenco_goes_live_with_autostore_solution_at_dc_in_jeffersonville_indiana

[138] M. Allinson, "Footwear company SoftMoc adopts AutoStore warehouse system for Ontario, Canada facility," Robotics & Automation News, May 4, 2025. Available: https://roboticsandautomationnews.com/2025/05/04/footwear-company-softmoc-adopts-autostore-warehouse-system-for-ontario-canada-facility

[139] AutoStore, "AutoStore Tech Powers Next-Gen Robotic Vertical Farm," May 14, 2025. Available: https://www.autostoresystem.com/news/autostore-onepointone-unveil-worlds-first-robotic-vertical-farm-utilizing-supplying-select-whole-foods-market-stores

[140] S. Francis, "Veloq and AutoStore unveil 'world's most advanced grocery fulfilment centre' in Vienna as part of global automation partnership," Robotics & Automation News, Oct. 20, 2025. Available: https://roboticsandautomationnews.com/2025/10/20/veloq-and-autostore-unveil-worlds-most-advanced-grocery-fulfilment-centre-in-vienna-as-part-of-global-automation-partnership/

[141] S. Francis, "Veloq and AutoStore unveil 'world's most advanced grocery fulfilment centre' in Vienna as part of global automation partnership," Robotics & Automation News, Oct. 20, 2025. Available: https://roboticsandautomationnews.com/2025/10/20/veloq-and-autostore-unveil-worlds-most-advanced-grocery-fulfilment-centre-in-vienna-as-part-of-global-automation-partnership/95647/

[142] SBS Swiss Business School, "From Fairy Tales to Factory Floors: How Odense Became Europe's Robotics Hub," SBS Foresight News, May 12, 2025. Available: https://www.sbs.edu/foresight-news/from-fairy-tales-to-factory-floors-how-odense-became-europes-robotics-hub/

[143] V. Turk, "Welcome to robot city," MIT Technology Review, Feb. 26, 2025. Available: https://www.technologyreview.com/2025/02/26/1111796/robotics-robots-denmark-odense-collaboration/

[144] Teradyne Robotics, "Teradyne Will Expand Its Global Robotics Presence with New US Operations Hub in Metro Detroit, Michigan," Business Wire, Dec. 9, 2025. Available: https://www.businesswire.com/news/home/20251209403077/en/Teradyne-Will-Expand-Its-Global-Robotics-Presence-with-New-US-Operations-Hub-in-Metro-Detroit-Michigan

[145] S. Devi, "Universal Robots are a big hit," Financial Times, May 19, 2014. Available: https://www.ft.com/content/44e83f24-dc4e-11e3-a33d-00144feabdc0

[146] Universal Robots, "Linatex – UR cobots provide cost efficient and flexible solution," Universal Robots Case Stories. Available: https://www.universal-robots.com/case-stories/linatex/

[147] Universal Robots, "Engmark Meteor AS – UR cobots take on repetitive work and optimize production," Universal Robots Case Stories. Available: https://www.universal-robots.com/case-stories/engmark-meteor-as/

[148] Universal Robots, "Böco Böddecker – Optimizing processes with UR cobots," Universal Robots Case Stories. Available: https://www.universal-robots.com/case-stories/boco-boddecker/

[149] Universal Robots, "Minitüb – Minitüb runs at maximum efficiency with UR cobots," Universal Robots Case Stories. Available: https://www.universal-robots.com/case-stories/minitub/

[150] Universal Robots, "Mjolkursamsalan Akureyri – UR5 cobots relieve workers of repetitive tasks," Universal Robots Case Stories. Available: https://www.universal-robots.com/case-stories/mjolkursamsalan/

[151] Control Engineering Staff, "Packaging robots plug and play," Control Engineering, Sep. 24, 2013. Available: https://www.controleng.com/articles/packaging-robots-plug-and-play/

[152] Universal Robots, "Universal Robots to significantly expand production capacity," Universal Robots News Center, Apr. 24, 2014. Available: https://www.universal-robots.com/it/news-e-media/news-center/universal-robots-to-significantly-expand-production-capacity/

[153] Universal Robots, "PLC Industries – Cobot automation addresses labor shortage and boosts output by 40%," Universal Robots Case Stories. Available: https://www.universal-robots.com/case-stories/plc-industries/

[154] Swiss Startup Association, "Meet Péter Fankhauser, CEO & Co-Founder at ANYbotics," Swiss Startup Association Blog, Jul. 27, 2023. Available: https://swissstartupassociation.ch/2023/07/27/meet-our-premium-member-peter-fankhauser-founders-of-anybotics/

[155] N. Tyler, "AI-driven robot inspection," New Electronics, Jun. 18, 2025. Available: https://www.newelectronics.co.uk/content/features/ai-driven-robot-inspection

[156] ANYbotics, "Case Studies," ANYbotics. Available: https://www.anybotics.com/news/category/case-studies/

[157] ANYbotics, "ANYbotics Update August 2017," ANYbotics News, Aug. 17, 2017. Available: https://www.anybotics.com/news/anybotics-update-august-2017/

[158] ANYbotics, "World's First Autonomous Offshore Robot," ANYbotics Case Studies, Oct. 25, 2018. Available: https://www.anybotics.com/news/worlds-first-autonomous-offshore-robot/

[159] ANYbotics, "ANYmal C – The Next Step in Robotic Industrial Inspection," ANYbotics News, Aug. 20, 2019. Available: https://www.anybotics.com/news/the-next-step-in-robotic-industrial-inspection/

[160] ANYbotics, "ANYbotics Announces CHF 20M Series A," ANYbotics News, Dec. 3, 2020. Available: https://www.anybotics.com/news/anybotics-announces-series-a/

[161] ANYbotics, "ANYmal Robots Inspecting PETRONAS' Offshore Platform," ANYbotics Case Studies, Jan. 22, 2021. Available: https://www.anybotics.com/news/anymal-robots-inspecting-petronas-offshore-platform/

[162] ANYbotics, "Automation & Digitalization at Scale: ANYmal Makes the Case at BASF," ANYbotics Case Studies, Mar. 25, 2021. Available: https://www.anybotics.com/news/anymal-case-basf-chemical-plant/

[163] ANYbotics, "ANYmal Helps Vale to Prioritize People in Mining," ANYbotics Case Studies, Aug. 17, 2021. Available: https://www.anybotics.com/news/anymal-helps-vale-to-prioritize-people-in-mining/

[164] ANYbotics, "ANYmal X complete Five-Week Test Program at Equinor Lab," ANYbotics Case Studies, Jun. 1, 2023. Available: https://www.anybotics.com/news/anymal-x-completes-five-week-test-program-at-equinor-lab/

[165] ANYbotics, "This is how Outokumpu wants to make work safer," ANYbotics Case Studies, Nov. 13, 2023. Available: https://www.anybotics.com/news/this-is-how-outokumpu-wants-to-make-work-safer/

[166] ANYbotics, "Pioneering Autonomous Inspection for Enhanced Nuclear Safety at Onkalo," ANYbotics Case Studies, Feb. 19, 2024. Available: https://www.anybotics.com/news/autonomous-inspection-for-enhanced-nuclear-safety-at-onkalo/

[167] H. J. Kamps, "Pitch Deck Teardown: ANYbotics' $50M Series B deck," TechCrunch, Aug. 11, 2023. Available: https://techcrunch.com/2023/08/11/sample-series-b-pitch-deck-anybotics-ag/

[168] D. Leichner, "Romain Moulin of Exotec On The Future Of Robotics Over the Next Few Years," Authority Magazine (Medium), Jan. 12, 2023. Available: https://medium.com/authority-magazine/romain-moulin-of-exotec-on-the-future-of-robotics-over-the-next-few-years-eced9bf99895

[169] A. Chowdhry, "Exotec: Interview With Co-Founder & CEO Romain Moulin About The Warehouse Robotics Company," Pulse 2.0, Jan. 5, 2026. Available: https://pulse2.com/exotec-profile-romain-moulin-interview/

[170] DTC Team, "From Spark to Innovation at Exotec: 'We're not doing some copycat.'," Dell Technologies Capital, Sep. 30, 2023. Available: https://www.delltechnologiescapital.com/resources/exotec-robotics

[171] F6S Network Limited, "Exotec – F6S Profile," F6S. Available: https://www.f6s.com/exotec

[172] Exotec, "Exotec Reaches $1 Billion in Systems Sold," Exotec News, Mar. 12, 2024. Available: https://www.exotec.com/news/exotec-reaches-1-billion-in-systems-sold/

[173] Exotec, "Exotec Celebrates 10 Years of Innovation: Driving A New Era of Warehouse Technology," PR Newswire, Oct. 15, 2025. Available: https://www.prnewswire.com/news-releases/exotec-celebrates-10-years-of-innovation-driving-a-new-era-of-warehouse-technology-301000000.html

[174] Exotec, "Exotec Celebrates 10 Years of Innovation: Driving A New Era of Warehouse Technology," Exotec News. Available: https://www.exotec.com/news/exotec-celebrates-10-years-of-innovation-driving-a-new-era-of-warehouse-technology/

[175] Exotec, "Inside Carrefour's 6-Week Grocery Micro-Fulfillment Transformation," Exotec Case Studies. Available: https://www.exotec.com/case-studies/carrefour-reims-france/

[176] Exotec, "Carrefour Powers 2-Hour Grocery Delivery to Meet Surging Demand," Exotec Case Studies. Available: https://www.exotec.com/case-studies/carrefour-le-plessis-pate-france/

[177] Exotec, "Cdiscount Picks Smarter, Stores Higher to 5x Throughput and Density" Exotec Case Studies. Available: https://www.exotec.com/case-studies/cdiscount-paris-france/

[178] Fast Retailing Co., Ltd., "Fast Retailing Announces Expansion of Strategic Global Partnership to Strengthen Supply Chain Transformation," Fast Retailing Press Release Archive, Nov. 13, 2019. Available: https://www.fastretailing.com/eng/about/news/2019.html

[179] Exotec, "A Successful Launch for Exotec in North America," Exotec News, Jan. 5, 2021. Available: https://www.exotec.com/news/a-successful-launch-for-exotec-in-north-america/

[180] Exotec, "GEODIS' Strategy for Urgent and Ever-Changing Fulfillment Demand," Exotec Case Studies. Available: https://www.exotec.com/case-studies/geodis/

[181] Télécom Paris, "Exotec, first French industrial unicorn!," Télécom Paris. Available: https://www.telecom-paris.fr/en/div/exotec-unicorn

[182] Exotec, "Exotec and POSCO DX Strike a Partnership to Bring Exotec Solutions to the South Korean Market," Exotec News, Aug. 3, 2023. Available: https://www.exotec.com/news/exotec-and-posco-dx-strike-a-partnership-to-bring-exotec-solutions-to-the-south-korean-market/

[183] Exotec, "Exotec Launches Next Generation of Skypod System, an All-in-One Robot-Based AS/RS, that Addresses the Majority of Processes Within a Warehouse," Exotec News, Feb. 6, 2025. Available: https://www.exotec.com/news/exotec-launches-next-generation-of-skypod-system-an-all-in-one-robot-based-as-rs-that-addresses-the-majority-of-processes-within-a-warehouse/

[184] iRobot Corporation, "History," iRobot. Available: https://about.irobot.com/history

[185] J. Biggs, "iRobot Acquires Evolution Robotics, Makers Of The Mint Robot Vacuum, For $74 Million," TechCrunch, Sep. 17, 2012. Available: https://techcrunch.com/2012/09/17/irobot-acquires-evolution-robotics-makers-of-the-mint-robot-vacuum-for-74-million/

[186] C. Loizos, "How iRobot lost its way home," TechCrunch, Dec. 14, 2025. Available: https://techcrunch.com/2025/12/14/how-irobot-lost-its-way-home/

[187] J. Pattison Tuohy, "How Roomba invented the home robot — and lost the future," The Verge, Dec. 15, 2025. Available: https://www.theverge.com/tech/844964/how-irobot-invented-the-roomba-and-lost-the-future-of-home-robotics

[188] J. Hughes, "Lessons from Roomba: sometimes being first mover sucks," Financial Times, Dec. 15, 2025. Available: https://www.ft.com/content/44e0d194-d33e-4eaa-b149-ad0c75e167a8

[189] K. Johnson, "The iRobot Deal Would Give Amazon Maps Inside Millions of Homes," WIRED, Aug. 5, 2022. Available: https://www.wired.com/story/amazon-irobot-roomba-acquisition-data-privacy/

[190] Institute for Health Metrics and Evaluation, "The Lancet: Dramatic declines in global fertility rates set to transform global population patterns by 2100," IHME News Release, Mar. 20, 2024. Available: https://www.healthdata.org/news-events/newsroom/news-releases/lancet-dramatic-declines-global-fertility-rates-set-transform

[191] P. Boyle, "U.S. physician shortage growing," AAMCNews, Association of American Medical Colleges, Jun. 26, 2020. Available: https://www.aamc.org/news/us-physician-shortage-growing

[192] Telefónica Communication Team, "How much time do we spend on our digital lives?" Telefónica Blog, Oct. 22, 2025. Available: https://www.telefonica.com/en/communication-room/blog/how-much-time-do-we-spend-on-our-digital-lives/

[193] M. Kan, "Amazon Astro Is a $1,449 Home Robot," PCMag, Sep. 28, 2021. Available: https://www.pcmag.com/news/amazon-astro-is-a-1499-home-robot

[194] M. Price, "Meet Jibo, the $899 social robot," CNET, Nov. 10, 2017. Available: https://www.cnet.com/pictures/meet-jibo-the-899-social-robot/

[195] E. Ackerman, "Mayfield Robotics Announces Kuri, a $700 Home Robot," IEEE Spectrum, Jan. 3, 2017. Available: https://spectrum.ieee.org/mayfield-robotics-announces-kuri-a-700-mobile-home-robot

[196] J. Brandon, "How an Adorable Robot Became a $2 Million Crowdfunded Wonder," Inc., Mar. 25, 2015. Available: https://www.inc.com/john-brandon/startup-success-jibo-creates-a-sense-of-robotic-wonder.html

[197] ROBOTS Guide, "Anki Drive," RobotsGuide.com. Available: https://robotsguide.com/robots/anki

[198] IEEE Spectrum, "Vector," IEEE Robots Guide. Available: https://robots.ieee.org/robots/vector/

[199] A. Robertson, "Within Seconds Cozmo Supersedes Other Robot Toys," Forbes, Oct. 7, 2016. Available: https://www.forbes.com/sites/andyrobertson/2016/10/07/cozmo-review/

[200] B. Heater, "Anki has sold 1.5 million robots," TechCrunch, Aug. 8, 2018. Available: https://techcrunch.com/2018/08/08/anki-has-sold-1-5-million-cozmo-robots/

[201] S. Captain, "What failed startup Anki did for robotics," Fast Company, Apr. 29, 2019. Available: https://www.fastcompany.com/90341936/what-failed-startup-anki-did-for-robotics

[202] E. Guizzo, "Jibo Is as Good as Social Robots Get. But Is That Good Enough?," IEEE Spectrum, Dec. 23, 2015. Available: https://spectrum.ieee.org/jibo-is-as-good-as-social-robots-get-but-is-that-good-enough

[203] A. Carman, "They welcomed a robot into their family, now they're mourning its death: The story of Jibo," The Verge, Jun. 19, 2019. Available: https://www.theverge.com/2019/6/19/18682780/jibo-death-server-update-social-robot-mourning

[204] Voicebot.ai, "RBC analyst says 52 million Google Home devices sold to date and generating $3–4 billion in 2018 revenue," Voicebot.ai, Dec. 24, 2018. Available: https://voicebot.ai/2018/12/24/rbc-analyst-says-52-million-google-home-devices-sold-to-date-and-generating-3-4-billion-in-2018-revenue/

[205] Bosch Business Innovations, "Saying goodbye to Mayfield Robotics," Bosch Business Innovations. Available: https://www.bosch-business-innovations.com/stories/mayfield-robotics/

[206] E. Ackerman, "Mayfield Robotics Cancels Kuri Social Home Robot," IEEE Spectrum, Jul. 25, 2018. Available: https://spectrum.ieee.org/mayfield-robotics-cancels-kuri-social-home-robot

[207] B. Heater, "Mayfield Robotics ceases production of Kuri robot amid a questionable future," TechCrunch, Jul. 25, 2018. Available: https://techcrunch.com/2018/07/25/mayfield-robotics-ceases-production-of-kuri-robot-amid-a-questionable-future/

[208] D. Lee, "The company behind the adorably doomed robot Kuri is shutting down," The Verge, Aug. 21, 2018. Available: https://www.theverge.com/circuitbreaker/2018/8/21/17765330/mayfield-robotics-kuri-robot-shutting-down

[209] P. Olson, "Amazon's New Robot Astro Is a Privacy Minefield," Bloomberg Opinion, Sep. 30, 2021. Available: https://www.bloomberg.com/opinion/articles/2021-09-30/amazon-s-astro-is-cute-and-smart-but-it-might-yet-be-a-privacy-minefield

[210] M. Gurman, "Amazon's Astro Home Robot Remains Elusive Six Months After Debut," Bloomberg, Feb. 22, 2022. Available: https://www.bloomberg.com/news/articles/2022-02-22/when-can-i-buy-amazon-s-astro-home-robot-amazon-astro-hands-on-impressions

[211] K. Wilder and P. Mackun, "Older Population Grew in Nearly All U.S. Metro Areas," U.S. Census Bureau, Jun. 27, 2024. Available: https://www.census.gov/library/stories/2024/06/metro-areas-population-age.html

[212] Starship Technologies, "Starship Technologies Raises $50M Series C to Scale Autonomous Delivery Across U.S. Cities," Oct. 15, 2025. Available: https://www.starship.xyz/press/starship-technologies-raises-50m-series-c/

[213] Coco Robotics, "Coco Robotics Launches Next-Gen Autonomous Robots for Urban Deliveries," PR Newswire, Feb. 26, 2026. Available: https://www.prnewswire.com/news-releases/coco-robotics-launches-next-gen-autonomous-robots-for-urban-deliveries-302698399.html

[214] Sodexo, "Sodexo strengthens its partnership with robotic delivery start-up Kiwibot to accelerate the transformation of its food model in US universities," Feb. 20, 2022. Available: https://www.sodexo.com/news/newsroom/2022/kiwibot-roll-out-USA

[215] Serve Robotics, "DoorDash and Serve Robotics Partner to Bring Serve's Delivery Robots to the DoorDash Platform," Oct. 9, 2025. Available: https://investors.serverobotics.com/news-releases/news-release-details/doordash-and-serve-robotics-partner-bring-serves-delivery-robots

[216] Capital One Shopping, "Package Delivery Statistics," Aug. 29, 2025. Available: https://capitaloneshopping.com/research/package-delivery-statistics/

[217] DoorDash, "Dash Report 2025: Delivery in America," DoorDash Dasher Central, 2025. Available: https://dasher.doordash.com/en-us/dash-report-2025

[218] A. Perez Pintado, "U.S. short 80,000 drivers; number could surpass 160,000 by 2030," The Gazette, Jan. 30, 2022. Available: https://www.thegazette.com/business/u-s-short-80-000-drivers/article_8672c988-be37-51b4-b7a5-1355f54de0ec.html

[219] World Economic Forum, *Transforming Urban Logistics: Sustainable and Efficient Last-Mile Delivery in Cities*. Geneva, Switzerland: World Economic Forum, Dec. 2024. Available: https://reports.weforum.org/docs/WEF_Transforming_Urban_Logistics_2024.pdf

[220] Starship Technologies, "About Starship – Our Journey and Autonomous Delivery Operations," Starship Technologies. Available: https://www.starship.xyz/about/

[221] V. Woollaston-Webber, "Starship partners with Just Eat to bring its delivery robots to the UK," WIRED, Jul. 6, 2016. Available: https://www.wired.com/story/starship-robots-just-eat-hermes-pronto/

[222] Starship Technologies, "Swiss Post joins Starship Technologies' Robotic Delivery Testing Programme in Switzerland," Aug. 23, 2016. Available: https://www.starship.xyz/press/swiss-post-joins-starship-technologies-robotic-delivery-testing-programme-in-switzerland/

[223] F. Searle, "Tesco added to robot buggy delivery service," Fresh Produce Journal, Jan. 11, 2019. Available: https://www.fruitnet.com/fresh-produce-journal/tesco-added-to-robot-buggy-delivery-service/177564.article

[224] M. R. Dickey, "Starship deploys autonomous delivery bots on a college campus," TechCrunch, Jan. 22, 2019. Available: https://techcrunch.com/2019/01/22/starship-deploys-autonomous-delivery-bots-on-a-college-campus/

[225] Starship Technologies, "Starship Technologies Celebrates 5 Years of Autonomous Robot Deliveries at George Mason University," Jan. 23, 2024. Available: https://www.starship.xyz/press/starship-technologies-celebrates-5-years-of-autonomous-robot-deliveries-at-george-mason-university/

[226] A. Feldman, "Starship Technologies Raises $40 Million To Expand Its Food-Delivery Robots On College Campuses," Forbes, Aug. 20, 2019. Available: https://www.forbes.com/sites/amyfeldman/2019/08/20/starship-technologies-raises-40m-to-expand-its-food-delivery-robots-on-college-campuses/

[227] IEEE Spectrum, "Starship," IEEE Robots Guide. Available: https://robotsguide.com/robots/starship

[228] M. Paris, "Last Mile Robot Launches Food Delivery Service In Snow Storm," Forbes, Dec. 15, 2019. Available: https://www.forbes.com/sites/martineparis/2019/12/15/last-mile-robot-launches-snowy-pilot-in-quest-to-own-food-delivery-market/

[229] B. Heater, "Serve's CEO on Why Roaches Beat Unicorns," Association for Advancing Automation (A3), Jan. 29, 2026. Available: https://www.automate.org/robotics/industry-insights/serves-ceo-on-why-roaches-beat-unicorns

[230] Serve Robotics, "Serve Robotics spins out of Postmates with seed funding from Neo to build the next generation of sidewalk delivery robots," Mar. 2, 2021. Available: https://investors.serverobotics.com/news-releases/news-release-details/serve-robotics-spins-out-postmates-seed-funding-neo-build-next

[231] Serve Robotics, "Serve Robotics Expands to Chicago, Bringing Autonomous Deliveries to the Windy City with Uber Eats," Sept. 30, 2025. Available: https://investors.serverobotics.com/news-releases/news-release-details/serve-robotics-expands-chicago-bringing-autonomous-deliveries

[232] Serve Robotics Inc., "Serve Robotics Inc. Announces Closing of $40 Million Public Offering and Uplisting to the Nasdaq Capital Market Under New Ticker 'SERV'," Apr. 22, 2024. Available: https://investors.serverobotics.com/news-releases/news-release-details/serve-robotics-inc-announces-closing-40-million-public-offering

[233] DoorDash, "DoorDash and Serve Robotics Partner to Bring Serve's Delivery Robots to the DoorDash Platform," Oct. 9, 2025. Available: https://about.doordash.com/en-us/news/doordash-serve-launch-partnership

[234] C. McDonald, "Forget the Robot Apocalypse. Order Lunch," California Magazine, Feb. 26, 2018. Available: https://alumni.berkeley.edu/california-magazine/online/forget-robot-apocalypse-order-lunch/

[235] N. Alarcon, "Kiwibot, Jetson-Powered Robot, Arrives in San Jose with Shopify and Ordermark," NVIDIA Developer Blog, Jul. 29, 2020. Available: https://developer.nvidia.com/blog/kiwibot-arrives-in-san-jose/

[236] C. Wood, "Why yes, those are robot delivery minions at UC Berkeley," Berkeleyside, May 23, 2017. Available: https://www.berkeleyside.org/2017/05/23/robots-delivery-comes-to-cal

[237] W. Kane, "Those four-wheeled robots on campus, explained," UC Berkeley News, May 31, 2018. Available: https://news.berkeley.edu/2018/05/31/those-four-wheeled-robots-on-campus-explained/

[238] Segway Robotics, "How Segway's Robotics Solutions Can Help Meet Food Delivery Demands," Segway Robotics News. Available: https://robotics.segway.com/news/how-segways-robotics-solutions-can-help-meet-food-delivery-demands/

[239] Sodexo, "Kiwibot Finishes Pre-Series A Funding and Closes the Biggest Expansion Deal With Sodexo," Sodexo Newsroom, Feb. 14, 2022. Available: https://us.sodexo.com/newsroom/2022/kiwibot-sodexo

[240] Kiwibot, "Kiwibot Launches Next-Generation Robot 4.0 in Partnership With Knight Foundation, Chick-Fil-A, and Announces Manufacturing Cooperation With Segway," PR Newswire, Apr. 13, 2021. Available: https://www.prnewswire.com/news-releases/kiwibot-launches-next-generation-robot-4-0-in-partnership-with-knight-foundation-chick-fil-a-and-announces-manufacturing-cooperation-with-segway-301267974.html

[241] University of Southern Indiana, "Kiwibot, campus dining delivery robots, arrive at USI for start of Fall Semester," Aug. 18, 2023. Available: https://www.usi.edu/news/releases/2023/08/kiwibot-campus-dining-delivery-robots-arrive-at-usi-for-start-of-fall-semester

[242] B. Heater, "Kiwibot acquires an ad startup to turn its delivery robots into mobile billboards," TechCrunch, Sept. 19, 2024. Available: https://techcrunch.com/2024/09/19/kiwibot-acquires-ad-startup-to-turn-its-delivery-robots-into-mobile-billboards/

[243] P. Heath and C. Petrow-Cohen, "We may not have flying cars, but more food delivery bots are coming to L.A.," Los Angeles Times, Jul. 24, 2025. Available: https://www.latimes.com/business/story/2025-07-24/food-delivery-robots-are-becoming-more-commonplace-in-los-angeles-neighborhoods

[244] B. Heater, "Last-mile robotic delivery firm Coco raises $36M," TechCrunch, Aug. 25, 2021. Available: https://techcrunch.com/2021/08/25/last-mile-robotic-delivery-firm-coco-raises-36m/

[245] T. Kaiser, "C3 Partners with Coco Robot Delivery," Food On Demand, Nov. 4, 2021. Available: https://foodondemand.com/11042021/c3-partners-with-coco-robot-delivery/

[246] DoorDash, "DoorDash and Coco Expand Global Partnership with U.S. Sidewalk Robot Delivery Launch," Apr. 10, 2025. Available: https://about.doordash.com/en-us/news/doordash-and-coco-expand-global-partnership

[247] R. Szkutak, "Sam Altman-backed Coco Robotics raises $80M," TechCrunch, Jun. 11, 2025. Available: https://techcrunch.com/2025/06/11/sam-altman-backed-coco-robotics-raises-80m/

[248] Bloomberg, "Humanoid Market to Reach $200B by 2035 Barclays Says," Bloomberg, Jan. 16, 2026. Available: https://www.bloomberg.com/news/videos/2026-01-16/humanoid-market-to-reach-200b-by-2035-barclays-says-video

[249] National Center for Health Workforce Analysis, "Physician Workforce: Projections, 2020–2035," Health Resources and Services Administration (HRSA), Nov. 2022. Available: https://bhw.hrsa.gov/sites/default/files/bureau-health-workforce/Physicians-Projections-Factsheet.pdf

[250] A. Kayser, "We're Nearing a Doctor Retirement Cliff. Can Health Care Survive the Fall?," Newsweek, Apr. 3, 2025. Available: https://www.newsweek.com/doctor-physician-retirement-cliff-health-care-2054258

[251] National Association of Manufacturers, "Facts About Manufacturing: The Top 18 Facts You Need to Know," National Association of Manufacturers. Available: https://nam.org/mfgdata/facts-about-manufacturing-expanded/

[252] W.-Y. Lee, "Foxconn to deploy humanoid robots at Houston AI server plant," *Reuters*, Oct. 28, 2025. [Online]. Available: https://www.reuters.com/business/foxconn-deploy-humanoid-robots-houston-ai-server-plant-2025-10-29/

[253] *F.02 Contributed to the Production of 30,000 Cars at BMW*, Figure.ai, Nov. 19, 2025. [Online]. Available: https://www.figure.ai/news/production-at-bmw

[254] Hyundai Motor Company, "Uniting humans and robots – Hyundai Motor Group and Boston Dynamics' vision for the future of work and HR evolution," *Hyundai Worldwide – Newsroom*, Jul. 14, 2025. [Online]. Available: https://www.hyundai.com/worldwide/en/newsroom/detail/uniting-humans-and-robots--hyundai-motor-group-and-boston-dynamics-vision-for-the-future-of-work-and-hr-evolution-0000000983

[255] I. Fujino, "China's BYD, Nio do more with AI-powered humanoid factory robots," *Nikkei Asia*, Dec. 19, 2024. [Online]. Available: https://asia.nikkei.com/business/china-tech/china-s-byd-nio-do-more-with-ai-powered-humanoid-factory-robots

[256] Porsche Consulting GmbH, "The 'real' colleague," *Porsche Consulting – The Magazine*, Dec. 11, 2025. [Online]. Available: https://newsroom.porsche.com/en/2025/company/porsche-consulting-the-real-colleague-41063.html

[257] D. Milmo, "Amazon 'testing humanoid robots to deliver packages'," *The Guardian*, Jun. 5, 2025. [Online]. Available: The Guardian. https://www.theguardian.com/technology/2025/jun/05/amazon-testing-humanoid-robots-to-deliver-packages

[258] N. Khan, "Tesla's Optimus robots to enter mass production in 2025, aims for 100,000 by 2026," *Yahoo Finance*, Jan. 10, 2025. [Online]. Available: https://finance.yahoo.com/news/teslas-optimus-robots-enter-mass-101736202.html

[259] Business Wire, "Global humanoid robots market report 2026–2040: Humanoid robot pricing drops below $10,000 as market expansion accelerates," *The AI Journal*, Oct. 29, 2025. [Online]. Available: https://aijourn.com/global-humanoid-robots-market-report-2026-2040-humanoid-robot-pricing-drops-below-10000-as-market-expansion-accelerates-researchandmarkets-com/

[260] Agility Robotics, "RoboFab: The first humanoid robot factory," *Agility Robotics*, 2025. [Online]. Available: https://www.agilityrobotics.com/content/opening-robofab-worlds-first-factory-for-humanoid-robots

[261] J. Biba, "Tesla's Optimus robot: Uses and release timeline," *Built In*, updated Nov. 17, 2025. [Online]. Available: https://builtin.com/robotics/tesla-robot

[262] K. Walia, "A humanoid robot is now on sale for under US$6,000 – what can you do with it?," *The Conversation*, Aug. 18, 2025. [Online]. Available: https://theconversation.com/a-humanoid-robot-is-now-on-sale-for-under-us-6-000-what-can-you-do-with-it-262183

[263] B. Heater, "Figure raises $70M to build its humanoid robots," TechCrunch, May 24, 2023. Available: https://techcrunch.com/2023/05/24/2546360/

[264] Figure AI, "F.02 Contributed to the Production of 30,000 Cars at BMW," Nov. 19, 2025. Available: https://www.figure.ai/news/production-at-bmw

[265] H. Sproule, "AI is up and walking," Warrington College of Business, University of Florida, Nov. 3, 2023. Available: https://warrington.ufl.edu/news/ai-is-up-and-walking/

[266] J. Covert, "Meet Wall Street's busiest and most selective recruiting firm," New York Post, Sep. 3, 2015. Available: https://nypost.com/2015/09/03/meet-wall-streets-biggest-and-most-selective-recruiting-firm/

[267] B. Adcock, "Bio," Brett Adcock Official Website. Available: https://www.brettadcock.com/bio

[268] S. Huff, "Figure Humanoid Robot Startup Comes Out of Stealth," Entrepreneur, Mar. 3, 2023. Available: https://www.entrepreneur.com/business-news/figure-humanoid-robot-startup-comes-out-of-stealth/446847

[269] B. Heater, "Figure rides the humanoid robot hype wave to $2.6B valuation," TechCrunch, Feb. 29, 2024. Available: https://techcrunch.com/2024/02/29/figure-rides-the-humanoid-robot-hype-wave-to-2-6b-valuation-and-openai-collab/

[270] Figure AI, "Figure exceeds $1B in Series C funding at $39B post-money valuation," Figure AI News, Sep. 16, 2025. Available: https://www.figure.ai/news/series-c

[271] J. Davalos and Bloomberg, "UPS in talks with startup Figure AI to deploy humanoid robots," Fortune, Apr. 28, 2025. Available: https://fortune.com/article/ups-robotics-figure-ai-humanoid-robots-shipping/

[272] IEEE Spectrum, "EVE," IEEE Robots Guide. Available: https://robotsguide.com/robots/eve

[273] Halodi Robotics, "Halodi Robotics inks contract with ADT Commercial for delivery of 140 humanoid robots," Business Wire, Mar. 23, 2022. Available: https://www.businesswire.com/news/home/20220322005398/en/CORRECTING-and-REPLACING-Halodi-Robotics-Inks-Contract-with-ADT-Commercial-for-Delivery-of-140-Humanoid-Robots

[274] R. Szkutak, "1X struck a deal to send its 'home' humanoids to factories and warehouses," TechCrunch, Dec. 11, 2025. Available: https://techcrunch.com/2025/12/11/1x-struck-a-deal-to-send-its-home-humanoids-to-factories-and-warehouses/

[275] 1X Technologies, "NEO home robot," 1X Technologies. Available: https://www.1x.tech/order

[276] NVIDIA AI Podcast, "The future of humanoid robots with 1X's Bernt Børnich," Apple Podcasts, Episode 259, Jun. 4, 2025. Available: https://podcasts.apple.com/lu/podcast/the-future-of-humanoid-robots-with-1xs-bernt-bornich/id1186480811?i=1000711175455

[277] EQT Ventures, "Empathetic robots: The vision of 1X founder Bernt Øivind Børnich," Medium, Jan. 27, 2025. Available: https://medium.com/eqtventures/empathetic-robots-the-vision-of-1x-founder-bernt-øivind-børnich-5bb17e4a7639

[278] T. Haugstad, "The robot from Moss becomes a security guard in the USA," TU.no (Teknisk Ukeblad), Jul. 3, 2022. Available: https://www.digi.no/artikler/roboten-fra-moss-blir-vekter-i-usa/520711

[279] 1X Technologies, "Announcing rebranding: From Halodi Robotics to 1X," Jul. 2, 2023. Available: https://www.1x.tech/discover/announcing-rebranding-from-halodi-robotics-to-1x

[280] B. Heater, "OpenAI-backed 1X raises another $100M for the race to humanoid robots," TechCrunch, Jan. 11, 2024. Available: https://techcrunch.com/2024/01/11/openai-backed-1x-raises-another-100m-for-the-race-to-humanoid-robots/

[281] S. Lundeberg, "Bipedal robot developed at Oregon State makes history by learning to run, completing 5K," Oregon State University Newsroom, Jul. 25, 2021. Available: https://news.oregonstate.edu/news/bipedal-robot-developed-oregon-state-makes-history-learning-run-completing-5k

[282] S. Lundeberg, "OSU spinout Agility Robotics, maker of legged workplace robots, secures $20 million in new investment," Oregon State University Newsroom, Oct. 15, 2020. Available: https://news.oregonstate.edu/news/osu-spinout-agility-robotics-maker-legged-workplace-robots-secures-20-million-new-investment

[283] L. Lee, "These robots are coming for the jobs no one wants — and could fill workforce gaps," Business Insider, Mar. 8, 2026. Available: https://www.businessinsider.com/agility-robotics-humanoid-robots-labor-shortage-aging-workforce-2026-3

[284] "Agility Robotics," Wikipedia, The Free Encyclopedia. Available: https://en.wikipedia.org/wiki/Agility_Robotics

[285] Ford Motor Company, "Ford acquires Digit, world's first commercially available humanoid robot," Ford Media Center, Jan. 28, 2020. Available: https://media.ford.com/content/fordmedia/img/za/en/news/2020/01/28/ford-acquires-digit--world-s-first-commercially-available-humano.html

[286] Agility Robotics, "Digit deployed at GXO in historic humanoid RaaS agreement," Agility Robotics, Oct. 3, 2024. Available: https://www.agilityrobotics.com/content/digit-deployed-at-gxo-in-historic-humanoid-raas-agreement

[287] Agility Robotics, "Agility Robotics broadens relationship with Amazon," Agility Robotics Press Release, Oct. 24, 2023. Available: https://www.agilityrobotics.com/content/agility-robotics-broadens-relationship-with-amazon

[288] R. Drew, "Humanoid robotics startup Agility is raising $400 million," The Information. Available: https://www.theinformation.com/articles/humanoid-robotics-startup-agility-raises-400-million

[289] "UBtech Robotics," Wikipedia, The Free Encyclopedia. Available: https://en.wikipedia.org/wiki/UBtech_Robotics

[290] C. Li, "China's UBTech Robotics eyes US$307 million in Hong Kong share placement," South China Morning Post, Jul. 22, 2025. Available: https://www.scmp.com/business/markets/article/3319084/chinas-ubtech-robotics-eyes-us307-million-hong-kong-share-placement

[291] UBTECH, "UBTECH humanoid robot Walker S2 begins mass production and delivery, with orders exceeding 800 million yuan," PR Newswire, Nov. 17, 2025. Available: https://www.prnewswire.com/news-releases/ubtech-humanoid-robot-walker-s2-begins-mass-production-and-delivery-with-orders-exceeding-800-million-yuan-302616924.html

[292] W. Langley, "Robots only half as efficient as humans, says leading Chinese producer," Financial Times, Jan. 25, 2026. Available: https://www.ft.com/content/0f831781-b450-4644-9f83-b3f76968a4af

[293] J. Malayil, "Audi integrates Walker S1 robot for smarter, safer quality inspections in China plant," Interesting Engineering, Mar. 28, 2025. Available: https://interestingengineering.com/innovation/audi-integrates-walker-s1-robot

[294] F. Master and Reuters Beijing Newsroom, "UBTech agrees Airbus deal to expand robot use in aviation manufacturing," Reuters, Jan. 21, 2026. Available: https://www.reuters.com/world/asia-pacific/ubtech-agrees-airbus-deal-expand-robot-use-aviation-manufacturing-2026-01-21/

[295] D. Seeley, "Texas Instruments to deploy humanoid robots on its chip production lines," Dallas Innovates, Jan. 27, 2026. Available: https://dallasinnovates.com/texas-instruments-to-deploy-humanoid-robots-on-its-chip-assembly-lines/

[296] M. Williams, "Carmakers working with robot suppliers to test flexible humanoids for lineside logistics operations," Automotive Logistics, Feb. 24, 2025 (updated Jun. 25, 2025). Available: https://www.automotivelogistics.media/digitalisation/carmakers-working-with-robot-suppliers-to-test-flexible-humanoids-for-lineside-logistics-operations/338738

[297] J. Malayil, "World's first humanoid robot swarm tackles complex factory tasks at China's car plant," Interesting Engineering, Apr. 30, 2025. Available: https://interestingengineering.com/innovation/china-robots-tackle-car-factory-tasks

[298] UBTECH, "Zhou Jian, 'Father of Robots' in China and Founder of UBTECH, 'Appears' at Times Square (New York)," PR Newswire, Feb. 10, 2015. Available: http://prnewswire.com/news-releases/zhou-jian-father-of-robots--in-china-and-founder-of-ubtech-appears-at-times-square-new-york-300032705.html

[299] Apptronik, "About Us," Apptronik. Available: https://apptronik.com/about-us

[300] Apptronik, "Austin-based Apptronik inks partnership with NASA for humanoid robots," Apptronik Press Release. Available: https://apptronik.com/news-collection/austin-based-apptronik-inks-partnership-with-nasa-for-humanoid-robots

[301] Apptronik, "Apptronik closes over $935 million Series A with new $520 million extension round," Apptronik Press Release, Feb. 11, 2026. Available: https://apptronik.com/news-collection/apptronik-closes-over-935-million-series-a

[302] V. Waldersee, "Mercedes-Benz takes stake in robotics maker Apptronik, tests robots in factories," Reuters, Mar. 18, 2025. Available: https://www.reuters.com/business/autos-transportation/mercedes-benz-takes-stake-robotics-maker-apptronik-tests-robots-factories-2025-03-18/

[303] I. Deng, "China's AgiBot targets US$142 million revenue as march of humanoid robots gathers pace," South China Morning Post, Dec. 24, 2025. Available: https://www.scmp.com/tech/big-tech/article/3337477/chinas-agibot-targets-us142-million-revenue-march-humanoid-robots-gathers-pace

[304] C. Feng, "Robot makers AgiBot, Unitree and UBTech land more orders from Chinese enterprises," South China Morning Post, Jul. 14, 2025. Available: https://www.scmp.com/tech/tech-trends/article/3318152/robot-makers-agibot-unitree-and-ubtech-land-more-orders-chinese-enterprises

[305] Reuters, "Chinese robot maker AGIBot completes new round of financing," Reuters, Aug. 1, 2025. Available: https://www.reuters.com/technology/chinese-robot-maker-agibot-completes-new-round-financing-2025-08-01/

[306] J. Zhu and L. Mo, "Exclusive: Chinese robot maker AgiBot plans Hong Kong IPO next year, sources say," Reuters, Oct. 10, 2025. Available: https://www.reuters.com/world/china/chinese-robot-maker-agibot-plans-hong-kong-ipo-next-year-sources-say-2025-10-10/

[307] AgiBot, "AgiBot makes Singapore market debut at AI Gravity 2025," PR Newswire, Nov. 26, 2025. Available: https://aapnews.aap.com.au/aapreleases/cision20251125AE32701

[308] AgiBot, "AGIBOT launches in Malaysia – marks the start of a new regional expansion," AgiBot, Available: https://www.agibot.com/article/231/detail/34.html

[309] GadgetMatch, "AgiBot robots can now learn skills on the factory floor," GadgetMatch, Nov. 3, 2025. Available: https://www.gadgetmatch.com/agibot-robots-learn-skills-factory-floor-real-world-reinforcement-learning-industrial-robotics/

[310] A. From, "AGIBOT completes European debut in Milan," Gasgoo, Feb. 2, 2026. Available: https://autonews.gasgoo.com/articles/news/agibot-completes-european-debut-in-milan-2018267456819011585

[311] D. Pokidaev, "China's AgiBot to produce robots in Kazakhstan," *The Times of Central Asia*, Feb. 3, 2025. Available: https://timesca.com/chinas-agibot-to-produce-robots-in-kazakhstan/

[312] J. P. Tomás, "China Mobile's revenues slightly up y-o-y in Jan–Sep," *RCR Wireless News*, Oct. 21, 2025. Available: https://www.rcrwireless.com/20251021/5g/china-mobile-2

[313] D. Shicong, "China's AgiBot debuts first unified world foundation model for robotic manipulation," *Yicai Global*, Aug. 14, 2025. Available: https://www.yicaiglobal.com/news/chinas-agibot-debuts-first-unified-world-foundation-model-for-robotic-manipulation

[314] W. Knight, "Meet the Chinese startup using AI—and a team of human workers—to train robots," *WIRED*, Nov. 5, 2025. Available: https://www.wired.com/story/agibot-robots-manufacturing/

Notes on Sources, Author Experience, and Educational Use

This book brings together lessons learned from the field with independent analysis of publicly available material. Certain chapters draw on the author's first-hand professional experience in building, deploying, and working with robots in real-world environments, and reflect practical observations and insights developed through that work.

Other chapters examine the approaches adopted by leading robotics companies and present technical descriptions derived from publicly available sources, including company disclosures, technical presentations, product documentation, academic publications, and industry analyses. These materials have been independently interpreted, synthesised, and contextualised by the author. The resulting analysis represents the author's understanding of the subject matter and does not constitute official documentation, endorsement, or representation by the companies or organisations referenced.

The content of this book is intended solely for educational and informational purposes. It is not intended to provide professional, engineering, legal, or commercial advice, and should not be relied upon as a substitute for independent evaluation, validation, testing, or consultation with qualified professionals.

About the Author

Prathamesh Khedekar is the founder of Boring Sage, an ed-tech company focused on bringing deep professional expertise in emerging technologies to engineers and founders. He has built mission-critical systems for the governments of the U.S., U.K., Norway, and the Netherlands, and autonomous robotic systems for Fortune 100 companies globally. An Einstein Visa recipient, his work has been featured in The San Diego Union-Tribune, Security Magazine, and American Banker. He writes and teaches at the intersection of emerging technologies and business applications, helping founders and engineers bring next-generation products to life.

Printed by Libri Plureos GmbH in Hamburg,
Germany

9 798994 296622